(*continued on back*)

Trace Analysis

CHEMICAL ANALYSIS

A SERIES OF MONOGRAPHS ON
ANALYTICAL CHEMISTRY AND ITS APPLICATIONS

Editors

P. J. ELVING, J. D. WINEFORDNER

Editor Emeritus: **I. M. KOLTHOFF**

VOLUME 84

A WILEY-INTERSCIENCE PUBLICATION

JOHN WILEY & SONS

New York / Chichester / Brisbane / Toronto / Singapore

Trace Analysis
Spectroscopic Methods
for Molecules

Edited by
GARY D. CHRISTIAN and JAMES B. CALLIS

Department of Chemistry
University of Washington
Seattle, Washington

A WILEY-INTERSCIENCE PUBLICATION

JOHN WILEY & SONS

New York / Chichester / Brisbane / Toronto / Singapore

Library of Congress Cataloging in Publication Data:

Main entry under title:

Trace analysis.

(Chemical analysis; v. 84)
"A Wiley-Interscience publication."
Includes bibliographies and index.
1. Trace elements—Analysis. 2. Spectrum analysis.
I. Christian, Gary D. II. Callis, James B.
III. Series.

QD139.T7T725 1986   543   85-26603
ISBN 0-471-87583-X

Printed in the United States of America

10 9 8 7 6 5 4 3 2 1

CONTRIBUTORS

Robert J. Hurtubise
Chemistry Department
University of Wyoming
Laramie, Wyoming

Dallas L. Rabenstein
Department of Chemistry
University of California
Riverside, California

Thomas T. Nakashima
Department of Chemistry
University of Alberta
Edmonton, Alberta, Canada

Kenneth L. Ratzlaff
Instrumentation Design Laboratory
Department of Chemistry
University of Kansas
Lawrence, Kansas

A. Lee Smith
Dow Corning Corporation
Midland, Michigan

PREFACE

As recently as 1979 (1), the field of organic trace analysis was heralded as a "new frontier." Its development as an analytical subdiscipline has been driven by the demands of applied scientists such as toxicologists and environmentalists for increasingly sensitive and selective elucidation of tiny amounts of species (pesticides, toxins, pollutants) in complex matrices (soil, blood, plants). In turn, increasing capabilities in instrumentation have led to new discoveries concerning the importance of organic trace materials and consequent demands for even greater sensitivity and selectivity. This upward spiral seems likely to continue until the ultimate capability is reached of detecting and characterizing one molecule, even if it is the only one of its kind, present at some arbitrary location in the volume of the known universe!

A quantitative definition of trace analysis will not be given in this preface, because its meaning in the past has varied widely with the background, interest, and intent of various authors. The definition must take into consideration the context of the sample, the analytical technique, and the particular analysis. Nevertheless, it is widely agreed that the term trace signifies a constituent which is present in minor concentration in another material referred to as the matrix. It is important to not confuse trace methods with micromethods. The latter refers to the processing and analysis of very small (microscopic) amounts of matter. Frequently, trace analysis is performed on microsamples using micromethods. Also, a trace analytical procedure may involve a concentration step followed by a micromethod of determination.

Spectroscopic instruments are an important analytical tool in trace analysis because they not only provide the means for quantitative determination, but they also provide a spectrum which serves as a "fingerprint" for confirmation of the identity of the trace constituent. However, in a complex matrix where the spectra of major and minor components are similar, the presence of the trace component may be unrecognized or even lost in the "noise" of the major constituents. This has led to the development and widespread use of hyphenated instruments in which a chromatographic technique is used to separate the components of a mixture and present them, one at a time, to a spectroscopic instrument which then provides their fingerprints. An excellent appraisal of the possibilities of these hyphenated techniques is given by Hirschfeld (2).

In this book we have chosen to concentrate on four spectroscopic techniques and their hyphenated combinations for trace analysis. Many readers will be surprised that we have not included the workhorses of trace analysis—mass spectrometry and its hyphenated combinations. We felt that a proper treatment of this instrument should be carried out in a separate monograph. In contrast, UV–vis absorption, fluorescence, infrared, and nuclear magnetic resonance (NMR) have recently seen vast improvements in their sensitivity and versatility to the point where they deserve consideration as trace analytical tools, especially with rapid development of related hyphenated methods for each. Accordingly we are pleased to have found five experts to review the use of these four instruments for trace analysis.

Chapter 1, by Professor Ratzlaff of the University of Kansas, deals with advances in ultraviolet–visible absorption spectrophotometry. The applicability of this technique to trace analysis, like the others treated in this book, has been broadened by dramatic advances in technology which have led to improved sensitivity. As Ratzlaff points out, the current scanning spectrophotometers are virtually photoelectron noise-limited down to absorbances of 10^{-5}. Gathering of more light requires further advances in light sources and high throughput monochromators. However, his review of novel laser based methods of absorbance detection shows that absorbance differences of 10^{-7} are attainable. This is especially true now that a method for reducing flicker noise in gas lasers has been discovered. (3).

Luminescence spectroscopy is described in Chapter 2 by Professor Hurtubise of the University of Wyoming. In some ways, this technique represents the ultimate in sensitivity and selectivity. With regard to the former, individual molecules (4) and atoms (5) have been detected and even imaged (6). With regard to the latter, low temperature spectroscopies have allowed determination of specific polyaromatic hydrocarbons in complex matrices with almost no sample preparation. Recently the technique of supersonic jet spectroscopy has been developed as a method for obtaining high resolution fluorescence spectra. The interfacing of this type of spectroscopy to a capillary GC provides an unusually powerful hyphenated method (7). Even for liquid samples at room temperature, impressive work has been done taking advantage of temporal gating (8) to reduce Raman interference. Also, multidimensional analysis holds promise as a technique to deconvolute overlapping spectra (9).

The advent of the Fourier transform spectrometer has made infrared spectroscopy a viable technique for trace analysis. Commercial instruments are available which can attain baseline noise levels of less than 10^{-4} absorbance units (peak to peak) after averaging 100 scans. As Dr. Smith of Dow Corning, the author of Chapter 3, points out, of equal importance to signal-to-noise ratio is the capability to measure diverse samples through various types of accessories. For example, Griffiths showed some years ago that diffuse reflectance provided a very convenient way to measure solid samples with many advantages over the conventional approach of pelleting or mulling. Most recently, the introduction of the circle cell has greatly simplified the handling and measurement of aqueous solutions (10). A final development of importance concerns the possibility that Raman spectroscopy can be done using the FT-IR's interferometer to measure the emission spectrum stimulated by a near infrared laser (11).

The final chapter in this book concerns the use of NMR spectroscopy as a trace analysis tool. As Professors Rabenstein and Nakashima explain, major advances in magnet technology (superconducting solenoids yielding field strengths of up to 11.75 tesla) and the advent of the Fourier-transform technique have combined to allow the routine determination of submicrogram quantities of small molecules. With such sensitivity, it may prove feasible to measure the proton NMR spectrum of liquid chromatographic effluents on the fly. The selectivity of NMR has been increasing lately as new pulse sequences have been developed. The two dimensional NMR spectroscopies NOESY and COSY greatly improve the resolving power of NMR and increase its structural information content as well.

Finally, we want to emphasize the key role that proper analysis of the data from these sophisticated instruments can play. The rapidly emerging field of chemometrics should be kept in mind (12). Excellent progress is being made on methods for solving the challenging problems of (a) the quantitation of a limited number of known components in a variable background of unknowns (13), and (b) the application of multivariate statistics and pattern recognition of spectroscopic data to properties of a sample other than chemical composition, for example, the cancer causing potential of a specific fraction of a petroleum sample.

REFERENCES

1. H. S. Hertz and S. N. Chester, *Trace Organic Analysis: A New Frontier in Analytical Chemistry*. NBS, Washington, D.C. (1979).
2. T. Hirschfeld, *Anal. Chem.*, **52**, 270A (1980).
3. R. E. Synovec and E. S. Yeung, *Anal. Chem.*, **57**, 2606 (1985).
4. T. Hirschfeld, *Appl. Opt.*, **15**, 2965 (1976).
5. D. J. Wineland, W. M. Itano, and R. S. Van Dyck, Jr., *Adv. At. Mol. Phys.*, **19**, 135 (1983).
6. S. Matsumoto, K. Morikawa, and M. Yanagida, *J. Mol. Biol.*, **152**, 501 (1981).
7. B. V. Pepich, J. B. Callis, J. D. S. Danielson, and M. P. Gouterman. *Rev. Sci. Instr.*, in press (1986).
8. N. Furata and A. Otsuki, *Anal. Chem.*, **55**, 2407 (1983).
9. I. M. Warner, G. Patonay, and M. P. Thomas, *Anal. Chem.*, **57**, 463A (1985).
10. J. S. Wong, A. J. Rein, D. Wilks, and P. Wilks, Jr., *Appl. Spectrosc.*, **38**, 32 (1984).
11. T. Hirschfeld and B. Chase, *Appl. Spectrosc.*, **40**, 133 (1984).
12. M. Sharaf, D. Illman, and B. R. Kowalski, *Chemometrics*, Wiley, New York (1986).
13. E. Sanchez and B. R. Kowalski, *Anal. Chem.*, **58**, 499 (1986).

CONTENTS

xi

Trace Analysis

CHAPTER

1

TRACE ANALYSIS BY ULTRAVIOLATE-VISIBLE SPECTROPHOTOMETRY

KENNETH L. RATZLAFF

Instrumentation Design Laboratory
Department of Chemistry
University of Kansas
Lawrence, Kansas

1. INTRODUCTION

A variety of techniques for trace analysis exist, and this text addresses several of the more important ones. Some of the fundamental definitions of trace analysis are discussed in Chapter 3 (Section 1): limits of detection (LOD), limits of quantitation (LOQ), and the problems of sampling and contamination that apply to all trace analyses.

Spectrophotometry in the visible region of the spectrum is one of the oldest analytical techniques, and it is still the most commonly used. The concept is simple: When a solution exhibits a color, the color results from the selective absorbance of certain wavelengths of light, and the concentration of analyte in that solution can be related to the amount of light absorbed. At one time, that light intensity was measured by visual comparison with standards, but now absorbance is measured by sophisticated instruments which provide greatly increased *sensitivity* and *selectivity*, the keys to trace analysis.

In this chapter, we begin by defining fundamental physical and instrumental parameters which are part of the measurement of absorbance in condensed-phase measurements (Section 1). Then methods for increasing sample signal and reducing interferent signals are discussed (Section 2). The chapter concludes (Section 3) with promising methods involving UV and visible absorption measurements which have not yet become standard.

1.1. The Origin of Ultraviolet–Visible Spectra

The energy of a photon in the UV–visible range of the electromagnetic spectrum lies between about 1.6 and 8 eV. This energy range corresponds roughly to the energy of outer-electron transitions, that is, a change of one in the electronic quantum number n. For a photon striking the molecule to be absorbed, it first must have the same energy as the energy level transition; then, the probability of its absorption is determined by quantum mechanical selection rules. Therefore, the electronic transition energy determines the region of the spectrum, and the probability determines the intensity of the absorption.

The energy of the transition depends on the type of bond and the environment of the bond. For example, the wavelength of maximum absorbance, λ_{max}, for a variety of $\pi \rightarrow \pi^*$ transitions in organic molecules depends on the molecule; the transitions of C=C and C≡C bonds have higher energy, and the absorbance occurs well into the UV region, at about 170 nm, compared with the absorbance of a C=N bond, which occurs at about 190 nm. Conjugated bonds absorb at significantly longer wavelengths.

If the transitions excited by ultraviolet and visible photons involved electronic energy levels exclusively, the spectrum would appear the same as atomic absorption spectra, a series of lines in the plot of absorbance versus wavelength corresponding to the changes in n. However, the total energy of a system depends on electronic, vibrational, and rotational energy. (We can neglect the translational energy E_{tran}, which is very small.) Following the Born–Oppenheimer approximation, they may be considered independently so that the total energy, E_T, is simply

$$E_T = E_{el} + E_{vib} + E_{rot} \qquad (1)$$

where E_{el}, E_{vib}, and E_{rot} are the electronic, vibrational, and rotational energies of the system. The energy of an absorbed photon must correspond to a change in E_T, ΔE_T, which is the sum

$$\Delta E_T = \Delta E_{el} + \Delta E_{vib} + \Delta E_{rot} \qquad (2)$$

Typical values of ΔE_{vib} are an order of magnitude smaller than ΔE_{el}, and the energy of a rotational transition is smaller by an additional one to two orders of magnitude.

These changes in energy are illustrated in Fig. 1 (1). At the time that a change occurs in the electronic energy level (quantum number n), an increase or decrease in vibrational quantum number v and/or rotational quantum number J may also take place. From that result, we would expect a series of lines with a symmetrical envelope: a line corresponding to the energy of the electronic transition alone ($\Delta n = 1$, $\Delta v = 0$, $\Delta J = 0$), and a symmetrical set of lines corresponding to the electronic transition plus or minus the energy change of some vibrational transitions; each of those lines would then be divided into more lines resulting from changes in rotational levels, $\Delta J = \pm 1$.

However, this fine structure is not normally observable. Except in low-pressure

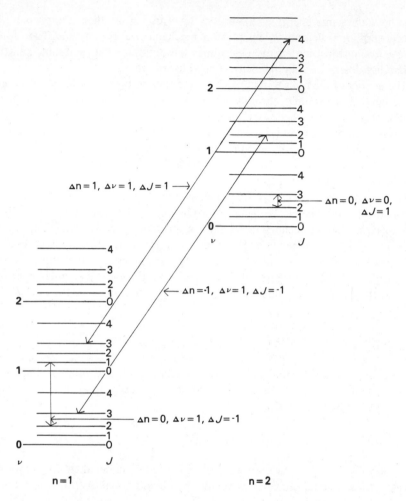

Fig. 1. Partial energy-level diagram for a typical molecule showing some allowed transitions. Transitions in which $\Delta n = 1$ result in absorbance in the UV–visible range.

vapors, the rotational structure is obliterated by the effects of the molecule's environment (the solvent or neighboring molecules), so that free rotation is obstructed, and the rotational energies are randomly perturbed. The consequence is the nearly total loss of rotational structure in the spectrum under normal analytical circumstances.

The vibrational structure suffers a similar fate. The example shown in Fig. 2 (2) shows the effect of solvent on the vibrational/rotational structure of *sym-*

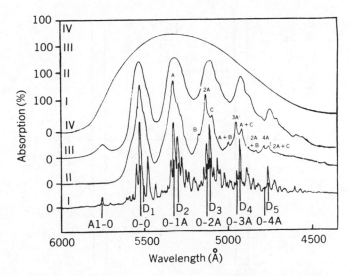

Fig. 2. Absorption spectrum of *sym*-tetrazine. Curve I, vapor at room temperature; curve II, at 77 K in a glass; curve III, in cyclohexane at room temperature; and curve IV, in aqueous solution at room temperature. (Reprinted from reference 2 by permission of The Chemical Society.)

tetrazine. The vibrational transitions are clearly visible in the vapor, and rotational transitions can also be detected; when placed in cyclohexane, the rotational structure is obliterated. A polar solvent, water, smooths even the remaining vibrational bands as a result of the stronger solvent interactions.

1.2. Beer's Law

As shown in Fig. 3, the radiant power of the incident beam may be diminished by absorption, scattering, and reflective losses (3). The diminution in the beam power due to scattering and reflective losses can contribute to errors, which will be considered later. In order to be quantitative, the measurement should respond only to absorption of radiation, and that response is described by a relationship variously attributed to Beer, Bouguer, Bunsen, and Lambert. The most useful component is the relationship of absorption to concentration, which was pointed out by Beer.

1.2.1. Quantitative Relationship

If an absorbing medium is uniform and nonscattering, the power after passing through thickness db will be decreased so that the ratio of the change in radiant power $d\phi$ to the radiant power incident on db, ϕ, is a constant. That is,

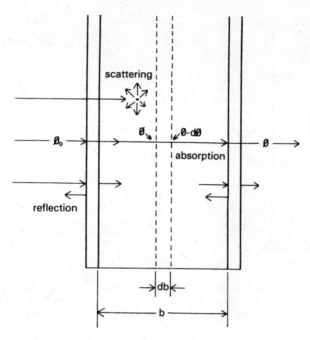

Fig. 3. Illustration of the phenomena that diminish the intensity of a transmitted light beam. In addition to absorption, scattering and reflection also take place.

$$\frac{d\phi}{\phi} = -k \, db \tag{3}$$

If an integration is performed over the length of the cell, from 0 to b, the attenuation in the incident radiant power ϕ_0 resulting in the transmitted radiant power ϕ is

$$\ln \frac{\phi}{\phi_0} = -kb \tag{4}$$

The transmittance T is defined as that quotient;

$$T = \frac{\phi}{\phi_0} \tag{5}$$

The relationship between the path length and attenuation in radiant power is sometimes known as Lambert's law, although it was first formulated by Bouguer in 1729 (1).

Beer, in 1852, stated the relationship between the attenuation in incident power and the number of molecules present. When a photon passes through a sample containing molecules capable of absorbing the photon, the photon may be absorbed. First, it must strike the molecule. Then, absorption requires both that the energy of the photon match the energy of a transition, and that the transition be allowed. The extent to which this may occur is measured as the molar absorptivity, ϵ. It can be expressed in terms of the area of the molecule, a, which helps determine the probability of a photon striking the molecule, and P, the probability of a transition determined by quantum mechanical selection rules. The result is

$$\epsilon \cong 9 \times 10^5 \, Pa_m \tag{6}$$

where a_m is expressed in square nanometers (4). The area of a typical molecule is about 0.01 to 0.1 nm^2; therefore, if the transition is allowed so that the probability is high ($P = 0.1$ to 1), ϵ may be as high as 10^5, resulting in a very intense absorption. Weak absorptions result from forbidden transitions where $P \leq 0.01$.

The constant in the equation is chosen so that $\epsilon = 2.303k$ when b is expressed in centimeters and c is expressed in moles per liter. The consequence is commonly called Beer's law:

$$-\log \frac{\phi}{\phi_0} = \epsilon bc \tag{7}$$

The value $-\log(\phi/\phi_0)$ is defined as absorbance and is assigned the symbol A so that $A = -\log T = \epsilon bc$.

1.2.2. Nomenclature

Although other terms enjoy common usage, the term absorbance is used by analytical chemists, the recommendation of the American Society for Testing Materials (5,6), having been endorsed by the journal *Analytical Chemistry*. It is preferred to the terms optical density or extinction, which are used in some quarters. Similarly, the ratio T (Eq. (5)) is transmittance, not transmission or transmittancy. The term molar absorptivity for ϵ is preferred to the term "extinction coefficient." When the concentration is expressed in grams per liter, the "absorptivity" is used in Beer's law with the symbol a; that is, $A = abc$. The symbol b, not l or d, is used for the path length. Finally, the term spectrophotometry is reserved for the measure of the *ratio*, or function thereof, of radiant power.

1.2.3. Deviations from Beer's Law

The deviations from Beer's law are generally grouped into two groups, real and apparent. Those real exceptions, where the true absorbance/actual concentration relationship is not linear, are found primarily at extremely high concentrations and are seldom important in trace analysis. Details can be found in standard analytical spectroscopy texts (e.g., reference 1).

Apparent deviations are produced when the experimental requirements are not met; for example, Beer's law is defined for the condition in which the incident radiation is monochromatic and the path length through the sample is constant for the entire beam cross section. These requirements are only met approximately. Consequently, a brief review of the effects of these approximations is useful. The factors which contribute to apparent deviations, considered separately below, include multiple internal reflections, multiple internal scattering, stray light, bandpass errors, and chemical errors.

1.2.4. Multiple Internal Reflections

As illustrated in Fig. 4, the transmitted beam suffers diminution due to reflection before entering and after leaving the sample solution. These losses can be determined from the Fresnel equation. Whenever light passes from one medium of refractive index n_1 to a second medium of refactive index n_2, some of the light is reflected. If the light is incident normally, the fraction reflected, f, is determined by the refractive indices of the media:

$$f = \{(n_1 - n_2)/(n_1 + n_2)\}^2 \tag{8}$$

Considering the possible interfaces through which the beam travels in a cell containing water or a common solvent, only the glass–air interface is significant. If $n_{air} = 1.00$ and $n_{glass} = 1.5$, then $f = 0.04$.

By following the path of the beam through the cell, as in Fig. 4, the effect of reflection at the glass–air interface on the transmitted power ϕ can be obtained. First, after the light has passed through the solution once, about 4% of the light (f_1, the reflection at the solution–glass interface at the exit face of the cell) is reflected back through the solution. If the absorbance of the solution is A, the light beam upon reaching the exit face of the cell is attenuated by the factor 10^{-A}. After the 4% (f_1) is reflected back through the solution, 4% of that power

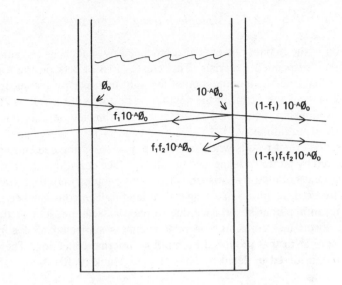

Fig. 4.   Light path of the energy reflected within the cell.

(f_2) is reflected back through the solution a third time. Neglecting any further reflections, the power transmittance is

$$\phi = (1\text{-}f_1) \, [10^{-A} + f_1 f_2 10^{-3A}] \, \phi_0 \qquad (9)$$

The value of ϕ_0 in Eq. (7) is usually determined from a reference (blank) solution, which experiences the same reflections except that $A = 0$. Therefore, the measured value of the absorbance, A', is related to the real absorbance, A by the equation

$$A' = -\log \frac{(1 - f_1) \, [10^{-A} + f_1 f_2 10^{-3A}] \phi_0}{(1 - f_1) \, [1 + f_1 f_2] \, \phi_0}$$

$$= -\log \frac{[10^{-A} + f_1 f_2 10^{-A}]}{[1 + f_1 f_2]} \qquad (10)$$

Typical values of f_1 and f_2 are 0.04. The relative error is largest at trace concentrations, but at a maximum of less than 0.3%, it is normally insignificant for most trace determinations. However, it should not be ignored; much higher f values have been suggested when reflections take place off other optics, apertures, and slits (7,8).

1.2.5. Multiple Internal Scattering

Diminution in the radiant power is also produced by scattering of light by either molecules or suspended materials. This creates two effects on the transmitted beam. The first effect is a diminution of the light intensity by scattering photons *out of* the path of the beam in the solution. The diminution follows a Beer's-law-like relationship; where suspended material is the analyte, this phenomenon is the basis of turbidimetry.

The second effect results when some photons seen by the detector are repeatedly scattered by particles suspended in solution. The conditions for the Lambert relationship are not entirely satisfied since some photons may be scattered back and then forward, resulting in a longer path length. The consequence is that, in addition to an apparent absorbance due to photons scattered out of the optical path, the absorbance/concentration relationship is subjected to a nonlinearity which is similar to that produced by multiple internal reflection. These effects have been considered in detail by Kubelka and Munk (9,10).

1.2.6. Stray Light

The error due to stray light results from light reaching the detector whose wavelength is outside the absorption band being measured *or* from light reaching the detector which failed to pass through the sample cell. For example, this light may due to higher-order diffractions in a grating monochromator, or due to less than perfect optics or light leaks. Imperfections in the grating produce "ghosts"; the occurrence of these imperfections has been dramatically reduced by the use of gratings ruled holographically rather than by a mechanical ruling engine (11).

The consequence for the absorbance measurement can be seen by assuming that a quantity of stray light reaches the detector, unabsorbed by the sample. That quantity can be considered to be some fraction α of the reference radiant power. Therefore, the transmittance with stray light error T' would be

$$T' = \frac{\phi + \alpha\phi_0}{\phi_0 + \alpha\phi_0} \tag{11}$$

If $\alpha \ll 1$, then the measured absorbance A' is related to the real absorbance by the equation

$$A' = -\log(10^{-A} + \alpha) \tag{12}$$

From the equation, and from a plot of the error shown in Fig. 5, we find that

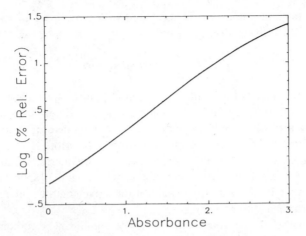

Fig. 5.   Error due to unabsorbed stray light as a function of absorbance. The curve is based on a 0.1% stray-light level.

the error is small when A is small, but that as A becomes large, A' tends toward $-\log \alpha$.

Clearly, stray light will *not* often be significant in trace analysis unless the absorbance of the analyte is to be measured in the presence of a large absorbance due to an interferent.

1.2.7.   Bandpass Errors

Beer's law is defined for the condition that either the incident radiation is monochromatic or that the molar absorptivity ϵ is unchanged over the wavelength range of the incident radiation. These ideals are never perfectly achieved, and consideration of the error produced is necessary.

A reasonably general approach is to consider the effect of a monochromator's bandpass on the measurement of the maximum absorbance of a Gaussian peak. The output of a monochromator can be approximated as a triangular function of intensity with respect to wavelength (12). The observed absorbance is obtained by determining the values of ϕ which are obtained by taking the integral of the product of the triangular slit-width function and the transmittance.

For the case where the bandpass of the incident radiation is symmetrically located on the peak *only*, the relative error is approximately constant with absorbance. Where the ratio of the half-width of the slit function to the half-width of the peak ranges from 0.1 to 0.85, the error ranges from about 1.5% to about 18%. Since in most cases the absorbance peaks of molecules in solution in the

UV–visible range are not narrow compared to the bandpass of even low-cost instruments, the error tends to be small.

1.2.8.   Chemical Errors

This is possibly the most difficult classification of error since there are so many unique chemical sources of error for each possible analyte. However, there are several categories which may be important in a given determination.

Interferents can be easily the greatest sources of error. If the analyte is to react with a reagent to form the colored species, the interferent may also react with the same reagent to form a compound with a similar spectrum. In some cases, the native interferent may absorb at the wavelength of interest. In either case, the error is additive.

Slight changes in the *equilibria* can make significant absorbance changes. These shifts may be caused by such factors as errors in pH buffering or insufficient excess of reagent.

Where the reaction product is slow to develop and/or unstable, measurement at the correct time is necessary. This time must be after sufficient color has developed but before the color decays or interferent reactions develop. When the measurement is time dependent, such factors as temperature and age of the reagent may also become important.

1.3.   Basic Instrumentation for Absorbance Measurement

The basic instrumentation for absorbance spectrophotometry must meet the requirements of Beer's law; that is, the radiant power of monochromatic light must be measured both after transmission through the sample (ϕ), and without the sample (ϕ_0). A block diagram is shown in Fig. 6. A light source and a means of producing monochromatic light are required, as well as a means of mounting

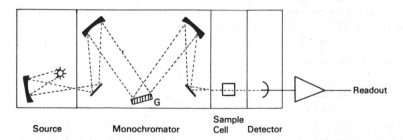

Fig. 6.   Block diagram of a simple spectrophotometer. The reflective element marked G is a plane grating; the detector is a photomultiplier tube.

the sample and a light detector. Finally, the detector output must be converted to user-readable form by the readout subsystem.

1.3.1. Light Sources

The primary characteristics of the ideal light source for general purpose spectrophotometry are stability and wide wavelength range. The intensity should remain constant over the period of an individual measurement. Any variations which do occur must be restricted either to a much shorter time scale than that of the measurement so that they can be filtered out, or to a much longer time scale (drift) so that they are cancelled out in the ϕ/ϕ_0 ratio. The wide wavelength range of the ideal source would allow measurements in all spectral ranges to be made with the same source.

Another characteristic of an ideal source would be constant intensity over the full wavelength range. Unfortunately, common sources cannot provide high intensity over the entire UV–visible range. Furthermore, sources based on arcs exhibit sharp peaks to which the light detector system must react as the wavelength is changed, and discrimination against high-intensity wavelengths when using a spectral region of low intensity can cause difficulties: the detector must react to a very wide range of power levels, and the likelihood of stray light is increased when a low power level is passed by a monochromator while rejecting a peak.

An ideal source would be monochromatic but tunable, eliminating the monochromator. A well-defined spatial distribution, preferably a collimated, narrow beam, would eliminate the need for most of the optics. Presently, these ideals would limit sources to lasers which fail to meet other usual criteria: low cost, long life, easy use, and small size.

Commonly used sources can be divided according to the type of emission process. The first type of source, *thermal*, is governed by the laws of blackbody radiation so that the emission spectrum is smooth and well-defined; the wavelength of maximum intensity is determined by temperature and is nearly independent of the material. The temperatures required to achieve significant UV output are prohibitive for most work, so that thermal sources are generally limited to visible (and infrared) spectrophotometry.

The tungsten-filament lamp is a thermal source which serves the visible range of the spectrum. At a temperature at 2600K to 3000K, the wavelength range is 350 to 2000 nm, the visible and near infrared regions, with the majority of the output in the infrared. A spectral distribution curve is included in Fig. 7. The wavelength range of the tungsten lamp can be enhanced toward shorter wavelengths by raising the operating current and therefore the operating temperature. In the quartz–halogen lamp, the addition of halogen to the lamp helps to reduce deterioration of the lamp at higher temperatures. During the life of the lamp,

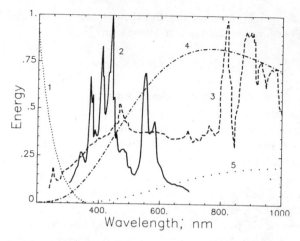

Fig. 7. Spectral energy distributions for various light sources: (1) hydrogen arc, (2) mercury arc, (3) xenon arc, (4) blackbody at 4000 K, and (5) blackbody at 3000 K.

tungsten sublimes from the filament and deposits on the envelope; when halogen gas is added, the sublimed tungsten forms gaseous tungsten halide, which is decomposed back on the hot filament. Consequently, the temperature can be increased without damage. Further increases in temperature can be traded for lamp life, the increase in output being approximately equal to the decrease in lamp life (13). The increases in temperature require a quartz rather than glass envelope.

Variations in the output intensity are derived from several sources. First, the intensity is a nonlinear, often exponential, function of the power supply voltage; since the intensity is related to temperature, which is determined by the current, a good current-related power supply will often suffice. However, intensity may also change with age and even with room temperature. Consequently, a photodetector is often employed to feed back intensity information to the supply in order to maintain constant intensity.

A second cause of variations in the source is noise generated by the filament itself. The magnitude of the noise decreases with frequency; hence the frequent reference to 1/*f noise*, also called *flicker*.

In *electric discharge sources*, electrical current is passed through gases, and radiation results from the electronic, vibrational, and rotational excitation of the gas molecules and ions. At very low pressures (<<1 atm), only a few discrete emission lines are observed, the simple line spectrum of the excited atom or molecule. As the pressure is increased to atmospheric pressure and then to high pressure, the lines are broadened so that a continuum is produced, usually with some discrete lines superimposed on it. For most molecular absorption spectro-photometry, high-pressure lamps are required to continuously cover the UV and/ or visible spectral ranges.

The hydrogen-arc and deuterium-arc lamps are often used in the ultraviolet range to complement the tungsten lamp. The output (Fig. 7) is limited at the short-wavelength end by the window material, and its usefulness extends to about 350 nm. The output of the deuterium-arc lamp is 3–5 times that of the hydrogen-arc lamp (1).

The xenon-arc lamp usefully covers both the UV and visible ranges with high-intensity output. However, it has several drawbacks which limit its use. First, the high-intensity lines (Fig. 7) superimposed on the continuum create serious stray-light problems. Second, there have been problems with stability; the arc tends to wander on the electrodes, and since the light source is a small point, this creates optical difficulties. Some relief has been obtained by better lamp design and subsystem regulation (14). Finally, a significant inconvenience is the intense radio-frequency noise pulse which may accompany start-up if proper grounding, shielding, and sequencing procedures are not employed. Otherwise, if accompanying instrumentation is not turned off, electronic circuits can be damaged, and computers will malfunction.

The final type under consideration, the *laser*, has many of the ideal charac-teristics. Laser beams are intense, collimated, and easily focused. However, most lasers have only fixed discrete lines, making them useful only at the pre-determined wavelengths.

The dye laser, however, is tunable; consequently, the entire visible range can be covered with no additional monochromating device. However, the high cost, low reliability, and inconvenience (relative to tungsten and deuterium lamps) limits laser use to specialized applications.

1.3.2. Detector Subsystems

The primary characteristic of an ideal light detector subsystem is stability; uncer-tainty should not be added to the measurement by the detector. Second, it should have adequate responsivity at all wavelengths, preferably equal response. Third, it should respond rapidly to changes in radiant power. Finally, it should have high sensitivity, small size, and low cost.

Compared with the characteristics of detectors for other types of spectroscopy, sensitivity is ranked relatively low since for the measurement of ϕ and ϕ_0, the latter is usually large enough to be easily measured (or can be made so with the right light source and optics) and most trace measurements involve values of ϕ which do not differ greatly from ϕ_0.

Detectors can be categorized in two fundamental ways, by the basis of oper-ation and by the presence or absence of gain. First, there are photoemissive and semiconductor detectors. The former, phototubes and photomultiplier tubes, can detect photons of sufficient energy to eject an electron from the surface, and consequently have greatest sensitivity in the UV and blue end of the spectrum. Semiconductor detectors have poor sensitivity in the UV, but good red and near–infrared sensitivity.

Detectors can also be categorized as those with internal gain (photomultiplier tubes and semiconductor devices with intensifiers) and those without. The gain is generated by electron multipliers. Signals resulting from a small number of photons are amplified with less added noise when electron multipliers are employed than with external amplifiers; however, the measurement of such low levels may not be necessary in absorption spectrophotometry, as was previously noted.

Phototubes are photoemissive devices. The output is the product of the radiant power ϕ and the efficiency of the photocathode. The latter seldom exceeds 30%, and is dependent on the photocathode material and the wavelength. Few phototubes are effective in the near-infrared region; the best near-infrared-sensitive tubes exhibit lowered UV response. Maximum currents should not exceed 10^{-10} A. Large variations in sensitivity may occur over the photocathode surface, resulting in some vibration and implementation problems.

Photomultiplier tubes add electron multiplier stages (Fig. 8) to the phototube; the consequence is a detector capable of detecting single photons. The photoelectron which is emitted by the collision of a photon with the photocathode is then attracted to a dynode at a more positive potential, and the resulting collision produces several electrons. The process is repeated over 8–13 additional dynodes so that up to 10^8 electrons are produced for each photoelectron. Consequently, the output current is the product of not only the radiant power and the photocathode efficiency, but also the gain. The gain is an exponential function of the photomultiplier-tube power-supply voltage. As a rule of thumb, the supply voltage stability must be ten times the desired gain stability. Fig. 8 shows the typical S-20 response curve, along with response curves for two red-sensitive PMTs. (See also Fig. 12 in Chapter 2 for PMT response curves.)

The PMT is also prone to several other sources of error. First, the efficiency of the PMT photocathode in converting photons to electrons varies markedly with position, causing significant variation when any aberrations in the optical path occur. Second, stray magnetic fields, from chopper motors for example, distort the electron paths. Also, hysteresis results from electrostatic charge on insulators and leads to gain changes; this charge depends on the current, that is, the light level and gain voltage (15,16).

Photomultipliers remain the most popular detectors because of their enhanced sensitivity, which is necessary in high-absorbance applications, and because the sensitivity can be electronically controlled via the gain voltage.

Photodiodes are the most important of semiconductor spectroscopic detectors.

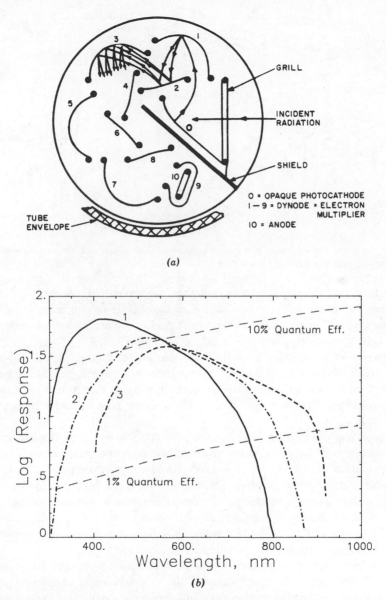

Labels in figure (a):
GRILL
INCIDENT RADIATION
SHIELD
O = OPAQUE PHOTOCATHODE
I — 9 = DYNODE = ELECTRON MULTIPLIER
IO = ANODE
TUBE ENVELOPE

(a)

Labels in figure (b):
Log (Response)
10% Quantum Eff.
1% Quantum Eff.
Wavelength, nm

(b)

Fig. 8. *a*. Diagram of the dynode structure for a side-on PMT. *b*. Typical (S-20) photocathode spectral response curve (1) and the response curves of two typical extended red photocathode tubes (2 and 3).

17

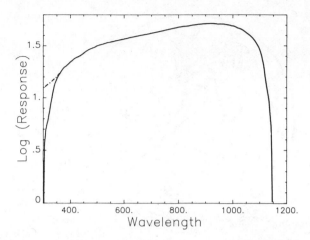

Fig. 9. Spectral response curve of a typical photodiode. The dashed line suggests the extended UV response available in some models.

They differ from the previous devices in that, as shown in Fig. 9, their optimum response lies in the near-infrared region of the spectrum, and usually the UV response is poor (although "blue-enhanced" devices are useful down to near 250 nm). Like phototubes, photodiodes exhibit no gain.

The primary advantages of photodiodes are low cost, small size, long-term stability, and the absence of a need for a medium- or high-voltage power supply. Some devices do have large active collection areas so that more light can be easily collected. However, poor UV response is the primary factor in limiting their application.

Photodetector arrays (PDAs) provide the opportunity for a fundamentally different approach to spectrometric instrumentation. A photodetector array offers a large number of discrete detectors (256–2048) at very close physical spacing. These separate detectors or *pixels* (picture elements) can be used *in parallel* to collect radiant power information. The device is typically mounted so that each pixel monitors a separate wavelength, and an entire spectrum is collected in parallel; however, the various elements can also be used at a common wavelength to monitor spatially separated sample characteristics.

There are a large variety of PDAs which have been developed for imaging tasks and considered for spectroscopy (17); a variety of electronics industry trade journals are devoted to electronic imaging technologies. Of these devices there are several which are useful; these PDAs may be categorized into two groups: solid-state arrays and silicon vidicon tubes. Certain features are held in common. First, they *integrate* the information in the detector unit; the signal which is finally obtained is the integral of the radiant power incident on the pixel over the time since the last readout. Second, although there is no

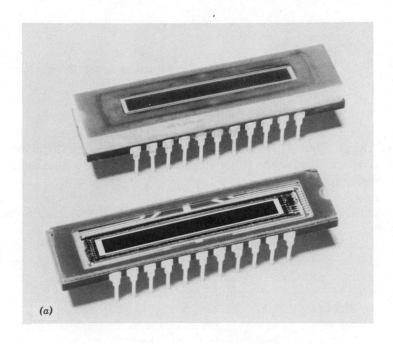

(a)

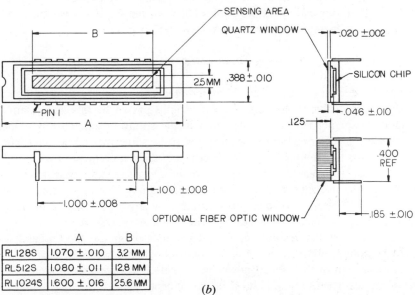

	A	B
RL128S	1.070 ± .010	3.2 MM
RL512S	1.080 ± .011	12.8 MM
RL1024S	1.600 ± .016	25.6 MM

(b)

Fig. 10. (a) Photograph of two 'S'-series photodiode arrays; the upper unit includes an integral fiber-optic faceplate. (b) Dimensional outline of an 'S' series photodiode array. (Courtesy of Reticon Corp., Sunnyvale, California.)

19

inherent gain in semiconductor detectors, they can be manufactured with accompanying devices which do provide gain.

Solid-state arrays (Fig. 10) are integrated circuits with the pixels integrated into the silicon substrate. They are normally self-scanned, that is, the electronic logic is provided to direct the signal for each pixel sequentially to one or two output pins. Recently, a technique has been developed to operate one type of PDA in a random-access, nondestructive mode (18).

Several technologies have been utilized, including charge-coupled devices (19) and charge-injection devices. However, the photodiode array (20) dominates both in commercial and noncommercial use primarily because one manufacturer has seen fit to prepare devices with large areas per pixel and with good UV response particularly for spectroscopic application; the "scientific" series of arrays from Reticon increases the height of each pixel up to 2.5 mm with a width of 0.025 mm.

The active surface of a vidicon tube is a silicon surface which collects the signal until interrogated by an electron beam. The beam is controlled outside the tube so that random access positioning is possible, and pixels can be scanned in any order. The price for this convenience is increased size, complexity, and somewhat reduced resolution.

More details of array and vidicon detectors and their properties are provided in Chapter 2, Section 3.7.3, as optoelectronic imaging devices for luminescence detection. That application differs somewhat from the usual absorption application since luminescence detection requires detector gain.

1.3.3. Wavelength Isolation Subsystems

In order for Beer's law to be satisfied, the light must be monochromatic, although the error is negligible if the band of wavelengths passed by this subsystem is substantially smaller than the peak width in the spectrum under study. Since UV–visible molecular absorption peaks tend to be somewhat broad, and since bandpass error is minimized at low absorbance, the requirements for the subsystem are less stringent than are found in other UV–visible spectroscopic methods. Among the characteristics of the ideal wavelength isolation subsystem would be continuously variable wavelength control so that any wavelength in the range of the instrument could be requested. The bandpass should be continuously selectable, even at high resolution, and all light not within the chosen bandpass should be completely rejected. Finally, the throughput should be near 100%; that is, nearly all of the radiant power at the selected wavelength entering the instrument also leaves at the exit of the instrument.

Several methods of obtaining a sufficiently narrow bandpass are available. First, a filter can be used which removes unwanted wavelengths by absorption, reflection, and/or interference. Second, dispersive devices can be used which spatially disperse the light by wavelength so that the desired wavelength can be isolated with a slit. Finally, a monochromatic light source (tunable laser) could be used, but this is seldom practical.

Interference filters (of the Fabry–Perot type) are assembled by "sandwiching" a transparent dielectric (typically calcium fluoride or magnesium fluoride) of thickness *d* between semireflecting metallized surfaces. This assembly is protected by outer layers of glass or quartz. The light incident on this filter is removed by reflection and destructive interference, except that of wavelength λ satisfied by the equation

$$\lambda = 2dn/m \tag{13}$$

where *n* is the refractive index and *m* is the order of the interference. Therefore, besides transmitting at wavelength λ, the higher orders allow this filter to also transmit at λ/2, λ/3, etc., and these bands must be removed by additional filtering. Filters are frequently employed in the second- and third-order modes since a narrower wavelength band is transmitted than in the first-order mode; however, the transmittance is lower, and more care is required to remove bands due to adjacent orders.

The primary advantages of interference filters are the high throughput and the low cost. Wavelength tunability is minimal, although the wavelength passed does change with the angle of incidence since light transmitted obliquely through the filter travels a longer length *d*; this is more often considered a source of error than an advantage. The bandpass, though not variable, is on the order of 10 nm, sufficiently small for most measurements. (See Chapter 2, Section 3.3.1 for a discussion of the use of absorption and interference filters in fluorescence instruments.)

Dispersive devices provide the flexibility of being able to select any wavelength in the instrument's range and, in most cases, the capability of determining the bandpass. Both grating and prism monochromators are used, although now, with recent strides in grating technology, the grating instrument is favored. The mounting methods for gratings have been adequately reviewed elsewhere (e.g., reference 13); variants of the Czerny–Turner mount are often chosen in part because of lower mirror costs and linear wavelength drive. Older ruled gratings have, in many cases, given way to holographically ruled gratings which, because of fewer imperfections, exhibit lower stray light and a flatter efficiency versus wavelength curve; these gratings in addition can be prepared readily on concave surfaces so that the grating also becomes a focusing element, and the optical path is simplified.

There are fundamental limits to the resolution which can be obtained by a monochromator, resulting from a diffraction of light at the slits; these limiting values are improved only by increasing the dispersion of the monochromator.

The actual resolution is also affected by the quality of the optical components, both mirrors and gratings or prisms.

Stray light in a monochromator comes from several sources. First, optical imperfections allow wavelengths of light to pass that are not predicted by the grating equation. Second, higher-order diffractions are passed unless separately removed, often by a filter with a bandpass small enough to discriminate against other grating orders, but large enough to pass the monochromator output over a sufficiently broad range that only a few filters (<10) are required to cover the instrument's range. In demanding situations, the stray light due to optical imperfections is removed by use of two monochromators in tandem.

Section 3.3.2 in Chapter 2 includes a discussion of monochromator systems for luminescence spectrometers. The grating equation and the resolving power of the grating are presented.

1.4. Signal-to Noise Ratio Theory for Absorbance Measurements

1.4.1. General Information

There are several types of uncertainty which are inevitably added to the information during the course of making an absorbance measurement. Some are developed *before* the instrumental measurement is made, and others are incurred as part of the measurement process. The former include uncertainties due to sampling, separations, mechanical sample manipulations, and chemical manipulations. The uncertainties incurred after the sample is brought to the spectrophotometer are the subject of this section.

The term uncertainty denotes the measurement *variations*, which will average to zero when measured with sufficient repetitions. Commonly, uncertainty which results from variations that occur on a time scale slower than that of the measurements is called *drift*, and the more rapid changes are called *noise*. The uncertainty affects the precision of measurement and may or may not limit the accuracy, depending on the magnitude of the determinate errors and the number of measurements averaged together. In much of the literature, the term *error* is used; that terminology should be discouraged and saved for systematic errors.

Following the nomenclature of Pardue et al. (21), the major sources of uncertainty are grouped according to the way in which they affect the experimental signal resulting from the measurement of radiant power. The absorbance is the result of two measurements of radiant power, previously denoted ϕ_0 and ϕ (Section 1.2.1.). The uncertainties that are normally experienced fall into the categories of those proportional to the signal, those proportional to the square root of the signal, and those independent of the signal magnitude. They are specified by the standard deviations, σ_P, σ_S, and σ_I, respectively. In addition, some uncertainty may be added if the readout device has insufficient resolution.

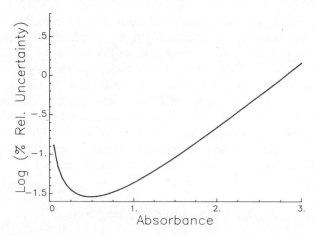

Fig. 11. Relative uncertainty in absorbance as a result of independent uncertainty which is 0.01% of the full-scale signal.

Knowing the standard deviation in the measurement of radiant power, the relative standard deviation in the absorbance, σ_A/A, is readily computed using propogation of error equations.

$$\sigma_A/A = (2.3A\ E_0)^{-1}\ [(\sigma 10^A)^2 + (\sigma_0)^2]^{1/2} \tag{14}$$

where A is the absorbance, and σ and σ_0 are the standard deviations in the signals for the measurements of ϕ and ϕ_0, respectively. E_0 is the voltage resulting from the measurement of ϕ_0.

In the following sections, the three categories of uncertainty will be considered in detail.

1.4.2. Independent Uncertainty

Sources of independent uncertainty are those in which the standard deviation in the measurements of ϕ and ϕ_0 is a constant σ_I, independent of the magnitude of ϕ. Substituting in Eq. (14),

$$(\sigma_A/A)_I = \sigma_I\ (2.3A\ E_0)^{-1}\ [10^{2A} + 1]^{1/2} \tag{15}$$

A plot of the curve of $(\sigma_A/A)_I$ as a function of A is shown in Fig. 11. A narrow minimum can be noted at an absorbance of about 0.4, and the precision is directly related to the magnitude of the reference signal E_0.

This curve is found in most classical instrumental analysis textbooks since it describes the uncertainty in the measurement of concentration that results from the reading of transmittance on the analog meter found on older spectrophotometers; since the meter cannot be read to better than about 0.2 divisions, σ_I/E_0 would be about 0.002. That source of independent uncertainty is becoming a thing of the past. However, other sources of independent uncertainty still exist: Johnson noise in resistors, a fundamental source; insufficient resolution in analog-to-digital converters when the radiant power signal is directly digitized (22); and noise in the detector power supplies and other electronics. While most sources of independent uncertainty can now be minimized and are usually inconsequential in trace analysis, the curve rises so rapidly at high absorbance that it can then become a factor.

1.4.3.   Square-Root Uncertainty

The random arrival of the discrete photons is described by Poisson statistics; the uncertainty in the resultant signal is proportional to the square root of the signal. This is true whether the signal is measured by discretely counting the photons or by measuring the photocurrent as an analog value (23). The resultant uncertainty is often called shot noise. In this case,

$$\sigma_S = \sqrt{n} = k\sqrt{E} \tag{16}$$

where n is the number of photons in the measurement interval and E is the electrical signal proportional to n. Substituting into Eq. (14),

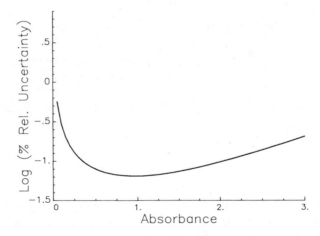

Fig. 12.   Relative uncertainty in absorbance as a result of square-root uncertainty when the full-scale signal results from the detection of 5×10^6 photons.

$$(\sigma_A/A)_S = k(2.3A\sqrt{E_0})^{-1}[10^A + 1]^{1/2} \tag{17}$$

The curve of $(\sigma_A/A)_S$ plotted as a function of A is illustrated in Fig. 12. A broad minimum can be noted with optimum precision in the region 0.5–1.5 absorbance units.

It should be noted that square-root uncertainty defines a fundamental limit to precision that is improved only by increasing the number of photons collected in the measurement interval. The increase can be effected by increasing light-source intensity or the measurement interval; the latter can be achieved in practice by applying an electronic filter with a longer time constant to the analog signal.

1.4.4. *Proportional Uncertainty*

Sources of proportional uncertainty are those which produce a standard deviation in the measurement of the radiant power that is proportional to the magnitude of the radiant power, in other words,

$$\sigma_P = \xi E \tag{18}$$

where ξ is a constant of proportionality.

When substituted into the equation above,

$$(\sigma_A/A)_P = \frac{\sqrt{2}\epsilon}{(2.3A)} \tag{19}$$

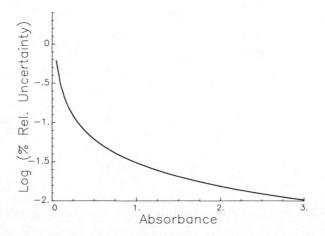

Fig. 13. Relative uncertainty in absorbance as a result of proportion uncertainty (flicker); the uncertainty in the signal is 0.04% of the signal.

Uncertainty that is proportional to the radiant power signal results in uncertainty in the absorbance measurement that is independent of the magnitude of the absorbance. The sources of these uncertainties are often termed flicker, so that ξ is often called the *flicker factor*. The relationship of the uncertainty to the absorbance is shown in Fig. 13.

Primary sources of flicker are the variations in light-source intensity, detector gain, and the positioning of the sample cell in the light beam (24). Light sources, particularly filament-type light sources, tend to exhibit a power spectrum (radiant power as a function of frequency) that follows a $1/f$ relationship, which is fundamental to hot radiant sources and cannot be completely eliminated with current regulation. However, in modulated systems, $1/f$ noise is not usually a dominant factor.

The work of Rothman et al. (24), and later Kaye (16), did, however, point to the very important effects of cell flicker. Even high-quality sample cells showed variations in transmission with the slight changes in position which occur when the sample is placed in the sample compartment. Temperature variations affect the beam focus and consequently the detector gain due to the spatial sensitivity of PMTs; apparent transmittance greater than one is possible. The scatter of the beam at the cell window makes it very difficult to match window losses when low absorbance measurements are made; the losses vary both between the members of the "matched" pair and across the surface of a single cell.

This result may argue for "sipper" attachments and flowthrough cells in which the sample cell is kept in a very rigid fixed position. Multiple wavelength measurements can also be used to remove the error by correlating the uncertainty (Section 2.4.2).

1.4.5. Absorbance Readout Uncertainty

Most spectrophotometers provide the user with a choice of transmittance or absorbance readout; the latter is usually more convenient. However, there is always some uncertainty in that readout, whether it is analog (analog meter or chart recorder) or digital (display or printer). Most digital readout devices provide more than enough resolution, but analog readout devices sometimes limit the overall precision of the measurement. The magnitude of the uncertainty is independent of absorbance unless the scale is changed. Therefore, if the uncertainty in the *readout of absorbance* is σ_A, the relative uncertainty is simply σ_A/A; the functional form is the same as that of proportional uncertainty, and the plot of relative uncertainty as a function of absorbance is the same as in Fig. 13.

1.5. Spectrophotometer Integration for Trace Analysis

A block diagram of the manner in which a spectrophotometer is usually organized is shown in Fig. 6. The main components are the source of radiation, the wavelength isolation system, the sample holder, and the detector. In principle, the wavelength isolation subsystem could be located either before or after the sample; however, it is usually placed before in order to reduce the danger of photochemical reactions induced by the source.

The block diagram illustrates a *single-beam instrument*. In this arrangement, the sample and reference measurements are made sequentially using the same sample holder; sample and reference are usually changed manually. The single-beam arrangement is dependent on the long-term stability of the light source, the detector, and all electronic components. Furthermore, manual operations are usually required to measure the value of ϕ_0 any time that the wavelength of the incident radiation is changed, and this makes wavelength scanning difficult.

In *double-beam instruments*, the beam is mechanically modulated between sample and reference cells. There are two advantages over single-beam instruments. First, if the wavelength is varied in the process of obtaining a spectrum, the measured value of ϕ_0 is rapidly and automatically updated. Second, whether or not the wavelength is scanned, any drift in source radiant power, detector gain, or electronic components, which occurs more slowly than the modulation frequency, will be correlated and will cancel in the ϕ/ϕ_0 quotient.

However, the mechanical operations required in double-beam instruments inherently add some measurement uncertainties due to the artifacts resulting from having different optical components in the two beams. First, each optical component involved in the modulation (mirrors, choppers, beam-splitters, and cells) has its own transmittance or reflectance spectrum, which is superimposed on the sample spectrum. Second, the modulation requires moving parts, which inevitably leads to positioning irreproducibility; changes in scattering or transmittance at optical surfaces and variations in beam focusing follow. In addition, square-root uncertainty is increased since a substantial portion of the measurement period is "wasted" during sample or beam movement (16).

Very rapid modulation was necessary in the pre-digital generation of instruments so that the ϕ/ϕ_0 quotient would be updated sufficiently often that both measurements could be assumed to have been made at the same wavelength; a scan rate that is too fast with respect to the modulation rate gives a "derivative effect" on the light source spectrum. However, when the scan motor (usually a digitally controlled stepper motor) is placed under the intelligent control of a microprocessor-based controller, measurement of both sample and reference radiant powers is possible between stepping sequences, and at precisely the same wavelength.

Consequently, two beam modulation methods have developed (25). The first, *double-beam-in-space*, utilizes a rapidly moving chopper/mirror to alternately steer the beam to the sample or reference cell. The second, *double-beam-in-time*, maintains a constant beam path, but alternately moves sample and reference cells into the path. The latter method reduces the baseline variations in the absorbance spectrum that result from slight absorption differences in the optical components used for the two paths. However, uncertainties may be introduced as a consequence of small differences in cell position during repeated cycles.

A *split-beam* approach is also possible (26).The monochromatic light emerging from the wavelength isolation device is split and directed to sample and reference cells simultaneously to be detected by a pair of detectors. Because of the additional error and complication of having *two* detectors rather than one, this approach is seldom used except in measurements of rapidly changing signals (26,27).

Three *feedback* techniques are often used to optimize the signal obtained from the detector. First, low-frequency variations in the source intensity can be substantially reduced by feeding back the signal from a photodetector mounted in the source housing to the source power supply (28). Second, when a PMT detector is used, the signal can be maintained in a reasonable range by feeding back the signal during the reference cycle to the PMT power supply that controls the gain. The signal resulting from ϕ_0 can be maintained at the full-scale level at all wavelengths (and slit widths) so that noise added to the signal in the readout electronics can be minimized. Third, the detector signal during the reference cycle can be used to vary the monochromator slit width in order to maintain a constant reference signal. This method has an advantage over the former, from the standpoint of the signal-to-noise ratio, since a constant, high photoelectron flux is obtained during the reference cycle. However, slit control is seldom used because (1) the wavelength resolution does not stay constant, and (2) the cost of implementation is greater.

Uncertainty in the signal due to spatial variations in the detector can be reduced by using frosted silica diffusers or frosted-surface PMTs, or by employing optical integrating spheres.

The single-beam spectrophotometer has made a comeback as a general purpose instrument (29). One might reason that the difference between single-beam and dual-beam instruments is simply that in the former the modulation is slower and is performed manually. The lower rate can be accomodated because progress in electronics has made possible computer-controlled instrumentation, highly stabilized power supplies, and detectors. An entire reference spectrum, ϕ_0 as a function of wavelength, can be obtained and stored in memory; then, when the sample is placed in the beam, the absorbance can be computed as the wavelength is scanned. The consequence can be reduced sample flicker and reduced square-root error since the photon flux is measured continuously. With the exception of instruments of the latter type, single-beam operation is limited to dedicated instruments in which the wavelength is seldom changed.

One instrumental parameter that can be altered, but is seldom changed, is the path length. Increasing path length from the common 1 cm to as much as 10 cm gains an order of magnitude increase in sensitivity without increasing uncertainty produced by the cell windows. However, if a larger increase in path length is desired, some fundamental changes in the optical system are required, since

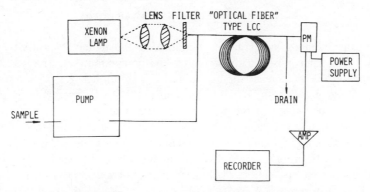

Fig. 14.   Spectrophotometer system using the hollow-fiber capillary cell. (Reprinted with permission from Fuwa et al., *Anal. Chem.*, **56,** 1640. Copyright 1984 American Chemical Society.)

it is not usually possible to maintain a well-characterized beam if the size of the sample chamber is enlarged.

Since it is well-collimated, a laser beam is ideal when the radiation is to be passed through a very long cell. One attractive possibility for cells with a very long path is the use of hollow-core fibers (30–32) (Fig. 14). Problems exist in filling and cleaning the fibers, and the determination of the actual path length would probably have to be done experimentally since the beam is repeatedly reflected internally off the fiber walls. Path lengths of 4.5–130 m were successfully tested.

2. METHODS FOR INCREASING A_{ANALYTE} AND DECREASING $A_{\text{INTERFERENT}}$

When carrying out a trace analysis, the first tendency is typically to find an instrumental technique capable of detecting the material at its native concentration with high selectivity. In order to measure the absorbance of an analyte detectably above that of uncompensated interference, the absorbance can be increased, the background (interferent) absorbance can be removed, or both. If one can raise the molar absorptivity sufficiently, the high sensitivity is less important.

If we look on the overall measurement as following the equation

$$A_{\text{total}} = (\epsilon b c)_{\text{analyte}} + (\epsilon b c)_{\text{interferent}} \tag{20}$$

we have the opportunity to enhance $(\epsilon b c)_{\text{analyte}}$ through $\epsilon_{\text{analyte}}$ or c_{analyte} or to suppress $(\epsilon b c)_{\text{interferent}}$ through $\epsilon_{\text{interferent}}$ or $c_{\text{interferent}}$.

Section 1 in Chapter 3 provides a discussion of concentrating analytes and removing interferents that is applicable to other forms of spectrophotometry.

2.1. Increasing Molar Absorptivity

Among the methods available for maximizing the molar absorptivity of the analyte, it is often possible to derivatize the compound, selectively adding a chromophore, or to increase the native molar absorptivity by selecting the optimum pH and wavelength. Finally, the sensitivity can be enhanced by placing the analyte in the proper solvent or in a micellar matrix.

Derivatization in inorganic analysis is quite common since reagents with high sensitivity have been found for most elements. In molecular analysis, one is often faced with probability of less specificity, but derivatization can provide two advantages: the possibility of enhancing the molar absorptivity and of optimizing the wavelength.

A large molar absorptivity at an inconvenient wavelength (e.g., the vacuum UV or the near infrared) does not lead to sensitive analyses. Derivatization may be required simply to obtain an absorbance in the UV–visible region. Optimization of wavelength has particularly great advantages in HPLC, where fixed wavelength (254-nm) detectors are the most sensitive, and in laser-based methods.

Sternson (33) and Lawrence (34) have listed a number of functionalities which can be added for increased sensitivity. An example of such an application is in the determination of the drug dianhydrogalactitol (DAH), which lacks a strong chromophore in the UV–visible range. However, derivatized with diethyldithiocarbamate ($\epsilon_{254nm} = 2.8 \times 10^4$), the product can be easily detected.

Derivatization schemes should, in the ideal case, involve reactions which (1) are rapid, (2) are quantitative or reproducible, (3) generate a single product, (4) utilize reagents that do not interfere with detection of the analyte product, and (5) do not require extreme conditions.

Among other variables to be optimized, the molar absorptivity is frequently a function of pH, and it always varies with wavelength. Normally, the latter suggests performing the analysis at the wavelength of maximum molar absorptivity. However, tradeoffs may be required: the interferent absorbance may be enhanced more than that of the analyte, or instrument performance may be poor at the absorbance maximum.

Absorption methods can in some cases be sensitized by the presence of surfactants, usually only if the critical micelle concentration is exceeded (35,36).

2.2. Increasing c—Preconcentration and Trace Enrichment

Typically, when measurements are made at the trace level, the sample volumes are relatively large. Expansion of the measurement range may be achieved by preconcentration techniques (37–39); the accuracy of the result is improved when this enrichment is also selective. Therefore, methods to raise the concentration should often be considered before the measurement is made.

The enrichment is defined by an *enrichment factor F*, defined as

$$F = \frac{Q_A/Q_M}{Q_A^0/Q_M^0} \tag{21}$$

where Q_A and Q_M represent quantities of analyte and matrix after enrichment, and Q_A^0 and Q_M^0 represent those quantities before the enrichment.

One should be aware that a method which will enrich the concentration of the analyte will also enrich the concentration of similar materials which are interferents.

Several techniques should be considered for enriching the trace material, and these are briefly listed.

(1)   Liquid–liquid extraction, although usually required primarily as a separation technique, is an important enrichment technique. Standard additions are usually required to calibrate the enrichment factor, or else standards are run through the same procedure.

(2)   Volatilization of the solvent with a rotary evaporator can enrich the concentration, but with no selectivity.

(3)   Dialysis, gel filtration, ultrafiltration, and many others are described by Mizuike (38) for inorganic analysis, but have general applicability in analysis for molecules.

(4)   Sorbent traps are valuable for concentrating trace organics from fluids. A tube packed with a sorbent material selectively retains the analyte, which is subsequently stripped by an appropriate solvent. Sorbents include polyurethane foam, activated carbon, and macroreticular XAD resins.

(5)   Several liquid chromatography variants have been described by Saner (37). Large sampling volumes are injected with a mobile phase of weak elution strength; the solvent passes through the column while the compound of interest is on the head of the column. Use of gradient elution or changing to a mobile phase which can displace the analyte leads to elution at a concentration greater than the injected concentration.

2.3.   Decreasing Interferent Concentration by Separation

Typically, the choice of a separation as a means of enhancing the integrity of a measurement entails a trade-off. The combined chromatographic/detector methods, usually the most useful, nearly always *decrease* the concentration of the analyte with the assumption that the measurement is improved more by removing the interferent than by maintaining the concentration.

Separations are a common preparatory step to most analytical techniques, and consequently it is not necessary that the separation techniques be discussed in detail. In this section, the discussion will be limited to the factors in the *detection* of the analyte after the separation, and here principally to high-performance liquid chromatography (HPLC).

HPLC places several additional constraints on the measurement of absorbance (40). First, the measurement is dynamic; the system records signal peaks rather than signal levels, and both shape and magnitude of the peak have meaning in the determination of the concentration. Therefore, fast response is required. Second, the detection cell volume must be small in order not to degrade the separation; typically, the volume is about 5 μL, so that the beam diameter must be quite small and/or the path length must be short. Third, the detector may be affected by a drift in the baseline due both to thermal effects and to solvent gradients. Fourth, the system typically is affected by pulsations in the cell pressure.

However, the absorbance detector for HPLC is quite valuable because of its low cost and nearly universal applicability. Having an inherent flow cell rather than a removable cell increases sensitivity. Therefore, significant attention has been paid to producing a design capable of high sensitivity.

There are two categories of detector systems, the variable wavelength and the fixed wavelength; the former typically employ a mercury lamp (about 5 W), and the latter require a continuum source, typically a 30-W deuterium lamp. Photodiodes are typically used as light detectors in HPLC absorbance detectors. Typically, matched dual photodiode detectors are employed since the light levels will remain high and differential drift can be controlled.

Recent figures presented by Abbot (41) provide good estimates of the limits of the detection system. The sources of noise that were considered are listed in Table 1.

The substantial differences in shot and Johnson noises for the two sources are due to the much higher output of the Hg lamp concentrated at the analytical

Table 1. Estimated Lower Limits for Noise from Sources in HPLC Absorbance Measurements

	Conditions		
	D_2 Lamp (30 W)	Hg Lamp (5 W)	PD Array Detection
Shot noise	9×10^{-6}	1.2×10^{-6}	2×10^{-5}
Pump pulsations	2×10^{-6}	2×10^{-6}	2×10^{-6}
Differential detector temperature coefficients	5×10^{-6}	5×10^{-6}	—
Mobile phase temperature coefficient	$<6 \times 10^{-6}$	$<6 \times 10^{-6}$	$<6 \times 10^{-6}$
Johnson or readout noise	8×10^{-6}	1.5×10^{-7}	4×10^{-5}
Dark current	1.8×10^{-7}	3.0×10^{-9}	$\sim1 \times 10^{-5}$
Total	1.4×10^{-5}	8.2×10^{-6}	4.6×10^{-5}

"Predicted values in absorbance units (reference 41).

wavelength and the higher throughput of a filter compared with a variable wavelength monochromator; the shot noise estimates are based on the collection of 1 and 57 μW, respectively.

The figure for noise due to pump pulsations is based on data where the actual variations in pressure have been minimized and the beam passes through the cell without touching the sides. Should the beam strike the sides, light piping occurs, which is very sensitive to pressure modulation.

The temperature coefficient for UV-enhanced photodiodes is about 0.02 to 0.2% quantum efficiency per degree celsius; use of a matched pair with a common heat sink can reduce the differential temperature coefficient to the order of the values listed. The variations due to the temperature coefficient of typical mobile phases will increase as the wavelength decreases, but the figure listed is valid at about 250 nm.

Estimates for an instrument using a photodiode array detector assume a D_2 lamp and a single-beam system. The higher figure for shot noise follows from the lower collection efficiency of each single diode. The readout noise results from a variable amount of charge injected into the signal by the inherent capacitance of the solid-state (jFET) switches when each diode is read. Although that source of noise is eliminated in the design of charge-coupled device arrays, commercially available versions of the latter have very poor UV response. One should quickly point out that these estimates do not allow for multiwavelength signal processing (e.g., integrating or curve fitting over wavelength) which could easily reduce the resultant noise by an order of magnitude.

This careful analysis suggests that where the separation is complete so as to provide a good baseline, HPLC can be very sensitive indeed.

2.4. Measurement and Compensation of Interferent
2.4.1. Blanks

An ideal blank is a solution whose preparation is identical to that of the unknown except that the species of interest in the determination is omitted. It can then be used as the reference in the absorbance measurement. The assumption then must be that any absorbing materials present in the sample other than the species of interest will also be present *in equal concentration* in the blank, and consequently the interferent absorbance will be cancelled. The difficulty lies in assuring that blank and sample differ only in the component of interest.

Although it may not always be possible, the best approach lies in a procedure which leads to a small blank because (1) the relative uncertainty in concentration due to instrumental factors will be lower since more signal is obtained from the detector, and (2) lower absorbance due to interferent concentration implies the *probability* of lower variation in interferent concentration. If the blank is normally low and is observed to increase on a given day, that change can be accomodated in principle, but it strongly suggests the presence of an uncontrolled parameter, which can lead to overall uncertainty.

2.4.2. Multiple Wavelength Measurement

If the spectral nature of the interferent is known, but the concentration varies, gathering information at more than one wavelength can readily lead to a multicomponent determination: a pair of simultaneous equations are solved for the concentrations when the absorbances and molar absorptivities are known at each wavelength. Three instrumental approaches have been suggested which can automate this possibility. The first two use information at two wavelengths; the last, based on the principle that if "two are good, more are better," uses the entire spectrum. These are considered separately below under the headings dual wavelength spectrophotometry, derivative spectrophotometry, and rapid-scan spectrophotometry.

2.4.3. Dual Wavelength Spectrophotometry

In its most common form, dual wavelength spectrophotometry (DWS) measures the *difference* in absorbance at two wavelengths.

$$\Delta A = A_1 - A_2 \tag{22}$$

The measured values are then

$$\Delta A = -\log(\phi_1/\phi_2) + \log(\phi_{0,1}/\phi_{0,2}) \tag{23}$$

where the subscripts 1 and 2 represent the two wavelengths. The second term represents a constant which can be determined with a reference solution. Then the first term can be determined from measurements involving the sample only; a ratiometric measurement, with its attendant advantages, is made without the need to modulate the beam between sample and reference cells, but instead between two wavelengths.

Substituting into Eq. (22) from Beer's law,

$$\Delta A = \epsilon_{1,a}bc_a + \epsilon_{1,b}bc_b - \epsilon_{2,a}bc_a - \epsilon_{2,b}bc_b \tag{24}$$

where subscripts a and b represent the species of interest and the interferent, respectively. If the wavelengths are chosen such that $\epsilon_{1,b} = \epsilon_{2,b}$, then Eq. (24) becomes

$$\Delta A = (\epsilon_{1,a} - \epsilon_{2,a})bc_a \tag{25}$$

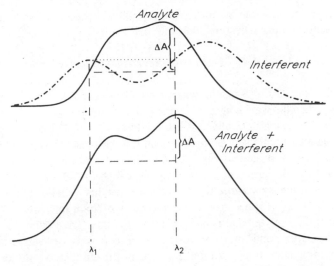

Fig. 15. Illustration of the choice of wavelength which allows a DWS measurement to be independent of interferent absorbance. The ΔA level is the same both with and without the interferent.

and the measured value is independent of the concentration of component b, as illustrated in Fig. 15.

One of the advantages of DWS measurement is that, since the beam path is nearly identical for both wavelengths, some of the errors related to cell quality and beam modulation are correlated to a large extent and are thereby nearly eliminated. Limitations to the identity of the two beams are due to chromatic aberrations in the optical path and to the wavelength dependence of scattering either on optical surfaces or in the sample due to a scattering interferent. The latter dependence ranges between λ^{-4} for particles much less than one wavelength in size and λ^{0} for very large particles.

DWS measurement is not without its attendant disadvantages. The uncertainties due to other factors besides cell flicker, as described in Section 1.4, are slightly elevated because both signals in the measurement of (ϕ_1/ϕ_2) are usually attenuated while the overall result, ΔA, is less than A. Therefore, a trade-off must be considered between wavelength-dependent errors such as the wavelength dependence for scattering which increases as $\Delta\lambda$ increases, and the measurement uncertainty which decreases as $\Delta\epsilon$ increases and therefore usually decreases as $\Delta\lambda$ increases (42,43).

This leads to the definition of conditions under which the application of DWS measurement for trace analysis might be useful. First, the interferent must be characterized sufficiently well that a pair of wavelengths of equal absorptivity can be chosen. Second, DWS is particularly valuable when the concentration of the interferent is variable so that a simple blank cannot be used, *and/or* the measurement circumstances are such that cell errors predominate.

2.4.4. Derivative Spectrophotometry

One special case of DWS results when the wavelength difference is sufficiently small that, when the wavelength is scanned, the result approximates a derivative.

The case for using derivative spectrophotometry can be easily seen by considering the measurement of an analyte band in the presence of a broader interferent band, as illustrated in Fig. 16 (44). In order to obtain a direct readout independent of the larger band, derivatives can be taken. A measure of the improvement is the *error ratio*, the ratio of the error that remains if the derivatives are used (1P or 2B) to the error if the uncorrected "total intensity" (TI) is used (Fig. 17). If the analyte band is narrower than the interferent band, the improvement is substantial, and improves further with increasing derivative orders.

This improvement is not without its cost; in the absence of further filtering, the SNR is degraded by roughly a factor of two for each stage of differentiation. Therefore, the trade-off between the error and the SNR would need to be carefully considered (45).

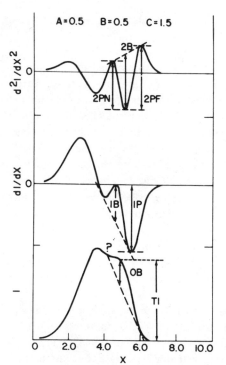

Fig. 16. From bottom to top, the zeroth, first, and second derivatives of a pair of overlapping peaks. Arrow OB shows the absorbance of the analyte peak compared with TI, the total intensity of analyte plus interferent. Arrows 1B and 1P in the first derivative curve, and 2B, 2PN, and 2PF, show ways in which the absorbance derivatives might be measured. (Reprinted from reference 44 by permission of Plenum Press.)

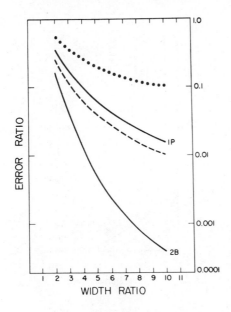

Fig. 17. The ratio of the error when the derivative measurement is made to the error when the total intensity (TI) measurement is made as a function of the interferent peak width to the analyte peak width. (Reprinted from reference 44 by permission of Plenum Press.)

Recently, the preferred method of taking the derivative has probably shifted to direct calculation (46). Earlier instruments relied on a myriad of mechanical techniques for wavelength modulation (44) or analog differentiators for differentiating the signal (47). The ubiquitous microprocessor, together with a modified least-squares algorithm (48), has simplified the computed approach; using simple weighting integers, the least-squares fit of a polynomial is calculated and differentiated, yielding a spectrum which is simultaneously smoothed and differentiated.

An example of the application of second-derivative measurements is illustrated in Fig. 18. In the determination of salicylic acid in aspirin powder, the salicylic acid is obscured by the overwhelming concentration of acetylsalicylic acid (49). However, the second derivative spectrum yields a distinct peak for the salicylic acid which can be used for the determination with good precision.

The use of derivative techniques in luminescence spectroscopy is discussed in Section 4.1 in Chapter 2.

2.4.5. Rapid-Scan Spectrophotometry

The objectives of DWS measurement are limited by access to only two wavelengths, largely an experimental problem during the development of the technique. If it were possible to obtain more information without loss of time, even more correlation of measurement errors would be possible.

The development of solid-state detector arrays has enabled this possibility by

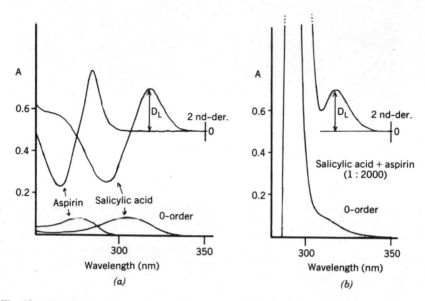

Fig. 18. Use of derivative spectra to determine salicylic acid in aspirin. (*a*) The zeroth and second derivatives of both materials are shown. (*b*) The equivalent spectra of the mixture in a 1:2000 ratio are shown. (Reprinted with permission from Kitamura and Majima, *Anal. Chem.,* **55,** 54. Copyright 1983 American Chemical Society.)

use of instruments such as those shown in Fig. 19. Using what is sometimes termed reverse optics, white light is passed through the sample to the dispersive device, a grating-based polychromator. The position of the detector array allows each pixel to obtain information at a separate wavelength *simultaneously*. Consequently, a spectrum with a width of 100 to 400 nm is obtained by the instrument in from 0.01 to 0.1 s. These spectra are normally signal-averaged for routine spectrophotometry so that the entire spectrum becomes available in about a

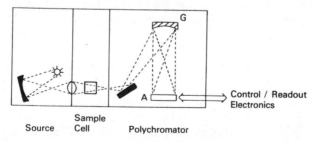

Fig. 19. A block diagram of a simple spectrophotometer based on an imaging array detector. The reflective element G is a concave grating; A is a linear imaging array detector.

second, roughly the time required for a single-wavelength absorbance value when a conventional instrument is used.

In contrast with other spectroscopies, absorption measurements can be made with high light levels since the level is determined primarily by the source and the polychromator throughput. Therefore, although the active area of a single pixel is much smaller than that of other detectors, higher source light levels and integration of the signal on the detector can compensate to minimize square-root uncertainty. The gain of the photomultiplier tube is not necessary for high-precision measurements. Hence, this instrument can easily have a signal-to-noise ratio comparable to that of conventional instruments.

Some loss in accuracy can be expected in a rapid-scan instrument, primarily at high absorbance; stray light becomes a more difficult problem since the conventional solution, double monochromators, cannot be used. However, stray light error is minimal at low absorbance (Section 1.2.6.).

Another source of error results from the optical arrangement. Since the entire source output is simultaneously incident on the sample, the possibilities of photodecomposition and fluorescence are greatly increased.

Simultaneously with the development of detector arrays came the development of intelligent signal processing, allowing sophisticated methods of error compensation to be automated, on-line with the experiment.

In principle, multicomponent measurements require an independent datum for each component. Considering, as in Eq. (24), the species of interest and the interferent to be two separate components, the minimum number of wavelengths is two. Having data from more wavelengths allows either the number of components to be increased or the precision of two-component measurements to be improved. In this system, Beer's law can be written

$$\mathbf{A} = \mathbf{X}\boldsymbol{\kappa} + \mathbf{E} \qquad (26)$$

where $\mathbf{A}$ is the matrix containing the absorbances at each wavelength, $\mathbf{X}$ is a two-dimensional matrix containing absorbances for each individual component at each wavelength, and $\boldsymbol{\kappa}$ is the concentration/path length product (ϵb). $\mathbf{E}$ is a matrix containing the random error in the measurement at each wavelength. The best estimate of the values for the $\boldsymbol{\kappa}$ matrix, $\mathbf{k}$, can be found by the method of least squares, which, written in matrix form, is

$$\mathbf{k} = (\mathbf{X}'\mathbf{W}\mathbf{X})^{-1}\mathbf{X}'\mathbf{W}\mathbf{A} \qquad (27)$$

$\mathbf{W}$ is the matrix of weights, proportional to the variances, which can be determined from the values of the absorbance at each wavelength using basic SNR theory (50).

The interferents to be included in the data treatment may either be molecular species whose reference spectra are available or scattering materials whose scattering coefficients are known (λ^{-n} where $0 < n \leq 4$). The wavelength dependence of uncertainty due to cell flicker can be approximated to be linear.

The consequence is that cell flicker uncertainty is reduced by at least an order of magnitude, and since other sources of uncertainty can be lowered by signal averaging, absorbance measurements can be made with very high precision at trace levels.

An interesting by-product of this computation is that since the variance of absorbance measurements becomes large at high absorbance, the weighting matrix in Eq. (27) causes the computation to discriminate against data prone to stray light error. Therefore, the dynamic range is increased at the high end of the calibration curve as well as at the low end.

A limitation of the technique described in the previous discussion results from the fact that, frequently, components exist for which the model spectrum is not available. Application of a linear least-squares method would yield negative values for some components in order to obtain a best fit.

In this case, it may still be possible to obtain an approximation of one or more of the components. One approach is the use of constrained least-squares and simplex optimization (51). However, Kowalski (52) and others (53) warn that serious errors can be masked by such a technique if applied before the unconstrained method.

Factor analysis (54) and its variants (55) can be applied without necessarily requiring the model spectra if the analyte spectrum can be at different concentrations relative to that of other components.

The use of rapid-scan detectors for multicomponent and total luminescence analysis and appropriate algorithms are presented in Chapter 2, Section 4.6.

2.4.6. *Reaction Rate Methods*

In conventional spectrophotometry, a reaction is usually required to produce the measured chromophore, and this reaction is allowed to move to equilibrium before the measurement is made. Reaction rate methods (56) depend on measuring the *rate* at which the reactions move toward equilibrium, and since these reactions for molecular analytes typically require an hour or more to equilibrate, a great savings in time can accrue.

Besides the time saved, there are other benefits which result from the use of rate methods: several types of interferences can be reduced. First, since the measurement is *relative*, monitoring the *change* in absorbance, any constant offset is eliminated; a constant absorbance or scattering error does not affect the accuracy. Second, if there is a competing reaction, either producing absorbance independent of the analyte or reducing absorbance as the product decomposes, it is likely to be slower or faster than the reaction of interest and will have only a small effect on the initial rate.

The application of rate methods is, unfortunately, not a general answer to spectrophotometric problems. A comparison of SNR for rate methods and equilibrium methods shows the latter to be greatly superior in most cases; the change in absorbance during the analytically useful initial portion of the reaction is too small for optimum precision to be possible (57). Catalytic methods involving metal ions have been found very sensitive in trace metal analysis (58), but these reactions are less common in molecular reactions. However, enzyme-linked assays, now being developed in clinical chemistry, are based on this principle.

2.4.7. Standard Additions

The method of standard additions (SA) (also called *known additions* or *spiking*) in its simplest form entails the measurement of a response for a sample after which, in lieu of reference or standard solutions, a known concentration increment is added; then, the response is measured again. The technique is of value when the solution conditions are not readily reproducible in reference media (59,60).

The SA technique in its simple form attacks a different class of interferent than is dealt with by previously considered techniques. While those methods are used when the interferent produces an *additive* error in the signal, the SA method finds its place when the error is *multiplicative*; the interferents generate a change in the *slope*. In particular, the SA method is necessary when the effect of competing equilibria involving the analyte cannot be eliminated. This type of interferent produces changes in the slope of the calibration curve without affecting the intercept. Other errors which may change slope might be a wavelength error (which changes ϵ) and pH or temperature errors.

The SA method requires that the following criteria be satisfied (61):

(1)   The technique must have a linear or otherwise known response with concentration over the concentration range encountered.

(2)   The chemical system must be free of deviations from linearity over the concentration range; such factors as ionic strength and pH must be sufficiently controlled so that interfering equilibria are not shifted by the addition.

(3)   The absorbance at zero concentration must be known; generally, it will be zero.

Within the constraints stated, the SA method represents an effective method for dealing with competing interferents. For trace methods, a large increment is necessary to obtain good precision; the increment should be large enough to place the resultant absorbance in a high precision range, usually about one. Care must be taken that condition (2) above is satisfied; to accomplish this, a series of additions is often useful.

The principal limitation of the SA method is clearly in condition (3). Like the other multiwavelength approaches mentioned previously for compensating *additive* interferents, an SA method with multiple sources of information is required. Kowalski and co-workers have suggested the generalized standard addition method and variants on it, providing methods of experimental design to determine how additions of various components should be made (62–64).

3. SPECIAL INSTRUMENTAL TECHNIQUES

Absorption spectrophotometry, largely unchanged over the past decade, has been augmented by new methods that are also based on the absorption of UV–visible radiation (65). These methods either require or are optimized by the use of lasers. The complications in using tunable lasers may be the limiting factor in their acceptance as general-purpose techniques at present, but spectacularly low levels of detection have been demonstrated.

These high-sensitivity methods can be divided into two groups: *transmission* methods and *calorimetric* methods. Transmission methods include methods in which the measured parameter is the light *not* absorbed by the sample, and include variants of conventional transmission measurements. In calorimetric methods, an effect is measured which results from the heat produced when the energy absorbed by the molecule is released to the solvent in a nonradiative (nonfluorescent) relaxation.

3.1. Laser Intracavity Absorption

The lasing action of a dye laser depends on the availability of enough radiant power in the resonating cavity to exceed a threshold. If the power falls below the threshold, stimulated emission is extinguished so that the slope of a plot of the relationship between input power and output power is very steep. Consequently, if the laser operates close to the threshold, a very small absorbance within the cavity will have a great effect on the output of the laser. Even above threshold level, where laser stability is improved, the change in radiant power is greatly enhanced over the change expected from simple absorption. For example, the change in output power when intracavity absorption changes from 9.9 to 10.1% can be nearly 100% (66).

Shirk and co-workers (66) have developed an analytical instrument capable of measuring an equivalent absorbance of 5×10^{-5} using the single-beam instrument shown in Fig. 20. The dye laser is pumped by an argon ion laser. Both

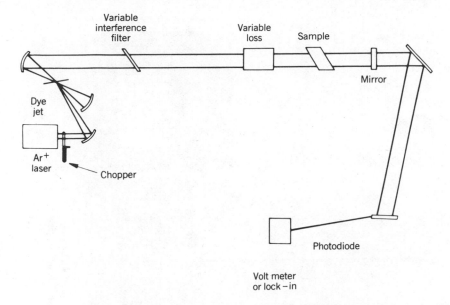

Fig. 20. Schematic diagram of the laser intracavity absorbance spectrophotometer. (Reprinted with permission from Shirk et al., *Anal. Chem.* **52**, 1701. Copyright 1980 American Chemical Society.)

the sample and an electrooptic cell are placed within the cavity, the latter acting simply as a programmable absorber. In order to maintain highest laser stability, the output power, detected by a photodiode, is kept constant by the optical null approach: the electrooptic device is controlled so that the attenuation of the sample and electrooptic device together remains constant.

Three factors finally determine the ultimate detection limit of this instrument. The first is the stability of the laser, and very careful attention must be paid to the optics and the dye jet in order to obtain reproducible operation. The second is the formation of a thermal lens; the slight local heating due to absorption of light energy leads to nonuniformity in the density of the solution, and the beam optics are affected. By chopping at about 130 Hz with a 40:1 off:on duty cycle, the thermal effect is not allowed to build up, and thermal lensing can be avoided. Where these factors are well controlled, the ultimate limit was said to be limited by the techniques for introducing the sample into the cell, as is often the case for other absorption methods.

3.2. Photoacoustic Spectroscopy

Photoacoustic spectroscopy (PAS) (67–69) is a *calorimetric* technique based on the detection of perturbations in the environment around a sample after absorbed energy is converted to heat. Consider the arrangement in Fig. 21. When a

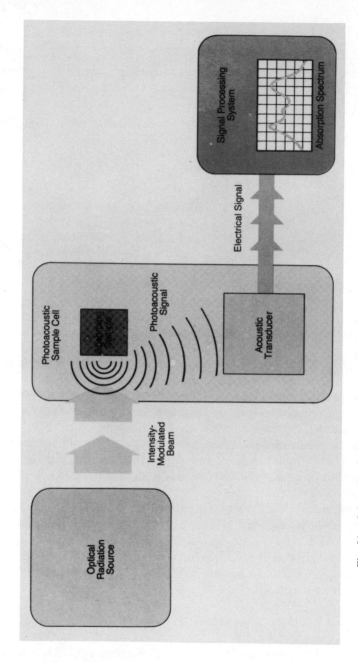

Fig. 21. Schematic of a photoacoustic spectrometer. (Reprinted with permission from Harris and Dovichi, *Anal. Chem.*, **52**, 695A. Copyright 1980 American Chemical Society.)

modulated light source, such as a chopped continuous source or a pulsed laser, is focused on a sample, the energy that is absorbed will subsequently be either re-emitted (fluorescence) or converted to heat. If the sample is itself a gas or is contained within an enclosed gas cell, those heat pulses generate sound waves as the gas expands at the chopping frequency of the incident light. An appropriate transducer, such as a microphone, generates the signal related to the absorbance spectrum. Hence the name *photoacoustic*.

Several experimental advantages have led to a rapid increase in PAS application over the past decade. The first is the simplification of certain parts of the instrumentation, feasible since the signal is acoustic, not optical; also, scattered light is less of a problem. Second, since the detected signal is acoustic, a single detector covers a wide wavelength range. Third, the measurement differs from that of absorbance in that it is not a measurement of a small difference between two large signals, but of a signal which increases with absorption. Fourth, this technique is suited to determinations in solids. Reduced sample preparation may also be a benefit.

The system diagrammed in Fig. 21 can be used for measurements both in gases and solids, although PAS is relatively insensitive in liquids (65). In solids, the gas serves as a transmitting medium for the detector; the thermal energy from the sample generates acoustic waves in the gas which are transmitted to a capacitative microphone.

In a liquid, the pressure waves can be measured directly using a pressure-transducing device: a piezoelectric detector (70). Significantly higher power is required in the source, and pulsed or modulated lasers are therefore commonly used. For a pulsed system, the phase of the pressure wave with respect to that of the incident beam can be used to distinguish a signal representative of the portion of the beam closest to the detector. The consequences are that (1) absorption at the cell windows can be rejected, and (2) the signal is independent of path length.

An improved method of coupling the cell to the piezoelectric disc detector (71) has led to detection limits comparable to an absorbance below 10^{-5}.

This system can be used in flowing systems (Fig. 22) by raising the modulation rate up to 4 kHz (72); at lower rates, the background acoustic noise in the flow would mask the photoacoustic signal. The detection limits for laser-induced PAS correspond to 8×10^{-6} absorbance units.

Some problems may arise when scattering is encountered; the scattered energy may strike the microphone surface or another surface acoustically coupled to it, dramatically increasing the signal. A platinum membrane over the detector surface (73) can reduce that type of error. However, the energy absorbed by the scattering material does generate its own signal, so that PAS can be used in place of turbidimetry with good results (74).

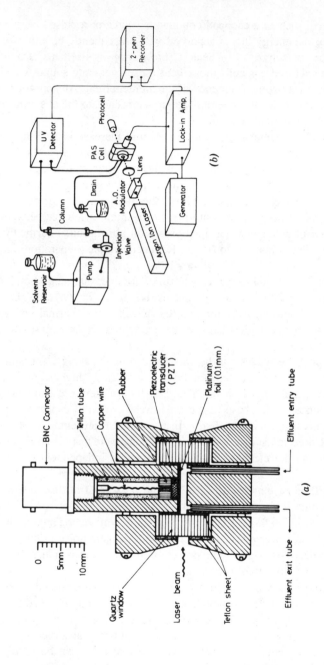

Fig. 22. (*a*) Schematic diagram of a flow cell and detector for photoacoustic spectroscopy. (*b*) Schematic diagram of the instrumentation arrangement for photoacoustic spectroscopy with a laser-induced signal in a flow cell. (Reprinted with permission from Oda and Sawada, *Anal. Chem.*, **53**, 471. Copyright 1981 American Chemical Society.)

46

3.3.   Thermal Lens Spectrophotometry

The thermal lens effect was previously mentioned as a source of error in laser intracavity methods. However, it has itself been shown to be a very sensitive probe of absorbance (64).

The thermal lens effect arises from the nonradiative relaxation of molecules which have absorbed the incident radiation, and as such is applicable to all substances which do not efficiently fluoresce. When the energy is released, local heating of the solution produces local changes in the refractive index of the solution.

Since the intensity of an ideal laser beam is distributed across the width of the beam in a Gaussian relationship, the induced profile of the refractive index approximates that of a lens which changes the focusing characteristics of the beam. Since the thermal expansion usually decreases the index of refraction, the effect is that of a concave lens. As illustrated in Fig. 23, the cell is normally placed past the narrowest position (the waist) of a focused beam, and the induced lens usually causes the beam to further diverge.

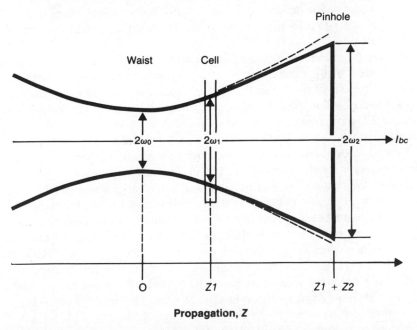

Fig. 23.   Optical arrangement for the thermal lens effect. The waist of the beam is at location O. The cell is located at Z1, one confocal distance from the waist. The pinhole through which the beam is measured is located at Z1 + Z2. (Reprinted with permission from Harris and Dovichi, *Anal. Chem.*, **52**, 695A. Copyright 1980 American Chemical Society.)

The amount of heat produced is dependent directly on the power of the beam and the energy absorbed. Also, in the absence of fluorescence, the heat generated is, at low absorbance, directly proportional to the absorbance.

Hu and Whinnery (75) have shown that for greatest precision and sensitivity, the cell is placed one confocal distance from the waist of the beam (Fig. 23). The intensity is typically sampled through a pinhole at the beam center where the intensity is denoted I_{bc}. Then the change in intensity due to the thermal lens, ΔI_{bc}, is related to the reference intensity I_{bc} by the equation

$$\Delta I_{bc}/I_{bc} = -\frac{2.303P(dn/dT)}{\lambda k} A \qquad (28)$$

where P is the laser power, dn/dT is the variation in refractive index with temperature, and k is the thermal conductivity. The latter two terms are defined by the solvent. This equation can, for a given system, be reduced to

$$\Delta I_{bc}/I_{bc} = 2.303PEA \qquad (29)$$

where E is defined as an enhancement factor. Typical values range from 0.09 mW^{-1} for water to 3.88 mW^{-1} for carbon disulfide (76) at 632 nm and room temperature.

Besides the enhancement factor, an advantage can be expected from the fact that the effect that is measured takes place away from the window, which is a source of serious error in conventional transmission measurement; the lens is formed totally within the cell and is independent of losses at the cell windows. Using a 175-mW argon-ion continuous-wave (CW) laser ($\lambda = 514.5$ nm), an equivalent absorbance of about 4×10^{-7} was measured.

Recently, several improvements have been made in the experiment to reduce measurement error and increase the signal. The first improvement entails separation of the heating function from the measurement function. In the earliest work, a CW laser was used as both the heating source to form the thermal lens and as the probe beam, whose intensity is measured. An improvement on this simple system is to use a pair of lasers. The first is the pump, a very strong heating source which creates the lens; the second is the probe, possibly a more stable laser (77). Since the probe beam is used only to measure the lens effect, it need not be at the absorption wavelength, and a common He–Ne laser is suitable. Such a system is shown in Fig. 24 (78). The heating source is a dye laser which can be turned to the absorption wavelength.

Second, a pulsed dye laser replaced the CW laser in order to obtain a modulated signal, which allowed better discrimination against noise sources; a significant advantage was seen in the enhancement factor (79). A reduced sensitivity

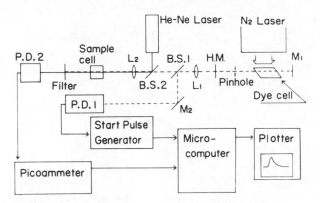

Fig. 24. Experimental arrangement for thermal lens spectrophotometry using a dye laser pump and a He–Ne laser probe. (Reprinted with permission from Miyaishi et al., *Anal. Chem.*, **54**, 2039. Copyright 1982 American Chemical Society.)

to sample flow was also observed, resulting in a broader range of applications (80).

A third improvement is obtained by monitoring the decay rate of the transient thermal lens produced by the modulated pump beam (81). In benzene, the time constant for the decay is about 50 ms, and a fit to the transient signal helps to reduce measurement noise.

Fourth is a technique which reduces a significant background signal which results from two-photon absorption in the solvent (82). The blank cell is placed before the waist and the sample cell after; any thermal lens effect common to both, as a result of the solvent, should be optically cancelled.

Improvement should be expected if a larger cross section of the beam is sampled. An optical imaging device is ideal for obtaining these data, and a linear photodiode array has been applied (83). Such an arrangement could make optimal use of the array since the detection system using the He–Ne laser always operates in the wavelength range for which the detector is most sensitive even when the absorbance takes place at much shorter wavelengths. Further improvements could be envisioned by using a two-dimensional array and correcting errors in the data treatment scheme (84).

Recently, Dovichi and co-workers have introduced a variant, dubbed laser-induced thermal refraction. As can be seen from Fig. 25, a pump beam is employed to create the lens that defocuses the beam of a probe laser (85,86). In this case, however, the two beams impinge on the sample at right angles; the thermal lens becomes a cylindrical rather than a simple lens. With this geometry, a spatial resolution on the order of 10^{-14} L is claimed.

Thermal lens spectrophotometry may well be the most sensitive of absorption

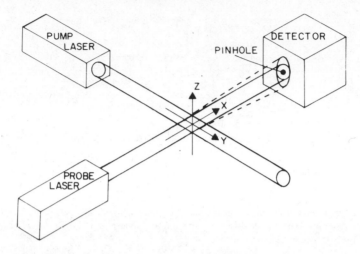

Fig. 25. Diagram illustrating the use of pump and probe lasers to make thermal lens measurements in small volumes. (Reprinted with permission from Dovichi et al., *Anal. Chem.*, **56,** 1700. Copyright 1984 American Chemical Society.)

spectrophotometric techniques presently on the horizon, with the realistic possibility of making measurements below 10^{-7} absorbance units.

REFERENCES

1. E. D. Olsen, *Modern Optical Methods of Analysis*, McGraw-Hill, New York, 1975.
2. S. F. Mason, *J. Chem. Soc.*, **1959**, 1263, (1959).
3. K. L. Cheng, in *Spectrochemical Methods of Analysis*, J. D. Winefordner, Ed., Wiley, New York, 1971.
4. E. A. Braude, *J. Chem. Soc.*, **1950**, 379 (1950).
5. *Pure Appl. Chem.*, **50**, 237 (1978).
6. *Pure Appl. Chem.*, **53**, 1913 (1981).
7. L. S. Goldring, R. C. Hawes, G. H. Hare, A. O. Beckman, and M. E. Stickney, *Anal. Chem.*, **46**, 181 (1953).
8. K. D. Mielenz, *J. Res. Natl. Bur. Stand., Sect. A.* **76**, 455 (1972).
9. P. Kubelka, *J. Opt. Soc. Amer.*, **38**, 448 (1948).
10. P. Kubelka and F. Munk, *Z. Tech. Phys.*, **12**, 593 (1931).
11. M. R. Sharpe, *Anal. Chem.*, **56**, 339A (1984).
12. S. Broderson, *J. Opt. Soc. Amer.*, **44**, 22 (1954).
13. H. A. Strobel, *Chemical Instrumentation*, second ed.; Addison-Wesley, Reading, Mass., 1973.
14. R. L. Cochran and G. M. Hieftje, *Anal. Chem.*, **49**, 2040 (1977).
15. W. Kaye, D. Barber, and R. Marasco, *Anal. Chem.*, **52**, 437A (1980).

16. W. Kaye, *Anal. Chem.*, **53**, 369 (1981).
17. Y. Talmi, *Anal. Chem.*, **47**, 658A (1975).
18. M. Denton, H. Lewis, and G. Sims, Analytical paper 98, *American Chemical Society Meeting*, Kansas City, September, 1982.
19. K. L. Ratzlaff and S. L. Paul, *Appl. Spectrosc.*, **33**, 240 (1979).
20. G. Horlick, *Appl. Spectrosc.*, **30**, 113 (1976).
21. H. L. Pardue, H. E. Hewitt, and M. J. Milano, *Clin. Chem.*, **24**, 1028 (1974).
22. P. C. Kelly and G. Horlick, *Anal. Chem.*, **45**, 518 (1973).
23. H. V. Malmstadt, M. L. Franklin, and G. Horlick, *Anal. Chem.*, **44(8)**, 63A (1972).
24. L. D. Rothman, S. R. Crouch, and J. D. Ingle, *Anal. Chem.*, **47**, 1226 (1975).
25. EU-700 Series Manual, GCA-McPherson, Acton, Mass., 1969.
26. J. D. Defreese and H. V. Malmstadt, *Anal. Chem.*, **48**, 1530 (1976).
27. K. R. O'Keefe, and H. V. Malmstadt, *Anal. Chem.*, **47**, 707 (1975).
28. H. L. Pardue and S. N. Deming, *Anal. Chem.* **41**, 987 (1969).
29. W. Kaye and D. Barber, *Anal. Chem.*, **53**, 366 (1981).
30. J. Stone, *Appl. Opt.*, **17**, 2876 (1978).
31. J. Stone, *Appl. Phys. Lett.*, **26**, 163 (1975).
32. K. Fuwa, W. Lei, and K. Fujiwara, *Anal. Chem.*, **56**, 1640 (1984).
33. L. A. Sternson, in *Chemical Derivitization in Analytical Chemistry*, R. W. Frei and J. F. Lawrence, Eds., Plenum Press, New York, 1981, Vol. 1, p. 127.
34. J. F. Lawrence, *Organic Trace Analysis by Liquid Chromatography*, Academic Press, New York, 1981.
35. W. L. Hinze, in *Solution Chemistry of Surfactants*, K. L. Mittal, Ed., Plenum Press, New York, 1979, Vol. 1, p. 79.
36. J. H. Callahan and K. D. Cook, *Anal. Chem.*, **56**, 1632 (1984).
37. W. A. Saner, in *Trace Analysis*, J. F. Lawrence, Ed., Academic Press, New York, 1982, Vol. 2.
38. A. Mizuike, *Enrichment Techniques for Inorganic Trace Analysis*, Springer-Verlag, New York, 1983.
39. F. W. Karasek, R. E. Clement, and J. A. Sweetman, *Anal. Chem.*, **53**, 1050A (1981).
40. P. C. White, *Analyst*, **109**, 677 (1984).
41. S. R. Abbot and H. Kelderman, Analytical paper 28, Philadelphia ACS meeting, August, 1984.
42. K. L. Ratzlaff and H. b. Darus, *Anal. Chem.*, **51**, 256 (1979).
43. K. L. Ratzlaff and D. F. S. Natusch, *Anal. Chem.*, **49**, 2170 (1977).
44. T. C. O'Haver, in *Contemporary Topics in Analytical and Clinical Chemistry*, D. M. Hercules, G. M. Hieftje, L. R. Snyder, and M. A. Evenson, Eds., Plenum Press, New York, 1978, Vol. 1.
45. T. C. O'Haver and G. L. Green, *Anal. Chem.*, **48**, 312 (1976).
46. T. C. O'Haver and T. Begley, *Anal. Chem.*, **53**, 1876 (1981).
47. G. Talsky, L. Mayring, and H. Kreuzer, *Angewandte Chemie*, **90**, 840 (1978).
48. A. Savitzky and M. Golay, *Anal. Chem.*, **36**, 1627 (1964).
49. K. Kitamura and R. Majima, *Anal. Chem.*, **55**, 54 (1983).
50. K. L. Ratzlaff, *Anal. Chem.*, **52**, 1415 (1980).

51. D. J. Leggett, *Anal. Chem.*, **49**, 276 (1977).
52. B. R. Kowalski, *Anal. Chem.*, **52**, 112R (1980).
53. J. B. Gayle and H. D. Bennett, *Anal. Chem.*, **50**, 2085 (1978).
54. P. C. Gillette, J. B. Lando, and J. L. Koenig, *Appl. Spectrosc.*, **55**, 630 (1982).
55. M. F. Delaney, *Anal. Chem.*, **56**, 261R (1984).
56. H. V. Malmstadt, C. J. Delaney, and E. A. Cordos, *CRC Crit. Rev. Anal. Chem.*, **2**, 559 (1972).
57. J. D. Ingle and S. R. Crouch, *Anal. Chem.*, **45**, 333 (1973).
58. E. B. Sandell and H. Onishi, *Photometric Determination of Traces of Metals: General Aspects*, fourth ed., Wiley, New York, 1978.
59. R. Klein, Jr. and C. Hach, *Am. Lab.*, **9(7)**, 21 (1977).
60. R. Klein, Jr. and C. Hach, *NBS Spcl. Publ.*, **464**, 61 (1977).
61. K. L. Ratzlaff, *Anal. Chem.*, **51**, 232 (1979).
62. B. E. H. Saxberg and B. R. Kowalski, *Anal. Chem.*, **51**, 1031 (1979).
63. C. Jochum, P. Jochum, and B. R. Kowalski, *Anal. Chem.*, **53**, 85 (1981).
64. J. H. Kalivas and B. R. Kowalski, *Anal. Chem.*, **54**, 560 (1982).
65. T. D. Harris, *Anal. Chem.*, **54**, 741A (1982).
66. J. S. Shirk, T. D. Harris, and J. W. Mitchell, *Anal. Chem.*, **52**, 1701 (1980).
67. A. Rosencwaig, *Photoacoustics and Photoacoustic Spectroscopy*, Wiley, New York, 1980.
68. Y. Pao, *Optoacoustic Spectroscopy and Detection*, Academic Press, New York, 1977.
69. J. F. McClelland, *Anal. Chem.*, **55**, 89A (1983).
70. W. Lahmann, H. J. Ludewig, and H. Welling, *Anal. Chem.*, **49**, 549 (1977).
71. E. Voigtman, A. Jurgensen, and J. Winefordner, *Anal. Chem.*, **53**, 1442 (1981).
72. S. Oda and T. Sawada, *Anal. Chem.*, **53**, 471 (1981).
73. S. Oda, T. Sawada, T. Moriguchi, and H. Kamada, *Anal. Chem.*, **52**, 650 (1980).
74. J. M. Harris and N. J. Dovichi, *Anal. Chem.*, **52**, 695A (1980).
75. C. Hu and J. R. Whinnery, *App. Opt.*, **12**, 72 (1973).
76. D. Solimini, *J. Appl. Phys.*, **37**, 3314 (1966).
77. M. E. Long, R. L. Swofford, and A. C. Albrecht, *Science*, **191**, 183 (1966).
78. K. Mori, T. Imasaka, and N. Ishibashi, *Anal. Chem.*, **54**, 2034 (1982).
79. A. J. Twarowski and D. S. Kliger, *Chem. Phys.*, **20**, 253 (1977).
80. N. J. Dovichi and J. M. Harris, *Anal. Chem.*, **53**, 689 (1981).
81. N. J. Dovichi and J. M. Harris, *Anal. Chem.*, **53**, 106 (1981).
82. N. J. Dovichi and J. M. Harris, *Anal. Chem.*, **52**, 2238 (1977).
83. K. Miyaishi, T. Imasaka, N. Ishibashi, *Anal. Chem.*, **54**, 2039 (1982).
84. J. T. Knudtson and K. L. Ratzlaff, *Rev. Sci Instrum.*, **54**, 856 (1983).
85. N. J. Dovichi, T. G. Nolan, and W. A. Weimer, *Anal. Chem.*, **56**, 1700 (1984).
86. T. G. Nolan, W. A. Weimer, and N. J. Dovichi, *Anal. Chem.*, **56**, 1704 (1984).

CHAPTER

2

TRACE ANALYSIS BY LUMINESCENCE SPECTROSCOPY

ROBERT J. HURTUBISE

Chemistry Department
University of Wyoming
Laramie, Wyoming

1. INTRODUCTION

As O'Haver (1) has discussed, the first recorded observation of fluorescence seems to have been in the sixteenth century by the Spanish physician and botanist Nicolas Monardes. Monardes mentioned a peculiar blue tinge in water which had been in cups made from a type of wood called *lignum nephriticum*. O'Haver (1) further traces the development of luminescence spectrometry as an analytical tool in terms of instrumentation and applications. Today molecular luminescence analysis is very widely used for organic trace analysis. It is not employed as extensively for inorganic species, but important advances have been made in the use of luminescent chelates for the trace analysis of inorganic ions.

Generally, luminescence is defined as radiation emitted when atoms or molecules undergo a radiative transition from an excited energy level to a lower energy level. The emission from an excited energy level produced by absorption of incident radiation is referred to as photoluminescence. Photoluminescence includes both fluorescence and phosphorescence. Another important type of luminescence is chemiluminescence. This type of luminescence arises from excited molecules or atoms that are produced from chemical reactions. Bioluminescence results if the reaction responsible for excitation is derived from a living system. Other types of luminescence include thermoluminescence (arises from the chemical reaction between reactive species trapped in a rigid matrix and released by raising the temperature), triboluminescence (mechanical excitation), radioluminescence (excitation results from radioactive decomposition), sonoluminescence (excitation by sound waves), and electroluminescence (electrical excitation). In this chapter, photoluminescence will be emphasized because it is most important for organic trace analysis. In addition, chemiluminescence and bioluminescence will be considered.

In the area of instrumentation, a variety of modern instruments and instrumental systems are used for luminescence analysis work. Capabilities of these instruments and systems range from simple collection of luminescence data to sophisticated data processing and handling. A variety of commercial instruments are available, from filter fluorometers to modular photon counting systems. Several research luminescence instruments have combined the latest laser and computer technology.

Applications in luminescence analysis abound. One finds applications in clinical analysis, drug analysis, air and water pollution analysis, biological and medical analysis, enzyme analysis, chromatographic analysis, industrial analysis, forensic analysis, agricultural analysis, immunochemical analysis, polymer analysis, and other areas. Luminescence analysis owes its popularity to several factors. Two notable factors are its sensitivity and selectivity. Also, one can obtain useful analytical luminescent signals under a variety of conditions, which adds to its versatility. For example, signals have been obtained for gas-phase components, materials in solution at room temperature and at liquid nitrogen temperature, compounds absorbed on solid surfaces, and components stabilized in micelle systems.

2. THEORY

2.1. General Considerations

The theory of molecular luminescence has been discussed comprehensively in the literature (2–7). In this section, only the fundamental theoretical aspects of molecular luminescence will be considered.

The majority of organic compounds that fluoresce or phosphoresce are aromatic types. However, a few highly unsaturated aliphatic compounds with extensive π systems also yield luminescence. The occupied orbitals of most organic molecules in the ground state each have a pair of electrons. The Pauli exclusion principle states that two electrons in an orbital must have opposing spins and thus the net spin for most ground-state molecules is zero. A molecule in this state, in which the electron spins are paired, is said to be in a singlet state. However, when an electron is promoted in a molecule, by absorption of ultra-violet or visible electromagnetic radiation, to an upper energy level, its spin is either in the same direction or opposed to the electron spin of the other electron in the orbital. For situations where the spins are in the same direction, the resulting state is known as a triplet state. The energy associated with electrons of opposing spins is greater than with electrons of parallel spins (Hund's rule).

2.2. Processes Involved in Excited States

The processes important in luminescence analysis are shown in Fig. 1 by a molecular energy-level diagram. When a molecule absorbs electromagnetic radiation in the ultraviolet or visible region, it is normally promoted to a vibrational

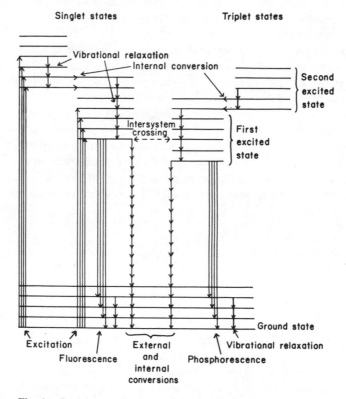

Fig. 1. Partial energy level diagram for a photoluminescent molecule.

level in an excited singlet state. The intensity of an electronic absorption band is a measure of the probability of an electronic transition between two states. The absolute intensity of an absorption band may be determined from the integrated absorption spectrum. However, molar absorptivity ϵ can be used as a measure of the probability of absorption (5). An ϵ value of 10^5 at a given wavelength, for example, implies a high degree of probability that the electronic transition will occur. The time for a molecule to pass from the ground state to an electronically excited state is about 10^{-15} s. Because this interval is so short, the atomic nuclei do not appreciably change positions. In other words, electron transitions occur without change of position of the nuclei. This is a statement of the Franck–Condon principle.

Immediately after excitation, a molecule has the same geometry and is in the same environment as it was in the ground state. Upper vibrational levels for an electronic state relax vibrationally in approximately 10^{-12} s, and within this time period the excited solute molecule relaxes to the lowest vibrational level of the lowest excited singlet state. In solution at room temperature, the solvent molecules reorient themselves to a state of equilibrium compatible with the new molecular polarity. The vibrational and solvent relaxation processes are accompanied by a loss of thermal energy. A radiative transition can then occur from the lowest vibrational level of the lowest excited singlet state, dropping the solute molecule into one of a number of possible vibrational levels of the ground electronic state. The radiative transition is called fluorescence and almost always occurs from the lowest vibrational level of the lowest excited singlet level in a molecule. The decay time of fluorescence is the same order of magnitude as the lifetime of an excited singlet state, namely, 10^{-9} to 10^{-7}. Based on spectroscopic selection rules, the transition from an excited singlet to a singlet ground state is spin-allowed and thus highly probable. After fluorescence occurs, vibrational relaxation and solvent reorientations occur and the molecule finally arrives in a ground-state equilibrium configuration.

As indicated in Fig. 1, other processes can compete with fluorescence. The excited molecule can lose energy by other mechanistic paths such as internal conversion and intersystem crossing. Internal conversion is a radiationless process whereby a molecule passes from a higher to a lower electronic state without emission of a photon. After internal conversion occurs, it is followed immediately by vibrational relaxation to the lowest vibrational state of the lower electronic level. A radiationless transfer occurring between the first excited singlet state to the lowest excited triplet state is known as intersystem crossing. Intersystem crossing involves a change in spin, and this means that spin selection rules are not obeyed rigorously. However, intersystem crossing can occur because of spin–orbital coupling between an excited singlet and an excited triplet state. If intersystem crossing is favored over fluorescence or internal conversion to the ground state, the molecule can pass from the lowest singlet state to a triplet state. Then the molecule will undergo vibrational relaxation to arrive at the lowest vibrational level of the lowest excited triplet state. From this state, electromagnetic radiation can be emitted, or internal conversion to the ground state can occur. The radiative transition is called phosphorescence. Because phosphorescence originates from the lowest triplet state, its decay time is similar to the lifetime of the triplet state, approximately 10^{-4} to 10 s. The transition from a triplet excited state to a singlet ground state is a forbidden transition because it involves a change in spin. Because phosphorescence is a spin-forbidden process, the rate of phosphorescence is relatively slow. This fact gives a triplet electronic state its long lifetime. For compounds in solution at room temperature, the long lifetime greatly increases the probability of collisional transfer to energy with solvent molecules or impurity molecules. This process is very efficient in solution at room temperature and is often the main pathway for loss of triplet-state excitation energy. Because of this, regular solution phosphorescence is rarely observed at room temperature unless special, and sometimes extreme, measures are taken to slow down the rates of competing processes. Analytically useful solution phosphorescence can be observed by dissolving the solute in a solvent that freezes to form a rigid glass at the temperature of liquid nitrogen (77 K). A solvent mixture consisting of ethyl ether, isopentane, and ethyl alcohol in a ratio of 5:5:2 can be used as a rigid glass medium. However, recent analytical advances have allowed phosphorescence to be obtained readily for several compounds at room temperature at the nanogram and subnanogram levels (8). These analytical breakthroughs will be discussed later.

2.3. Excitation and Emission Spectra

Fluorescence and phosphorescence emission spectra are both recorded by measuring the luminescence intensity as a function of wavelength at a fixed excitation wavelength. In contrast, an excitation spectrum is recorded by measuring luminescence intensity at a fixed emission wavelength while varying the excitation wavelength. Figure 2 shows examples of typical room-temperature excitation and fluorescence emission spectra and a low-temperature phosphorescence emission spectrum for benzo[f]quinoline. Because the energy transitions in fluorescence and phosphorescence are smaller than those involved in excitation, the luminescence spectra will occur at longer wavelengths than the wavelengths associated with the excitation spectrum. Because the lowest triplet state is at lower energy than the first excited singlet state, phosphorescence occurs at longer wavelengths than fluorescence (9).

The spectra in Fig. 2 are rather broad due to vibrational transitions superimposed on the electronic transitions and perturbations caused by the solvent. Generally, the more polar the solvent the greater the loss in vibrational detail. In the gas phase, more vibrational structure appears unless the gaseous material

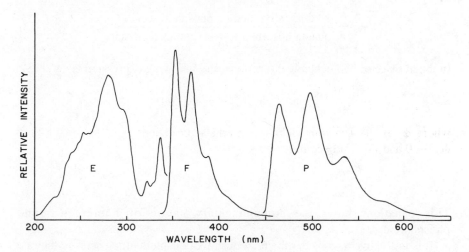

Fig. 2. Excitation and emission spectra for benzo[*f*]quinoline in ethanol: E, excitation; F, fluorescence; and P, phosphorescence.

is under high pressure. For solid phases which are highly ordered, spectra tend to be more structured than those from solution.

Normally, all excited states produced by absorption of ultraviolet or visible radiation relax to the lowest vibrational level of the first excited singlet state. This means that the excitation spectrum should match the electronic absorption spectrum for a compound. For the match to be observed experimentally, it is necessary to correct for the variation of source intensity with wavelength and change in excitation monochromator spectral characteristics with wavelength. Many of these aspects will be considered later.

2.4. Luminescence Efficiencies, Lifetimes, and Polarization

The efficiency of luminescence can be discussed in terms of quantum efficiency (quantum yield). The fluorescence efficiency ϕ_f is defined as follows[1]:

$$\phi_f = \frac{\text{quanta emitted as fluorescence}}{\text{quanta absorbed to a singlet excited state}} \tag{1}$$

Phosphorescence quantum efficiency is defined in a similar fashion:

[1]Note that in Chapter 1 we employed the symbol ϕ for radiant flux as recommended by the Commission on Spectrochemical Analysis of the International Union of Pure and Applied Chemistry. In this chapter, we employ the symbol I for luminescence intensity and ϕ represents luminescence efficiency, as commonly practiced.

$$\phi_p = \frac{\text{quanta emitted as phosphorescence}}{\text{quanta absorbed to a singlet excited state}} \tag{2}$$

In the absence of photochemical reactions, the following expression is valid (3):

$$\sum \phi_i = \phi_f + \phi_p + \phi_{ic} = 1. \tag{3}$$

where ϕ_{ic} is the quantum yield for internal conversion. If ϕ_{ic} is small, then $\phi_{ic} \to 0$ and Eq. (3) becomes

$$\phi_f + \phi_p = 1 \tag{4}$$

This equation shows the complementary nature of fluorescence and phosphorescence.

It is instructive to consider the relationship of quantum efficiencies to the rate constants of the various excited-state processes (3,5). Figure 3 shows a summary of the important rate constants for excited state processes and Eq. (5) and (6) define the fluorescence and phosphorescence quantum efficiencies in terms of the rate constants.

$$\phi_f = \frac{k_f}{k_f + k_c + k_x} \tag{5}$$

$$\phi_p = \left(\frac{k_p}{k_p + k_c'}\right)\left(\frac{k_x}{k_f + k_c + k_x}\right) \tag{6}$$

These equations illustrate the competitive nature of the excited processes. For example, if $k_f \gg k_c$ and k_x, then the fluorescence efficiency will be high and approach unity. The quantum efficiency of phosphorescence, as indicated in

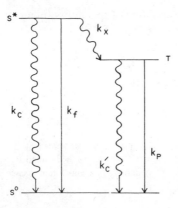

Fig. 3. Summary of rate constants for excited-state processes: S_0, ground state; S^*, lowest excited singlet state; T, lowest triplet state; k_c, internal conversion for lowest excited singlet state; k_c', internal conversion between lowest triplet state and ground state; k_p, phosphorescence; k_x intersystem crossing. Wavy lines indicate nonradiative transitions. (Reprinted with permission from D. M. Hercules, Ed., *Fluorescence and Phosphorescence Analysis*, Wiley, New York, 1966.)

Eq. (6), is defined by a more complicated equation. The quantum efficiency depends on competition between internal conversion from the lowest triplet state and phosphorescence emission, and it also depends on the rate of intersystem crossing relative to fluorescence and internal conversion from the excited singlet state (Fig. 3). Generally, k_f and k_p are dependent on molecular structure and are affected slightly by the environment. The term k_x is dependent on molecular structure, but is also a function of the environment. The k_c' and k_c terms are strongly dependent upon the molecular environment and to some extent on the nature of the electronic states in the molecule. It is important to realize that fluorescence and phosphorescence are only two of many processes that take place after molecular excitation. Thus, the structural and environmental changes that affect rate constants other than k_f or k_p can have a tremendous effect on ϕ_f or ϕ_p (5). Demas and Crosby (10) have reviewed extensively the measurement of photoluminescence quantum yields, and Bridges (11) has considered some recent developments in the determination of fluorescence quantum yields.

Another important parameter in describing luminescence phenomena is luminescence lifetime. Experiments have shown that the fluorescence intensity, for example, decays after the removal of the exciting source by a first-order rate equation, as indicated in Eq. (7),

$$I = I_0 e^{-t/\tau} \tag{7}$$

where I_0 is the initial intensity, I is the intensity at some time t, and τ is a constant. When t is equal to τ, the fluorescence intensity has fallen to $1/e$ of its initial value. By definition, τ is the mean decay time for the fluorescence process or mean lifetime of the excited state. Experimental lifetimes reflect the overall rate at which the excited state is deactivated which includes both radiative and nonradiative processes. The intrinsic or natural lifetime τ_0 is obtained if fluorescence is the only mechanism by which the excited state returns to the ground state. The relationship between τ and τ_0 for fluorescence is $\phi_f \tau^0 = \tau$. For phosphorescence, the decay also follows the first-order rate equation. The intrinsic or natural lifetime τ_0 for phosphorescence emission is also defined in the same way as for fluorescence. However, the relationship between the quantum yields and intrinsic and observed lifetimes is quite different for phosphorescence compared to fluorescence. The intrinsic lifetime of phosphorescence is related to the observed lifetime as follows (3):

$$\tau_0 = \tau \left(\frac{k_x}{k_c + k_x} \right) \left(\frac{1 - \phi_f}{\phi_p} \right) \tag{8}$$

Is is often assumed that internal conversion occurs only from the triplet state. If this is true, then Eq. (8) simplifies to

$$\tau_0 = \tau\left(\frac{1 - \phi_f}{\phi_p}\right) \qquad (9)$$

Fluorescence lifetimes are approximately 1–20 ns; however, as discussed previously, phosphorescence lifetimes are substantially longer, ranging from milliseconds to seconds. More detailed discussions on luminescence lifetimes can be found in the literature (7,12).

Emission polarization can be used to establish the relative orientation of the emission transition dipole to that of a particular absorption band. Polarization arises because electronic transitions are characterized by transition moments, which have unique orientations with respect to molecular structure. For aromatic hydrocarbons, it is known that allowed transitions are in-plane polarized (3). Therefore, it is possible to determine something about the direction of the emission. It has been found that for several hydrocarbons, phosphorescence is polarized perpendicular to the plane of rings. The previous information is helpful in determining state assignments and the nature of the perturbing states. Figure 4 shows a typical experimental setup used in the measurement of polarization. The exciting radiation is polarized in the XZ plane. The resulting emission is resolved into $I_{\parallel}$, the intensity of the emission in the YZ plane, and $I_{\perp}$, the intensity of emission in the XY plane. The polarization p of emission is defined by Equation 10.

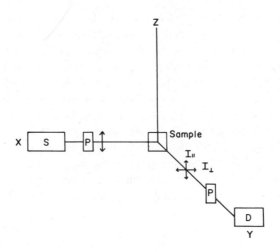

Fig. 4. Block diagram of apparatus for fluorescence polarization measurements: S, detection system; P, polarizer; $I_{\parallel}$, intensity of fluorescence polarized in same plane as excitation radiation; $I_{\perp}$, intensity of fluorescence in plane perpendicular to excitation radiation. (Reprinted with permission from W. R. Seitz, "Luminescence Spectrometry," in P. J., Elving, E. J. Meehan, and I. M. Kolthoff, Eds., *Treatise on Analytical Chemistry*, 2nd ed., Part I, Vol. 7, Sec. H, Wiley, New York, Chap. 4.)

$$p = \frac{I_\parallel - I_\perp}{I_\parallel + I_\perp} \tag{10}$$

The value of p can vary from 1 to -1. Natural light corresponds to $p = 0$ ($I_\parallel = I_\perp$). More detailed discussions on polarization can be found elsewhere (3,5,13).

2.5. Excited State Dissociation and Other Phenomena

The acid–base chemistry of the excited states in luminescent molecules is an important consideration in trace analysis for several reasons. For example, if a molecule loses a proton in the excited singlet state before fluorescence emission, then the emission properties of the deprotonated species could be quite different from those of the original molecule. It is well known that the excitation of a molecule from the ground electronic state to its lowest excited singlet state involves a change in the electronic dipole moment of the excited molecule. The excited molecule is usually very different from that of the molecule in the ground state, and thus the chemical properties of the molecule in the excited state often are very different from the chemical properties of the molecule in the ground state. The charge distribution between the ground and lowest triplet states of a molecule is the basis of most photochemistry. The triplet state is important because of its relatively long lifetime, which permits photoexcited molecules adequate time to encounter a potential reactant. The lowest excited singlet state has a much shorter lifetime than the triplet state, which lowers the probability of a fluorescent molecule encountering a reactant which would alter its fluorescent properties. However, chemical reactions such as proton exchange with acids and bases and association reactions with ions and molecules in solution can occur during the lifetime of the lowest excited singlet state. It is important for an analyst to be aware of the possible reactions of molecules in the excited state. The majority of luminescence methods in organic trace analysis do not involve complex organic photochemistry. Nevertheless, the more information one has about the excited state chemistry as it relates to such things as quenching and the effect of impurities on quantum yield, the smaller the possibility of erroneous data being obtained.

Schulman (14) has given a detailed discussion of the acid–base chemistry of excited singlet states. Three important situations can arise: (1) excited-state proton exchange is much slower than fluorescence in an acid or conjugate base, (2) excited-state proton exchange is much faster than fluorescence in an acid or conjugate base, and (3) excited-state proton exchange and fluorescence are of comparable rates. With the first situation, fluorescence will occur from the excited species unperturbed by excited-state proton transfer reactions. This means that the quantum yield of fluorescence from an excited species should be pH independent, and the intensity of the fluorescence will depend only upon the absorbance at the excitation wavelength and the concentration of the ground state species. Thus, the variation of fluorescence intensity of an acid or base will depend only on the thermodynamics of the ground state acid–base reaction. The variation of fluorescence intensity should parallel the absorption dependence on pH of each species.

In the second situation, the rate of dissociation of excited acid and protonation of excited conjugate base are much greater than the rates of their fluorescence, and prototropic equilibrium will be established in the lowest excited singlet state. In this case, it is the dissociation constant of the excited-state proton exchange which determines the fluorescence behavior of the species. Because the electronic distribution of an electronically excited molecule is different from the ground state, the excited state pK is usually very different from pK_a.

For the third situation, as an example, the rates of proton transfer between excited acid and solvent or between excited base and solvent can be comparable to the rates of deactivation of acid and conjugate base by fluorescence. The variations of the quantum yields of fluorescence of acid and conjugate base with pH will then be governed by the kinetics of the excited-state prototropic reaction.

The previous three cases have assumed that the only sources of proton donor and acceptor species in solution were H_3O^+ and OH^- from the solvent. However, many analytical measurements are performed with solutions containing buffer ions for pH control. The buffer ions can be proton donors or acceptors and thus may enter into proton-transfer reactions with excited molecules. The rate constants for the interactions of many common buffer ions with several fluorescent species have been shown to be comparable to the rate constants for the interaction of H_3O^+ and OH^- with the fluorescent species. Because the concentrations of buffer ions in buffered solutions are substantially greater than the concentrations of H_3O^+ and OH^-, the probability of reaction of excited species with buffer ions could be large. The presence of buffer ions in a sample could thus be a potential source of interference in a fluorometric analysis (14,15).

Jackson and Porter (16) determined the acidity constants of the lowest triplet state for several aromatic compounds by two methods. In the first method, the triplet state was populated by flash photolysis and the acid–base equilibrium was studied spectrophotometrically. In the second method, the difference between acidity constants in the triplet and ground states was calculated from the energy levels of the acid and base derived from phosphorescence spectra. Richtol and Fitch (17) have reported triplet state acidity constants for hydroxy and amino substituted anthraquinones and related compounds. Generally, there are far fewer reports of triplet-state acidity constants compared to the many reports of singlet-state acidity constants. As discussed earlier, the study of phosphorescence in fluid solutions is rare. The majority of phosphorescence work has been done in

Table 1. Excited-State pK_a Data^a

Compound	Reaction^b	pK(S$_0$)	pK(S$_1$)	pK(T$_1$)
2-Naphthol	Deprotonation	9.49	2.81	7.7
1-Naphthoic acid	Deprotonation	3.7	7.7	4.6
2-Naphthylamine	Deprotonation	—	12.2	3.1
Acridine	Protonation	5.45	10.3	5.6
9-Anthroic acid	Deprotonation	3.0	6.5	4.2

^{a}From Vander Donck (18).
^{b}Deprotonation means the pK_a is given for the ionization of the compound. Protonation means the pK_a is given for the protonated form of the compound.

rigid solutions at liquid-nitrogen temperature. Under these conditions, acid–base equilibrium cannot be quickly established after excitation, and phosphorescence has a relatively limited dependence upon pH in rigid solutions. Normally, the triplet-state acidity constants lie much closer to the ground-state acidity constant compared to an excited-singlet-state acidity constant. Vander Donckt (18) has reported several pK values for compounds in the ground state, lowest excited singlet state, and lowest excited triplet state. Several of these values are given in Table 1.

Apparently, there have been no reports of excited-singlet-state or excited-triplet-state acidity constants for compounds adsorbed on solid surfaces at room temperature. Also, it seems there have been no reports for triplet-state acidity constants for micelle-stabilized compounds at room temperature. This is an area of research that should be fruitful in yielding mechanistic information on these systems.

2.6. Delayed Fluorescence

The analyst should be aware that delayed fluorescence can occur in solution and may also occur on solid surfaces. Parker (19) has considered delayed fluorescence in solutions; however, very little work has been done on delayed fluorescence from compounds adsorbed on solid surfaces. Three different types of delayed fluorescence have been demonstrated. First, *E-type delayed fluorescence* occurs as follows. A molecule is excited to the first excited-singlet state followed by intersystem crossing to the triplet state. If the energy between the singlet and triplet states is small enough, it is possible that the singlet state can be repopulated by thermal excitation. Fluorescence emission can then occur from the singlet state. Because this process involves thermal excitation, the fluorescence emission is highly temperature dependent. Eosin and anthraquinone are examples of compounds which show E-type delayed fluorescence. A second type of delayed fluorescence has been designated *P-type delayed fluorescence*. Its intensity is proportional to the square of the intensity of the excitation source, indicating two photons are needed to produce the fluorescence. Two molecules are excited to the singlet state, each undergoing intersystem crossing to the triplet states. The two triplet states interact with each other to yield an excited singlet state and a ground state. The regenerated excited singlet state then gives the delayed fluorescence. This type of fluorescence has been observed for several aromatic hydrocarbons. The third type of delayed fluorescence occurs when the excited singlet state undergoes photooxidation with the ejection of an electron followed by recombination to regenerate an excited singlet state that emits fluorescence. This type of fluorescence has been called *recombination fluorescence*. An example of a compound showing this type of fluorescence is acriflavine (9,12).

2.7. Excited State Complexes

At relatively high concentrations, solute–solute interactions may occur in the lowest excited singlet state. The formation of a complex of an excited solute molecule with a ground-state molecule of the same type results in the formation of an excimer. The excimer can dissociate to two ground-state singlets accompanied by the emission of fluorescence.

$$^1M^* + {}^1M \rightarrow {}^1M_2^* \tag{a}$$

$$^1M_2^* \rightarrow {}^1M + {}^1M + h\nu \tag{b}$$

The excimer emission is observed at longer wavelengths than normal fluorescence because the excimer is lower in energy than the unassociated excited singlet state. In addition, the intensity of excimer emission is proportional to the square of the concentration of the fluorescent molecule. Pyrene is an example of a compound that gives excimer fluorescence. Parker (4) has considered excimers in some detail. Sawicki (20) has shown that excimer formation can occur on thin-layer chromatoplates. For example, the fluorescence spectra of pyrene (0.1 μg) and of benzo[*a*]pyrene (0.5 μg) on a cellulose chromatoplate showed excimer bands. When excited state complex formation occurs between two different solute molecules, it is said that an exciplex has been formed. Froehlich and Wehry (21) have given an interesting and detailed discussion of exciplexes. They also briefly considered triplet exciplexes and the use of phosphorescence spectroscopy to study these complexes. However, excimer and exciplex formations are usually observed with fluorescence phenomena because diffusion of the excited species is needed to form the excited state complex (15).

2.8. Other Phenomena

An analyst should be aware of the possibility of photochemical reactions in luminescence analysis work. Because relatively high energy sources are used, it is possible to induce photochemical reactions by overexposing the sample to the source radiation. A discussion of photochemistry is beyond the scope of this chapter; however, Turro (22) has discussed molecular photochemistry in considerable detail. Erratic results or changes in the excitation and emission spectra suggest the possibility of a photochemical reaction. Luminescence quenching is another important consideration, and quenching phenomena will be considered later in the chapter. Noncollisional energy transfer (sensitized luminescence) can occur over distances larger than the contact distances of molecular collision. Noncollisional energy transfer is not spin-forbidden and singlet–triplet, triplet–singlet, singlet–singlet, and triplet–triplet transfer can occur. Noncollisional energy transfer probably arises from a vibrational coupling interaction between the excited states of the donor and the acceptor. An example of noncollisional transfer is the triplet–triplet energy transfer between benzophenone as donor and naphthalene as acceptor. If both benzophenone and naphthalene are in solution at 77 K and excited at 365 nm, benzophenone absorbs the exciting radiation, undergoes intersystem crossing to its triplet state, transfers its excitation energy to naphthalene, and naphthalene phosphorescence is observed. The phosphorescence of naphthalene is not excited by 365-nm radiation (4).

2.9. Room-Temperature Phosphorescence

Analytical room-temperature phosphorescence (RTP) is a recent development in organic trace analysis. RTP can be obtained from several compounds by adsorbing the compound on certain solid surfaces, using micellar solutions, and in some cases by sensitized RTP in liquid solutions. The fundamental theoretical reasons for RTP have not been fully developed.

In 1967, Roth (23) listed the RTP detection limits of 18 compounds adsorbed on filter paper. The interest in RTP was most likely stimulated by reports from Schulman and Walling (24,25). In 1974, Paynter et al. (26) reported the range of linearity and detection limits for several compounds adsorbed on filter paper. The field of analytical solid-surface RTP has expanded substantially since their report. Recent reviews and reports on analytical solid-surface RTP can be consulted for details (27–30).

No general model has been developed to explain the interactions required for RTP from compounds adsorbed on solid surfaces. However, some general trends have emerged. In most cases, the adsorbed phosphor must be dried to enhance the RTP signal. The effects of moisture and oxygen on the RTP of sodium 4-

biphenylcarboxylate and sodium 1-naphthoate adsorbed on filter paper were investigated (31). From this work, it was proposed that hydrogen bonding of ionic organic molecules to hydroxyl groups of the filter paper is the primary mechanism of providing the rigid matrix for RTP. Presumably, the rigid matrix minimizes collisional deactivation of phosphor molecules in the triplet state. In addition, it was noted that moisture acts to disrupt hydrogen bonding and aids in the transport of oxygen, which is a quencher, into the sample matrix.

For p-aminobenzoic acid adsorbed on sodium acetate, strong RTP was observed (32). The main interactions proposed with this system were the formation of the sodium salt of p-aminobenzoic acid and hydrogen bonding of the amino group of the acid with the carbonyl group of sodium acetate. Niday and Seybold (33) have suggested that packing of filter paper with materials such as salts or sugars inhibits the internal motion of the phosphorescent compound. Further, the added compounds "plug up" the channels and interstices of the matrix and thus decrease the oxygen and moisture permeability. Hurtubise and Smith (34) reported relatively strong RTP from the dianion of terephthalic acid adsorbed on a mixture of the anion of polyacrylic acid and sodium chloride. With the previous system, hydrogen bonding was not possible. In addition, Bower and Winefordner (35) showed that several polycyclic aromatic hydrocarbons give RTP on sodium acetate and filter paper with a heavy atom present. Ramasamy and Hurtubise (36) used reflectance spectroscopy to provide evidence for the forms of benzo[f]quinoline and quinoline adsorbed on filter paper, silica gel chromatoplates, and a 0.5% polyacrylic acid–NaCl mixture. Infrared data suggested several hydrogen bonding interactions for benzo[f]quinoline and phenanthrene with polyacrylic acid–salt mixtures. With the reflectance and infrared data, a partial model for phosphor–solid surface interactions was developed for the compounds. Dalterio and Hurtubise (37) employed several spectral techniques to study the interactions of hydroxyl aromatics and aromatic hydrocarbons with polyacrylic acid–salt mixtures and filter paper and related the data to RTP. The results revealed hydrogen bonding of the model hydroxyl aromatics and aromatic hydrocarbons to the support materials. The individual phosphors acted as hydrogen donors, hydrogen accepting species, or as both hydrogen donors and hydrogen acceptors. Infrared data showed a net increase or decrease in the intermolecular and intramolecular hydrogen-bonding network of the solid supports with the addition of phosphors, suggesting specific geometric requirements for adsorption of the phosphor. Diffuse reflectance spectra and luminescence emission spectra indicated that RTP was emitted by the neutral hydroxyl aromatics adsorbed on the solid surfaces. It is clear from the literature that no single working model for solid-surface RTP has been developed.

Cline Love and co-workers (38,39) have developed the new analytical approach of solution micelle-stabilized RTP. Phosphorescence can be observed from many aromatic molecules in fluid solution at room temperature by incorporating the phosphor into a micellar system. Generally, a detergent concentration above the critical micelle concentration is used to ensure micelle formation, heavy atoms are employed, and an inert gas is used to deoxygenate the sample solution. Apparently the micelle can organize reactants on a molecular scale and increase the proximity of the phosphor and heavy-metal counterion. This would raise the effective concentration of the heavy metal and increase the probability of spin–orbit coupling. Other factors are also involved (38).

Phosphorescence intensities in liquid solutions are generally too low to be useful in analytical work. However, Donkerbroek et al. (40,41) recently investigated RTP of phosphors in liquid solutions without the use of micelles. They concluded that direct solution RTP of phosphors with intensities high enough to be useful in analytical chemistry is rather rare. A useful alternative is sensitized phosphorescence in solution at room temperature. As applied by Donkerbroek et al., sensitized phosphorescence can be described as follows: after excitation and before radiationless decay of the analyte, the analyte transfers its triplet-state energy to an acceptor molecule and then the acceptor molecule emits phosphorescence. 1,4-Dibromonaphthalene and biacetyl were investigated as acceptor molecules.

2.10.   Chemiluminescence and Bioluminescence

Certain chemical reactions lead to electronically excited products which emit photons or transfer their energy to emitting species. The resulting luminescence is called chemiluminescence or bioluminescence, depending on the nature of the reacting species. In chemiluminescence and bioluminescence, no external source excites the luminescence because the luminescence is generated by chemical reactions. For a reaction to yield chemiluminescence or bioluminescence, it must fulfill three conditions (9):

1.   There must be sufficient energy available to produce an electronically excited product.
2.   The reaction pathway must favor the formation of an electronically excited product.
3.   The electronically excited product must either itself luminesce or transfer its energy to another molecule that luminesces.

The actual luminescence process is the same as normal photoluminescence. The intensity of the luminescence is proportional to the rate at which the excited-state product is formed. For chemiluminescence and bioluminescence in chemical analysis, reactant concentrations are adjusted so that the reaction is first order in the material whose concentration is to be determined. Chemiluminescence and bioluminescence analyses are considered a form of kinetic analysis. The luminescence intensity is measured at a specified time after the reaction has been initiated or an intensity versus time curve can be integrated over a known time interval. It is important that the variables affecting reaction rate be controlled carefully and that the reaction be initiated by mixing the reactants in a controlled manner. The theory of chemiluminescence and bioluminescence as related to chemical analysis has been considered extensively in the literature (42,43).

2.11. Relationships between Luminescence and Concentration

Solution fluorescence intensity is proportional to the number of excited-state species. The fluorescence intensity in turn is proportional to the intensity of absorbed excitation radiation:

$$F = \phi_f(I_0 - I) \tag{11}$$

where F is the fluorescence intensity, I_0 is the intensity of incident excitation radiation, and I is the intensity of transmitted excitation radiation. The difference between I_0 and I is equal to the absorbed intensity. By substituting Beer's law for I in Eq. (11), one obtains

$$F = \phi_f I_0[1 - \exp(-2.3\epsilon bc)] \tag{12}$$

Expanding the exponential term as a series gives

$$F = 2.3\phi_f I_0 \epsilon bc[1 - \frac{2.3\epsilon bc}{2!} + \frac{(2.3\epsilon bc)^2}{3!} - \cdots + \frac{(2.3\epsilon bc)^n}{(n+1)!} \tag{13}$$

When $\epsilon bc < 0.05$, Eq. (13) reduces to

$$F = 2.3\phi_f I_0 \epsilon bc \tag{14}$$

In practice, the measured fluorescence will include a geometric factor depending on the effective solid angle viewed by the detector, and a conversion factor for the change in response of the detector with wavelength (44). Equation (14) describes the linear relationship between fluorescence intensity and concentration. Similar arguments apply to solution phosphorescence, and the simplified relationship between phosphorescence intensity and concentration is given by Eq. (15):

$$P = 2.3\phi_p I_0 \epsilon bc \tag{15}$$

If ϵbc becomes very large, Eq. (12) becomes

$$F = \phi_f I_0 \tag{16}$$

Equation (16) shows that F is independent of concentration under this condition. With ϵbc extremely large, essentially all the incident radiation is absorbed by the sample and none is transmitted. Thus, in this situation the fluorescence intensity would be independent of concentration. A similar situation would apply to solution phosphorescence.

In dilute solutions, the fluorescence and the phosphorescence are proportional to concentration. However, at relatively higher concentrations, deviations from linearity will occur because the higher terms in Eq. (13) become important. (This condition is different than the condition for Eq. (16).) In practice, the relationship between fluorescence and concentration becomes nonlinear because of the "inner filter" effect. It is normally assumed that every molecule in a sample cell is excited with equal probability and that the emitted radiation is not affected by the presence of solute molecules. In concentration regions where Eq. (13) is applicable, absorption of the fluorescence emission by some species in solution may occur or nonhomogeneous excitation of the sample can take place. The "inner filter" effect does not include those processes that affect the quantum

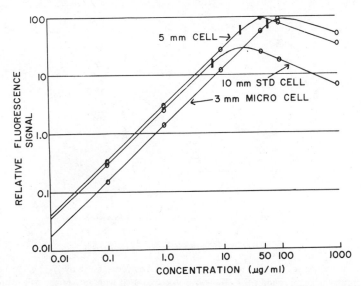

Fig. 5. Fluorescence analytical curves for quinine sulfate. (Reprinted with permission from J. D. Winefordner, S. C. Schulman, and T. C. O'Haver, *Luminescence Spectrometry in Analytical Chemistry,* Wiley, New York, 1972.)

efficiency of the molecule. "Inner filter" effects are indicated in several ways. Excitation and emission spectra may be altered or a calibration curve may become nonlinear at higher concentrations. If the luminescent component is the cause of the "inner filter" effect, dilution of the sample can often correct the problem (5). Figure 5 shows some typical fluorescence calibration curves for quinine sulfate. Figure 6 shows a phosphorescence calibration curve for benzaldehyde. The linear range for solution low-temperature phosphorescence calibration curves is generally greater than for solution room-temperature fluorescence calibration curves. The greater linear range can occur for several reasons. Phosphorescence is farther removed from the excitation spectrum and less self-absorption would occur compared to fluorescence. At 77 K, less collisional deactivation can occur compared to room temperature. Also, wider slits in the excitation mode can be used because the phosphoroscope minimizes the effect of scattered radiation (see discussion on phosphoroscopes in Section 3.6).

Zweidinger and Winefordner (45) have evaluated a rotating sample cell for quantitative low-temperature phosphorimetric measurements of organic compounds in clear, snowed, and cracked matrices. Phosphorescence intensity expressions were derived for the optically inhomogeneous matrices using the classical equations of Kubelka and Munk (46). Experimental data showed that the derived equations were valid. They showed that analytical curves (log I_p

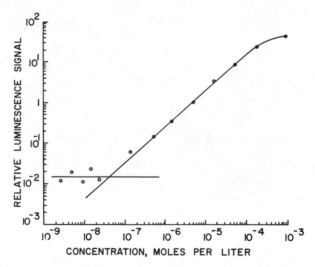

Fig. 6. Phosphorescence analytical curve for benzaldehyde excited by right-angle illumination measurement in 2-mm round cells. The curve was measured with an Aminco SPF with phosphoroscope attachment. Sample was at 77 K. (Reprinted with permission from J. D. Winefordner, S. C. Schulman, and T. C. O'Haver, *Luminescence Spectrometry in Analytical Chemistry*, Wiley, New York, 1972.)

versus $\log C$) in the optically inhomogeneous media should have a slope of unity at low concentrations, a slope of 0.5 at intermediate concentrations, and a slope of zero at high analyte concentrations. One very important consequence of their work in phosphorimetry is that snows at low temperature are essentially as useful analytically as clear rigid glasses.

Solid surface luminescence analysis involves the measurement of fluorescence and phosphorescence of components on solid materials (30). In various analyses, several solid materials have been used, such as silica gel, aluminum oxide, filter paper, silicone rubber, sodium acetate, and cellulose. Little work has been done in establishing the theoretical analytical aspects of luminescence reflected or transmitted from compounds adsorbed on solid surfaces. Pollak and Boulton (47) used the Kubelka and Munk theory to develop a theoretical basis for the fluorescence of adsorbed components on thin-layer chromatoplates. Pollak (48,49) further discussed this theory and its potential application to thin-layer chromatography. Goldman (50) also used the Kubelka and Munk theory of radiative transfer in scattering media to develop a theory of fluorescence for thin-layer chromatoplates. Neither Pollak and Boulton nor Goldman gave experimental data to support their theoretical conclusions. There is a definite need in solid-surface luminescence analysis for a theory that has been substantiated by experimental data. Presently, experimental conditions are usually obtained by trial and error, and empirical calibration curves are used without a sound theoretical basis. Hurtubise (51) obtained fluorescence experimental data from thin-layer chromatoplates using fluoranthene as a model compound to offer experimental support for two simplified equations obtained from the Goldman theory. The equations are

$$I^+/i_0\alpha = 1/3kX \, (1 - 7/30sX \cdot kX) \tag{17}$$

$$J^+/i_0\alpha = 2/3kX \, (1 - 4/30sX \cdot kX) \tag{18}$$

where I^+ is the transmitted fluorescence from the chromatoplate, i_0 the initial exciting intensity, α the proportion of absorbed radiation converted into fluorescence, k the absorption coefficient of exciting radiation, s the scattering coefficient of exciting radiation, X the thickness of the scattering medium, and J^+ the reflected fluorescence. In Eqs. (17) and (18), the scattering coefficient for fluorescence radiation does not appear because of mathematical approximations. The term kX is proportional to the amount of absorbing compound in the exciting beam. Once sX and kX values were obtained, $I^+/i_0\alpha$ values were calculated from Eqs. (17) and (18). With these values and the corresponding kX values, theoretical calibration curves were plotted for reflected and transmitted fluorescence. Also, experimental fluorescence calibration curves were obtained for fluoranthene

Table 2. Comparison of Theoretical and Experimental Approximate Points of First Slope Change in Terms of Micrograms of Fluoranthene[a]

	Developed Al_2O_3			Undeveloped Al_2O_3			Developed SiO_2			Undeveloped SiO_2		
Plate	Plate	Theor.	Exptl.	Plate	Theor.	Exptl.	Plate	Theor.	Exptl.	Plate	Theor.	Exptl.
Transmitted Fluorescence												
1	1	0.12	0.17	3	0.15	0.17	5	0.16	0.18	7	0.15	0.18
2	2	0.12	0.18	4	0.11	0.15	6	0.16	0.20	8	0.18	0.18
Reflected Fluorescence												
1	1	0.12	0.14	3	0.21	0.15	5	0.16	0.22	7	0.19	0.13
2	2	0.13	0.15	4	0.16	0.14	6	0.14	0.15	8	0.18	0.16

[a]Reprinted with permission from R. J. Hurubise, *Anal. Chem.*, **49**, 2160. Copyright 1977 American Chemical Society.

in the transmission and reflection modes. Table 2 shows a comparison of theoretical and experimental approximate points of slope change in terms of micrograms. Data were obtained for fluoranthene on developed and undeveloped aluminum oxide and silica gel chromatoplates to see if chromatographic conditions had any influence on the reflected and transmitted fluorescence. The data in Table 2 show fairly good comparison between theoretical and experimental values for both developed and undeveloped chromatoplates, whether the fluorescence was measured in the transmission or reflection modes. These results indicate that Goldman's simplified equations can be used to predict approximately when a calibration curve first changes slope. However, there is a need for more research under a variety of experimental and instrumental conditions to establish the validity of Goldman's equation and other theories. The equations developed by Zweidinger and Wineforder (45) discussed earlier should prove useful in solid-surface luminescence analysis. Vo-Dinh and Winefordner (52) have considered the application of the Zweidinger-Winefordner equations to room-temperature phosphorimetry.

2.12.   Environmental Effects on Luminescence

Environmental factors can influence one or more of the rate constants involved in excited state processes or alter the various energy levels in a luminescent molecule. Becker (3) has considered several aspects of environmental effects on fluorescence and phosphorescence, and Wehry (53) has discussed environmental factors that influence fluorescence. Hurtubise (30) considered environmental effects on the fluorescence and phosphorescence of compounds adsorbed on various solid surfaces.

Temperature is an important consideration because an increase in temperature causes an increase in the rate of collisions of excited molecules or an increase in the collisional rate with foreign molecules. Generally an increase in collisional rates causes a loss in luminescence efficiency. McCarthy and Winefordner (12) calculated the quantum efficiencies for several types of luminescence as a function of temperature for a hypothetical molecule. The results are shown in Figure 7. As indicated, the quantum efficiencies for fluorescence, phosphorescence and P-type delayed fluorescence decrease with temperature over the entire temperature range. For E-type delayed fluorescence the quantum efficiency increases with temperature, reaches a maximum value, and then slowly decreases.

Analytically useful luminescence signals have been obtained for molecules in the gas phase, in solution, in the crystalline form, and in various rigid media. The gas phase is suitable for luminescence measurements; however, many times the compounds need relatively high vapor pressures so sufficient sample can be obtained for luminescence measurements. At low pressure in the gas phase,

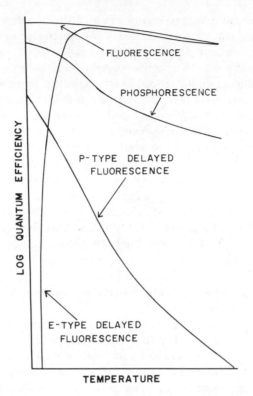

Fig. 7. Typical plots of quantum efficiencies of fluorescence (ϕ_f), phosphorescence (ϕ_p), E-type fluorescence (ϕ^E_{DF}), and P-type delayed fluorescence (ϕ^P_{DF}) versus sample temperature T for a hypothetical molecule. (Reprinted by permission from W. J. McCarthy and J. D. Winefordner, *J. Chem. Educ.*, **44**, 136 (1967).

molecular interactions are minimized and unperturbed luminescence spectra can be approximated.

The liquid phase is the most common medium for observing fluorescence. Solute–solvent interactions have the effect of altering vibrational energy levels in an irregular manner and broad luminescence spectra result. The polarity of the solute or solvent is probably the most important parameter influencing solution luminescence. For example, for many polar molecules the excited state is more polar than the ground state. With an increase in the polarity of the solvent, a greater stabilization of the excited state occurs for polar molecules relative to the ground state. Under these conditions, one would then expect to see a red shift in the fluorescence spectra as the solvent polarity increases. This approach has been used to determine if a $\pi \rightarrow \pi^*$ transition is involved. In the case of $\pi \rightarrow \pi^*$ transitions, the excited singlet state is usually more polar than the ground state. For $n \rightarrow \pi^*$ transitions, the excited singlet state is normally less polar than the ground state. Thus, increasing the solvent polarity stabilizes the ground state to a greater extent than the excited state and the emission spectrum is shifted to shorter wavelengths. Other factors can cause spectral shifts as a result of solvent polarity such as an increase in the energy difference between a Franck–Condon state and an equilibrium state. The above comments are an oversimplification of a relatively complex topic. Several authors have discussed the theory of solvent effects on electronic spectra in some detail (54–56).

Solvent viscosity is a factor that influences luminescence quantum yield. A higher viscosity can reduce the number of collisions an excited molecule undergoes. The net effect is a reduction in the rate of internal conversion to the ground state with a corresponding increase in luminescence quantum efficiency.

Hydrogen bonding interactions of luminescent molecules with a solvent can greatly influence the luminescence quantum yield. The effect of hydrogen bonding has been investigated more for fluorescence than for phosphorescence. Usually large alterations in the electronic charge distribution occur for electronically excited organic molecules. The probability that a photoexcited molecule will participate in proton-transfer reactions or hydrogen bonding interactions can be quite high compared to the ground state of the same molecule. Excited state hydrogen bonding can cause large changes in the shape of the fluorescence spectrum, the maximum fluorescence wavelength, and the fluorescence quantum yield. Wehry (53) has considered some aspects of the effects of hydrogen bonding on fluorescence spectra. For example, if it is assumed that a solute molecule C can hydrogen-bond with a molecule or solvent D, then generally an excited state hydrogen-bonded complex, (C–D)*, can be formed by two paths. Formation of the complex can occur by excitation of the species C–D already in the ground state:

$$\text{C–D} \xrightarrow{h\nu} \text{(C–D)}^*$$

Additionally, during the lifetime of an uncomplexed excited molecule C*, hydrogen bonding with D may occur:

$$\text{C} \xrightarrow{h\nu} \text{C}^*$$
$$\text{C}^* + \text{D} \rightarrow \text{(C–D)}^*$$

In the first situation, both the excitation and fluorescence spectra of C will be affected by the solvent molecule D. In the second case, hydrogen bonding occurs after excitation; thus only the fluorescence spectrum of C will be altered by the interaction. By comparing the effect of D on the excitation and fluorescence spectra of C, one can determine the relative abilities of C and C* to hydrogen-bond with D.

Many times it is observed that excited-state hydrogen bonding decreases the quantum efficiency of C. One example of this effect is the interaction of 5-hydroxyquinoline with hydrogen-bond-accepting solvents. A correlation was found between the ability of solvents to accept hydrogen from 5-hydroxyquinoline in hydrogen-bond formation and the quantum efficiency of fluorescence. For example, the quantum efficiencies of 5-hydroxyquinoline were 0.30 in isopentane and 0.07 in dimethyl sulfoxide (57).

As discussed earlier, with analytical RTP hydrogen bonding interactions have been presented as one explanation why RTP is observed from compounds adsorbed on solid surfaces (30,36,37).

Solution pH can cause considerable alteration in luminescence quantum efficiencies. As considered earlier, Schulman (14,15) has discussed the acid–base chemistry of excited singlet states. As an example of the effect that pH can have on ultraviolet absorption maxima, fluorescence emission maxima, and relative fluorescence intensity, Table 3 shows several examples for naphthols.

Heavy-atom solvents tend to enhance phosphorescence at the expense of fluorescence. Heavy-atom solvents usually induce a decrease in fluorescence quantum efficiency and an increase in the efficiency of intersystem crossing. The external heavy-atom effect usually increases the rates of both intersystem crossing and phosphorescence. There is still some uncertainty about the actual processes responsible for external heavy-atom effects (53). For halogen-containing solvents or alkali halide salts, there is strong evidence that a 1:1 complex is formed between an excited state of the solute and the heavy-atom-containing species. The exciplexes appear to be of a charge-transfer nature. However, the previous model does not explain all the observations from heavy-atom perturbations. The general consequences of the external heavy-atom solvents offer a means of enhancing sensitivity and selectivity in phosphorescence analysis. However, heavy-atom solvents should generally be avoided in fluorescence analysis.

The quenching of luminescence is a rather broad topic. Molecular oxygen is a well known quencher of fluorescence and phosphorescence. Some organic compounds are more susceptible to oxygen quenching than others. Some experimenters routinely bubble nitrogen or some other inert gas into the solution containing the luminescent component to displace dissolved oxygen prior to an analysis. Wehry (53) has listed five mechanisms for oxygen quenching. Which of the processes describes oxygen quenching of a particular fluorescent solute depends on the properties of both the solute and the medium. Some metal ions can also quench fluorescence and paramagnetic transition-metal ions usually produce the largest effects. Chen (59) has discussed the effects of silver and mercury ions on the luminescence of tryptophan and proteins and has emphasized that the mechanism of fluorescence quenching by these ions is complex. Sawicki

Table 3. Effect of pH on Ground and Excited State Disassociation of Naphthols[a]

Compound	State of Dissociation[b]	Medium	Ultraviolet Absorption Maxima (nm)	Fluorescence Emission		
				Medium[c]	Maxima (nm)[d]	Intensity[e]
	a	Ethyl alcohol	277(sh),288,298, 327,337	0.2 N H₂SO₄	385	0.75
	b	pH 10.4	255(sh),321	pH 6.1	453	3.30
	c	2 N NaOH	318,355	0.2 N NaOH	459	4.85
	a	Ethyl alcohol	276,287,300,323,336	Ethyl alchohol	365	4.00
	d	0.1 N NaOH	303,315(sh),350	ph 9.7	488	2.50
	a	Ethyl alcohol	263(sh),273,282,292, 306,312,320,327	0.2 N H₂SO₄	345	6.50
	d	0.2 N NaOH	286(sh),324(sh), 338,347	0.1 N NaOH	401	2.45
	a	Ethyl alcohol	272,282,293(sh), 320,339	0.2 N H₂SO₄	363	1.35
	d	0.2 N NaOH	268,301(sh),348	0.1 N NaOH	440	8.45
	a	Ethyl alcohol	287,306(sh),315, 321,329(sh)	0.2 N H₂SO₄	347	5.80
	d	0.2 N NaOH	329,343	pH 8.4	418	1.75

[a]From VanDuuren (58).
[b]a, undissociated; b, monoionic; c, diionic; d, dissociated.
[c]Concentration of solute: 5×10^{-5} mole/L^{-1}.
[d]Excitation at 313 nm.
[e]Uncorrected relative fluorescence intensities.

(20) has taken advantage of solvent quenching effects in the fluorescence analysis of air pollutants. Hurtubise (60) used selective fluorescence quenching with 80% ethanol/20% chloroform to analyze phenolic antioxidant mixtures. Generally, quenching is a process by which an electronically excited molecule gives up its energy by interacting with another solute molecule. Bimolecular luminescence quenching can be described by the following kinetic scheme (61):

$$D + h\nu \rightarrow D^* \qquad\qquad\qquad (a)$$

$$D^* \xrightarrow{k_e} D + h\nu \qquad\qquad\qquad (b)$$

$$D^* \xrightarrow{k_q} D + \Delta \qquad\qquad\qquad (c)$$

$$D^* + Q \xrightarrow{k_2} D + Q + \Delta \qquad\qquad (d)$$

Step (a) represents the excitation of ground state D to form an excited species D*, step (b) the radiative decay of D* with the emission of radiation, step (c) the first-order and pseudo-first-order quenching of D* by intramolecular interactions or solvent and impurity quenching, and step (d) the bimolecular quenching of D* by quencher Q. The mean lifetime of D* is given by Eq. (19).

$$\tau = \frac{1}{k_e + k_q + k_2[Q]} \qquad\qquad (19)$$

Equation (19) can be put in the following form:

$$1/\tau = 1/\tau_0 + k_2[Q] \qquad\qquad (20)$$

$$\tau_0 = \frac{1}{k_e + k_q}$$

where τ_0 is the decay time of D* in the absence of Q. A plot of $1/\tau$ versus [Q] should be linear with a slope of k_2. Alternately, the well known Stern–Volmer equation may be written

$$\tau_0/\tau - 1 = K_{sv}[Q] \qquad\qquad (21)$$

$$K_{sv} = k_2\tau_0$$

where K_{sv} is the Stern–Volmer quenching constant. In addition, the Stern–Volmer constant can be determined from fluorescence intensity measurements by varying the concentration of quencher, [Q]

$$\theta_0/\theta - 1 = K_{sv}[Q] \qquad\qquad (22)$$

where θ_0 is the intensity in the absence of quencher, and θ is the intensity with quencher present. The previous discussion concerns diffusional or dynamic quenching. There is an additional quenching mechanism possible, static or associational quenching. For static quenching, the quencher is chemically associated with the luminescent donor to form a nonluminescent association pair in the ground state. It can be shown when both static and dynamic quenching are present that Eq. (23) is valid (61),

$$\theta_0/\theta - 1 = (k_{sv} + \beta K_{eq})[Q] + K_{sv}\beta K_{eq}[Q]^2 \tag{23}$$

where K_{eq} is the association constant, K_{sv} is the Stern–Volmer constant, and β varies from 1 to ϵ_{DQ}/ϵ_D for measurements under optically dilute to optically dense conditions. The terms ϵ_{DQ} and ϵ_D are the molar absorptivity of the association pair and free D at the excitation wavelength, respectively. The term [Q] represents the free quencher concentration rather than the analytical quencher concentration. If only static quenching is present then $K_{sv} = 0$ and a linear relationship results from an intensity versus [Q] plot; however, the slope now equals βK_{eq}. In this situation, the data do not indicate that the kinetic scheme (a)–(d) is operative unless τ measurements are also made.

2.13. Luminescence Properties and Molecular Structure

Strictly speaking, the effects of molecular structure on luminescence properties cannot be considered without reference to the environmental effects because luminescence properties depend on both structural features and the environment. Fluorescence and phosphorescence always compete with other processes. After a molecule is promoted to an electronically excited state, it may lose the excess energy in a variety of ways. Several of these ways were discussed in the previous section. The effects of structure on luminescence have been reviewed by Wehry (53), Becker (3), and Seitz (9).

Generally, strongly fluorescent compounds possess the following characteristics, although there are certainly exceptions to these guidelines (53):

1. The lowest energy spin-allowed electronic absorption transition has a large molar absorptivity.
2. The energy of the lowest spin-allowed absorption transition should be fairly low because the larger the excitation energy, the more probable photodissociation may occur.
3. The electron that has been excited should be located in an orbital which is not strongly involved in bonding.
4. The molecule should not have structural features or functional groups that enhance the rates of radiationless transitions.

Aromatic hydrocarbons have the above characteristics and are usually intensely fluorescent. The π electrons in aromatic hydrocarbons are less strongly held than σ electrons and can be promoted to π* antibonding orbitals by absorption of electromagnetic radiation of relatively low energy. In addition, a π → π* transition in many aromatic hydrocarbons is strongly allowed. Things are more complicated in compounds with carbonyl groups and in heterocyclic compounds. In these compounds, nonbonding n electrons are available for excitation into π* orbitals. Frequently, the energyof n → π* absorption transition will be lower than the lowest π → π* absorption transition. This fact is significant because n → π* transitions are less intense than π → π* transitions. One would thus expect aromatic carbonyl compounds and heteroaromatics to fluoresce less strongly than aromatic hydrocarbons. Saturated hydrocarbons contain σ-bonding electrons, and transitions involving σ electrons, occur at very high energies and disrupt bonding in the hydrocarbon. Aliphatic carbonyl compounds can undergo n → π* transitions and can exhibit fluorescence; however, the quantum efficiencies are small. Highly conjugated compounds like β-carotene and vitamin A are fluorescent because of π → π* transitions. Nevertheless, the vast majority of fluorescent and phosphorescent organic compounds are aromatic.

Increasing the magnitude of a π electron system has the effect of increasing the rates of absorption and fluorescence and lowering the energy difference between the ground state and first excited singlet. For example, benzene, naphthalene, anthracene, and naphthacene have fluorescence quantum efficiencies in EPA glass at 77 K of 0.11, 0.29, 0.46, and 0.60, respectively. The maximum fluorescence emission wavelengths are 278, 321, 400, and 480 nm, respectively (9). Zander (62) has discussed the low-temperature phosphorescence characteristics of aromatic hydrocarbons.

Structural rigidity and steric hindrance are factors that can affect luminescence quantum yield. Frequently cited examples are phenolphthalein and fluorescein (63). Fluorescein gives very intense fluorescence in solution; however, phenolphthalein does not, although the two compounds are structurally similar. The main structural difference between the compounds is an oxygen bridge in fluorescein which is absent in phenolpthalein. Apparently, in phenolphthalein, rotations and vibrations of the aromatic rings occur with ease so that electronic excitation can be lost internally without leading to fluorescence. With a rigid

CIS   TRANS

structure like fluorescein, vibrational dissipation of electronic energy is much more difficult. In general, steric hindrance promotes radiationless deactivation of excited states. An example is *cis-* and *trans*-stilbene. As seen above, *cis*-stilbene involves steric interaction between two phenyl groups. A similar interaction cannot occur with the trans isomer. The cis isomer is not highly fluorescent, but the trans isomer is highly fluorescent (9). Kasha (2) has discussed the diminished luminescence of cyanine dyes because of steric crowding in the dyes.

The effect of heavy-atom solvents (external heavy atom) on fluorescence and phosphorescence quantum efficiencies was discussed earlier. Similar effects are observed for a molecule that contains a heavy atom (internal heavy atom). In general, the fluorescence efficiency decreases and the phosphorescence efficiency increases. The heavy atom perturbs electron spins to a greater extent than lighter atoms. The perturbation results in the mixing of electronic states. The net result is that the singlet states have greater triplet character and the triplet states have greater singlet character. This diminishes the spin forbiddenness of intersystem crossing and thus accelerates its rate. Kasha (2) has shown that the ratio of the quantum yield of phosphorescence to that of fluorescence increases in the series naphthalene, 2-chloro-, 2-bromo-, 2-iodonaphthalene; however, the total quantum yields were not affected.

A very important consideration in determining the luminescence behavior of a compound is the nature of the lowest lying singlet state. In most cases, absorbed photons arrive at the lowest excited singlet state in a molecule. This means that the behavior of a molecule after excitation will not depend on the initial excited state, but rather on the nature of the lowest excited state. If the lowest excited singlet state is a π,π^* state, the molecule will show the characteristics of a π,π^* state. In contrast, if the lowest excited singlet state is an n,π^* state, it will show the characteristics of an n,π^* state. Table 4 compares approximate values for maximum molar absorptivity, lifetime of the excited state, singlet–triplet split, and relative rate of intersystem crossing for n,π^* and π,π^* states (5). Generally, $n \rightarrow \pi^*$ transitions are less intense than $\pi \rightarrow \pi^*$ transitions, and thus n,π^* excited states have longer lifetimes. This factor tends to enhance intersystem

Table 4. Comparison of n,π^* and π,π^* Singlet States[a]

	n,π^* States	π,π^* States
ϵ_{max}	10–10^3	10^3–10^5
Lifetime	10^{-7} to 10^{-5}	10^{-9} to 10^{-7}
Singlet–triplet split	Small	Generally large
Rate of intersystem crossing	Greater than for fluorescence	Of the same order as fluorescence

[a]From Hercules (5).

crossing from n,π^* states. In addition, the smaller singlet–triplet split for n,π^* electronic states also tends to enhance intersystem crossing. Generally, if the lowest excited singlet state is n,π^*, intersystem crossing is favored over fluorescence.

Organic compounds with nonbonding electrons and specific geometric features can complex with a variety of metal ions. In many situations, the effect of complex formation is to change the nature of the first excited singlet state from a n,π^* to a π,π^* state. The complex can yield very efficient fluorescence, which is useful for trace analysis of metal ions. As a general rule, chelates with high-atomic-number metal ions fluoresce less efficiently but yield more efficient phosphorescence than do chelates with low-atomic-number metals (9).

The effects of substituents on the fluorescence of a parent compound are summarized in Table 5. Many exceptions to almost all the substituents in Table 5 have been documented, especially in the case of groups that can interact strongly with solvents (53). The general guidelines are that meta-directing (electron withdrawing) groups reduce fluorescence, and orthopara-directing (electron-donating) groups enhance fluorescence. Another general guideline is that substituents that interact with a solvent tend to reduce the vibrational structure in luminescence spectra. Hydrogen bonding substituents favor room temperature phosphorescence of compounds adsorbed on solid surfaces; however, room temperature phosphorescence has been observed from polycyclic aromatic hydrocarbons adsorbed on filter paper with a heavy atom salt (35). Below is a more detailed summary of the effects of substituents on luminescence given by Seitz (9).

Table 5. Effects of Substituent on the Fluorescence of Aromatics[a]

Substituent	Effect on Frequency of Emission	Effect on Intensity
Alkyl	None	Very slight increase or decrease
OH, OCH$_3$, OC$_2$H$_5$	Decrease	Increase
CO$_2$H	Decrease	Large decrease
NH$_2$, NHR, NR$_2$	Decrease	Increase
NO$_2$, NO	Large decrease	Large decrease
CN	None	Increase
SH	Decrease	Decrease
F		
Cl	Decrease	Decrease
Br		
I		
SO$_3$H	None	None

[a]Reprinted from Wehry (53), p. 92, by courtesy of Marcel Dekker, Inc.

Nitro Groups. Nitro groups introduce a low-lying $n \to \pi^*$ transition into a molecule. Thus, nitro compounds rarely fluoresce; however, some nitro compounds phosphoresce.

Carbonyl Groups. Compounds containing carbonyl substituents include carboxylic acids, ketones, esters, and aldehydes. These compound types generally have a low-lying $n \to \pi^*$ transition, which causes a decrease in the fluorescence frequency and a decrease in fluorescence quantum yield. Carbonyl compounds tend to phosphoresce efficiently (e.g., acetophenone, benzophenone, and benzaldehyde).

Alkyl Groups. Generally, alkyl groups do not significantly affect luminescence properties of a molecule unless steric hindrance is involved. With steric hindrance, internal conversion is increased and fluorescence quantum efficiency decreases.

Amino Groups. Amino groups tend to shift fluorescence to longer wavelengths and increase fluorescence quantum efficiency. Because amino groups are basic, the luminescence characteristics are dependent on pH. For example, aniline is fluorescent, but the anilinium cation is nonfluorescent (64). Also, amino groups can hydrogen bond to various solvents, which can affect the luminescence behavior of the compounds.

Hydroxyl Groups. Hydroxyl groups shift fluorescence to longer wavelengths and increase fluorescence quantum efficiency. Because hydroxyl groups are acidic and capable of hydrogen bonding, these properties can influence luminescence properties.

Cyano Groups. Cyano groups introduce an $n \to \pi^*$ transition, but it is usually higher in energy than the lowest $\pi \to \pi^*$ transition. Because of this, the cyano substituent does not affect fluorescence wavelength and may actually increase fluorescence.

Sulfonate Groups. Sulfonate groups do not greatly affect the luminescence characteristics of a compound. Thus, this functional group is useful in increasing water solubility of compounds with relatively high luminescence efficiencies.

Halo Groups. The main effect of a halogen substituent is the internal-heavy-atom effect.

Multiple Substituents. If multiple electron-donating substituents or multiple electron-withdrawing substituents are present, they appear to reinforce each other.

If both electron-donating and electron-withdrawing groups are present, the situation is more complex.

Table 6 lists several examples of various luminescent compounds and the variety of conditions that can be used to induce the luminescence. Numerous examples of luminescent compounds can be found in the biennial reviews on luminescence analysis in *Analytical Chemistry* (74).

Table 6. Luminescent Compounds and the Conditions for Luminescence

Compound	Excitation Wavelength (nm)	Luminescence	Conditions	Reference
2-Amino-anthracene	382.5	Fluorescence	In glycerol/HCl (2N) glass, 4.2 K	65
2-Naphthol	312	Fluorescence	In argon, matrix isolation	66
Methyltesto-sterone	470	Fluorescence	Solution at room temperature with 37% HCl and ascorbic acid	67
2-Chloroquino-line	313	Phosphorescence	In ethanol–water 10:90 (v/v) at 77 K	68
Pyrene	336	Phosphorescence	In thallium lauryl sulfate/sodium lauryl sulfate micelle system, room temperature	38
Propyl gallate	276	Fluorescence	Chloroform solution	69
Epinephrine	280	Phosphorescence	Sample spotted onto filter paper from 1M NaOH and 1M NaI, room temperature	70
Benzo[a]pyrene	Longwave UV radiation	Fluorescence	Sample adsorbed on 30% acetyated chromatoplate	71
2,7-Dintro-9-fluorenone	364	Phosphorescence	In ether–isopropanol (3:1, v/v) at 77 K	72
α-Methylene carbonyl compounds	374,375,378	Fluorescence	Compounds reacted with N'-methylnico-tinamide chloride	73

2.14. Signal-to-Noise Ratio Considerations

The ratio of signal-to-noise in luminescence spectroscopy is of fundamental importance to luminescence measurements (7,75–79). There are several factors which influence the measured signal and noise in fluorescence and phosphorescence experiments. In this section, the general theory and signal-to-noise (S/N) ratio equations for photodetector signals will be discussed. In Section 3.7, a comparison is given between S/N ratios for silicon intensified target (SIT) vidicon camera detectors and photomultiplier detectors (78,79).

St. John et al. (75) have discussed the application of signal-to-noise theory in molecular luminescence spectrometry. In discussing signal expressions, they made several assumptions in deriving the signal expressions. These assumptions are:

1. The entrance and exit slit widths and heights are equal for the excitation and emission monochromators.
2. The spectral bandwidth of the excitation monochromator, s, is appreciably less than the half-intensity width of the excitation spectral band, and that the spectral bandwidth of the emission monochromator, s', is appreciably less than the half-intensity width of the emission spectral band.
3. The luminescence emission band shape is given by a Gaussian function of wavelength.
4. The intensity of absorption and emission are approximately constant over the spectral bandwidths s and s'.
5. The measured emission band results from only a single electronic transition.

The usual experimental arrangement for the measurement of fluorescence or phosphorescence is with the excitation source and monochromator optically aligned perpendicularly to the emission monochromator and photodetector. St. John et al. (75) used these conditions in their detailed derivation of signal expressions. Equation (24) shows the intensity of a luminescent sample if the sample is being excited continuously and if the luminescence is being continuously measured. This is normally the case in fluorescence measurements.

$$I_{\mathrm{m}} = \frac{P_{\mathrm{abs}}\phi_s f_3}{4\pi A_s \Delta\lambda'} \tag{24}$$

Here, I_{m} is the intensity of emission (W cm^{-2} ster^{-1} nm^{-1}); P_{abs} the radiant power absorbed by the sample from b_1 to b_2 (W); b_1 the distance into sample cell before which no emission is observed (cm); b_2 the distance into sample cell after which no emission is observed; ϕ_s the energy yield of luminescent process for sample (ergs emitted s^{-1} per ergs absorbed s^{-1}) f_3 the fraction of emitted radiation reaching d_0 (no units); d_0 the exit point for luminescence from cell; A_s the area of emitting sample surface (cm^2); and $\Delta\lambda'$ the half-intensity width of the emission band (nm).

Phosphorescence measurements can be made with a dynamic shutter mechanism, such as a rotating can, to minimize the signal due to fluorescence and incident light scattering. The measured photodetector signal depends upon the time elapsed between initial excitation and measurement of the luminescent signal. O'Haver and Winefordner (80) have derived an expression which accounts for the loss of signal due to intermittent excitation and observation. A parameter α' was defined by them as the ratio of the observed power emitted when using intermittent excitation and observation to the power emitted if using continuous radiation. The parameter is unity for continuous excitation and observation. For phosphorescence signals, St. John et al. (75) assumed that the time dependency of the decay of phosphorescence could be described by a single exponential term. Thus, if t is the time at which the phosphorescence is measured after complete termination of the excitation radiation, then the emitted intensity and the resulting photoanodic current i is proportional to the exponential factor $\exp(-t/\tau_s)$, where τ_s is the decay time for phosphorescence (in seconds). The exponential factor then has to be taken into consideration when relating the photoanodic current to phosphorescence at time t.

The photoanodic current i due to radiant power reaching the photocathodic surface can be found by multiplying the radiant power in watts reaching the photocathode by γ. The term γ is the sensitivity factor and has units of amperes at the photodetector anode per watt of radiant power incident on the photocathode. The photoanodic current i is given by

$$i = \frac{P_{abs}\phi_s f_3 \gamma \alpha'}{4\pi A_s \Delta\lambda'}$$

St. John et al. (75) discussed two limiting cases for Eq. (25); however, only one limiting case will be discussed. This limiting case represents the luminescence of a component in a dilute solution of a solvent which does not absorb appreciably at the wavelengths of interest. This situation approximately describes many quantitative luminescence measurements. Making assumptions which are approximately obeyed for most quantitative measurements (75), the photoanodic signal i is given by

$$i = 2.3 I^0 k_a^2 a_s b C_s \phi_s \left[\frac{\gamma \alpha'}{4\pi A_s \Delta\lambda'} \right] \tag{26}$$

where I^0 is the intensity of source excitation (W cm^{-2} ster^{-1} nm^{-1}); k_a a function of slit height, reciprocal linear dispersion, transmission factor of monochromator, area of collimating mirror or lens, focal length of collimating mirror or lens, and slit width of monochromator; a_s the molar absorptivity of the luminescent sample at the excitation wavelength (L mole^{-1} cm^{-1}); b the cell length (cm); and C_s the concentration of luminescent sample (moles L^{-1}). One of the most important conclusions from Eq. (26) is that the concentration of the luminescing species, C_s, is proportional to i.

The total root-mean square (rms) noise current $\overline{\Delta i}_T$ in amperes, at the photoanode, is given by Eq. (27) (75),

$$\overline{\Delta i}_T = (\overline{\Delta i_p}^2 + \overline{\Delta i_s}^2 + \overline{\Delta i_c}^2)^{1/2} \qquad (27)$$

where Δi_p is the rms phototube noise due to the shot effect, Δi_s the rms effective luminescence flicker noise, and Δi_c the rms luminescence convection noise. St. John et al. (75) have discussed the noise terms in detail and have derived Eq. (28) for $\overline{\Delta i}_T$,

$$\overline{\Delta i}_T = [\Delta f k_D(i_d + i_B + i) + \xi^2[\sqrt{\Delta f_s}\, i + \sqrt{\Delta f_B}\, i_B]^2 + \epsilon^2 \Delta f[i + i_B]^2]^{1/2} \qquad (28)$$

where Δf is the frequency response bandwidth of electrometer–readout system (s^{-1}); k_D a function of the overall gain of the phototube and charge on an electron (coulombs); i_d the photoanodic dark current (A); i_B the background luminescence current (A); ξ the effective source flicker factor (s$^{1/2}$); Δf_s the effective frequency response bandwidth of electrometer readout system for sample (s^{-1}); Δf_B the effective frequency response bandwidth of electrometer readout system for background (s^{-1}) and ϵ the convection flicker factor (s$^{1/2}$). St. John et al. (75) combined Eqs. (26) and (28) to obtain an approximate signal-to-noise equation. The equation was used for predicting the variation of $i/\overline{\Delta i}_T$ as a function of a variety of experimental parameters. For example, by plotting the signal-to-noise ratio versus the value of an experimental parameter, it was possible to determine if there was an optimum value of the parameter. If an optimum value were not obtained, it was possible to predict the best experimental conditions for an analysis.

Signal and noise considerations have been discussed for the measurement of fluorescence from organic compounds adsorbed on solid surfaces such as thin-layer chromatoplates (49,81). The solid surfaces can extensively scatter both exciting radiation and fluorescent radiation. Even though direct fluorescence measurements can give a nearly ideally flat baseline, optical noise can be a problem. Optical noise is caused by random fluctuations of the optical transfer in the scattering medium. It is usually the principal factor which limits performance in solid surface luminescence analysis. Electrical noise becomes important at very low light levels and is generated in the photodetector and preamplifier stages. Several sources of optical noise are listed below for solid surfaces such as thin-layer chromatoplates (49,81).

1.   Any specular component of the scattered radiation at the surface of the medium varies randomly and as a result the intensity of the light entering the medium exhibits random fluctuations from point to point. These fluctuations appear as optical noise in both the transmission and reflection modes.

2.   In the reflection mode, it is possible for part of the specularly reflected luminescence to reach the photodetector and cause optical noise. This can be reduced by careful optical design.

3.   One serious source of optical noise is caused by local variations in the thickness of the medium.

4.   Density fluctuations of the medium and nonuniform particle size can cause optical noise.

5.   Special treatment of the medium such as with chemical reagents can cause nonrandom changes in optical parameters.

6.   Optical noise can also originate in the source radiation through lamp instability.

Pollak (49) stated that, theoretically, the signal-to-noise ratio has a definite value which does not depend on the amount of fluorescent component. He estimated the signal-to-noise ratio for transmitted fluorescence (TF) and reflected fluorescence (RF) of a component adsorbed on a chromatoplate. The approximate relationships are

$$(S/N)_{TF} \simeq \frac{1}{\delta_\gamma} \tag{29}$$

$$(S/N)_{RF} \simeq \frac{\gamma}{\delta_\gamma} \tag{30}$$

The term γ is the natural logarithm of the transmittance of a very thin sheet of medium, which has negligible reflectance. The term δ_γ represents the rms value of the fluctuations of the γ term. Equations (29) and (30) indicate that the signal-to-noise ratio is both constant and independent of the amount of the fluorescer. However, the output signal of the photodetector is dependent on the stability of the excitation source.

Pollak (49) has discussed double-beam scanning for fluorescence from solid surfaces. The goal of double-beam scanning in fluorometry is different from that in absorption work. In absorption work, the main goal is to obtain a smooth baseline. Noise produced by incremental changes in the amount of absorber in the sample beam (incremental zone noise) is not reduced. With fluorescence work, baseline noise can be very low and the main concern is reduction of incremental zone noise. Pollak (49) gives the signal-to-noise expression ratio for scanning the double-beam transmission mode by the following equation:

$$S/N \simeq \frac{1 + \rho_F}{\delta\rho_F} \tag{31}$$

The term ρ_F is the coefficient of fluorescence reflectance of a sheet of medium thick enough so that its transmission can be disregarded. The term $\delta\rho_F$ is the rms value of the fluctuations of the coefficient ρ_F. Pollak (49) has commented that a double-beam scanning system could yield a larger S/N ratio than a single-beam scanning system. However, there are apparently no double-beam scanning instruments in operation, and thus the previous hypothesis awaits experimental verification.

2.15.   Limit of Detection

Limits of detection in molecular luminescence analysis have been reported for numerous luminescent compounds. Long and Winefordner (82) have discussed, in general, limit of detection and the difficulties associated with various definitions of limit of detection. The limit of detection is a concentration or an amount which describes the lowest concentration or amount of a component that is statistically different from an analytical blank. As Long and Winefordner have emphasized, significant problems have occurred in expressing limits of detection because the approaches used for the term are statistically different. The International Union of Pure and Applied Chemistry (IUPAC) adopted a model for the limit of detection calculation, and the ACS Subcommittee on Environmental Analytical Chemistry has supported the model (83). The limit of detection is based on the relationship between the gross analyte signal S_t, the blank S_b, and the variability in the blank σ_b. The limit of detection is defined by the extent to which the gross signal exceeds a defined multiple K_d of S_b (83).

$$S_t - S_b \geq K_d\sigma_b \tag{32}$$

If blanks are not available or if a single sample is being analyzed for which there are no blank data, then the limit of detection is based on the peak-to-peak noise measured on the baseline close to the actual or expected analyte peak. It has been recommended that detection should be based on a minimal K_d value of 3 (83). Thus, the limit of detection is located at $3\sigma_b$ above the blank signal S_b. Long and Winefordner (82) have recommended that analysts report limits of detection using the IUPAC approach because errors in the analyte measurements can be incorporated into the detection limit.

In terms of signal-to-noise ratio theory, the smallest analytical signal that can be detected with a given level of confidence has been derived from small-sample statistical theory. Equation 33 defines the limiting signal-to-noise ratio at the limiting detectable sample concentration C_m (84,85).

$$\left(\frac{S}{N}\right)_{Cm} = \frac{t\sqrt{2}}{\sqrt{n}} \tag{33}$$

Here, t is the Student "t" and n is the number of combined sample and blank (or background) readings. The number of degrees of freedom is $2n-2$. The noise N is the total root-mean square noise due to all sources and is approximately equal to the peak-to-peak noise divided by five (76). Cetorelli et al. (85) have derived equations for the evaluation of C_m based on $(S/N)_{C_m}$ in Eq. (33). The equations are rather complex and contain several instrumental factors. Nevertheless, the equations are very important in relating the S/N ratio to important fundamental instrumental parameters in absorption, fluorescence, and phosphorescence spectrometry.

In practice, many recent papers on luminescence analysis have reported detection limits based on the concentration or amount of material that gives a S/N ratio of three (35,86–91). Aaron et al. (68) have reported limits of detection as the concentration giving a signal equal to three times the standard deviation of 10 blanks. In addition, Voigtman et al. (90) gave the limits of detection as that concentration which gives a signal three times the standard deviation of sixteen blanks. Lai et al. (91) reported the limit of detection as a signal three times the background noise, the latter being equal to $N_{p-p}/5$. The limits-of-detection definitions discussed in this paragraph are essentially within the IUPAC and ACS guidelines. The exact approach to use in reporting limits of detection will depend to some extent on what luminescence technique is being employed. However, the guidelines set down by IUPAC and ACS should be followed so that meaningful comparisons between analytical methods and instruments can be made (82,83).

3. INSTRUMENTATION

3.1. General Considerations

To obtain photoluminescence measurements, the following components are necessary: a source of excitation radiation, a device for selecting the excitation wavelength, a sample cell, a device for selecting the emission wavelength, and a detector. Figure 8 gives a schematic of a typical layout of components for

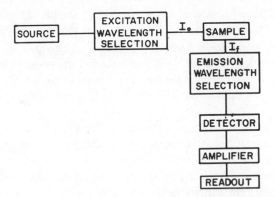

Fig. 8. General schematic for luminescence instrumentation (Reprinted with permission from W. R. Seitz, "Luminescence Spectrometry," in P. J. Elving, E. J. Meehan, and I. M. Kolthoff, Eds., *Treaties on Analytical Chemistry*, 2nd ed., Part I, Vol. 7, Sec. H, Wiley, New York, Chap 4).

photoluminescence measurements. For chemiluminescence and bioluminescence measurements a detector, a cell, provisions for mixing reactants, and electronics for readout and data processing are needed. Spectral resolution is rarely required (92). With fluorescence measurements from surfaces such as thin-layer chromatoplates and filter paper, spectrodensitometers are available and the arrangement of components is frequently different from that shown in Fig. 8. The different arrangement occurs because the luminescence is usually diffusely scattered from the surface and measured in the reflection mode or because the scattered luminescence is transmitted through a transparent solid surface and measured in the transmission mode (93). For dilute solutions or gases, the right-angle arrangement in Fig. 8 is normally used. This positioning is particularly advantageous because the detector "sees" the luminescent radiation and not the source radiation.

The components employed in commercial and many research luminescence instruments generally consist of the same type of components as in ultraviolet–visible spectrophotometers. In comparing luminescence measurements with absorption measurements, it is important to emphasize that the intensity reaching the detector tends to be much lower for luminescence measurements for several reasons (9):

1. Analytical luminescence measurements are generally made at low absorbances ($A < 0.05$).

2. The fraction of the emitted luminescence impinging on the detector system is frequently quite small.

3. To minimize spurious optical effects, luminescence instruments are nor-

mally designed to view a narrow interval of sample in the middle of the cell rather than viewing the entire length of the sample cell.

4. Frequently the fluorescence or phosphorescence quantum efficiency is considerably less than one; thus, luminescence instrumentation is designed to maximize instrumental factors that affect the signal. One of the main objectives is to obtain the highest possible signal-to-noise ratio.

An important factor affecting luminescence instrumentation and measurements is that most of the instrumental components affecting the measured luminescent signal are wavelength dependent. To obtain the true excitation spectrum and emission spectrum of a luminescent species, it is necessary to correct for the instrumental wavelength dependence.

3.2. Sources

3.2.1. Conventional Sources

High-pressure mercury and xenon arc lamps (gas discharge lamps) are the most common sources in luminescence instrumentation. A comparison of the spectral outputs of these types of lamps is given in Fig. 9. As seen in Fig. 9, most of the mercury emission lines are more intense than the continuum output of the xenon lamp. The high-pressure mercury lamp is advantageous in situations where a luminescent component has to be excited at one of the mercury emission lines.

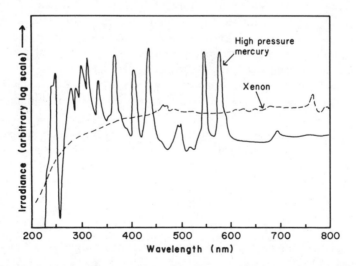

Fig. 9. Comparison of spectral outputs of arc light sources. (Reprinted by permission from T. D. S. Hamilton, I. H. Munro, and G. Walker, "Luminescence Instrumentation," in M. D. Lumb, Ed. *Luminescence Spectroscopy*, Academic, London, 1978, (p. 159.)

The xenon arc lamp is more commonly used because of the continuum output. The radiation from a gas discharge lamp originates with the excitation of the gas molecules to higher electronic stages by high-energy intermolecular collisions. This is accomplished by applying a high voltage across an enclosed gas. Ultraviolet and visible radiation are emitted by the excited molecules returning to the ground state. An example of a simple gas discharge lamp is the fluorescent lamp used for indoor illumination. At very low gas pressure in the gas discharge lamp, emission for the gas is monochromatic. As the gas pressure in the lamp is increased along with the current flowing through the lamp, the emission lines from the gas are broadened to bands. At very high pressures the emission of a gas discharge lamp broadens to such an extent that a continuum spectrum is obtained (15).

Other types of lamps are used to excite luminescence, namely, incandescent lamps and low-pressure mercury-vapor lamps (94). Tungsten lamps (incandescent) are stable and inexpensive and do not require elaborate power supplies. The main disadvantage of incandescent lamps is the limited useful output range. They cover the visible range and have no output below 300 nm. Thus, tungsten lamps are only useful for excitation in the visible range. They have found little use in luminescence work because many compounds are excited in the ultraviolet range. The low-pressure mercury-vapor lamp is used fairly widely in luminescence work. It is very stable and does not require a sophisticated power supply. The mercury-vapor lamp gives intense emission at 253.7, 296.5, 302.2, 312.2, 313.3, 365.0, 366.3, 404.7, 435.8, 546.1, 577.0, and 579.0 nm (15). Low-pressure mercury-vapor lamps find extensive use in filter fluorometers because specific lines can be used to excite samples with good efficiency. The low-pressure mercury-vapor lamp is used as a light source for the calibration of monochromators. However, because of its line emission, the mercury-vapor lamp is normally not employed with scanning spectroluminescence instruments.

Excitation source stabilization is important because luminescence intensity is proportional to source intensity. Intensity fluctuations occur because of power supply variation, changes in arc or discharge conditions, and geometry or sputtering of the quartz envelope with electrode material (95). Hamilton et al. (95) have discussed means for source stabilization.

3.2.2. Lasers

The use of lasers in analytical luminescence spectroscopy continues to grow (96–104). Most of the laser-excited luminescence spectrometry studies of analytical importance have involved fluorometry. However, recently laser excitation has been applied to phosphorimetry (98).

Wright (105) has compared the theoretical capability of a xenon arc lamp with a typical laser for excitation in fluorometry. Lytle (104) has further discussed the comparison developed by Wright and other aspects of laser-excited molecular fluorescence of solutions such as limit of detection. For example, a 150-W xenon lamp can deliver 10 μW at 500 $\pm$ 5 nm (2.5 $\times$ 10^{13} photons s^{-1}) to a monochromator slit. If the throughput efficiency is 30%, 7.5 $\times$ 10^{12} photons s^{-1} will be irradiating the sample (104). Lytle assumed the following: each of the photons absorbed would be subsequently reemitted at the specified fluorescence wavelength, the emission monochromator passed 10% of the signal, and there was a lens collection efficiency of 5%, a monochromator throughput of 30%, and a photomultiplier quantum yield of 20%. Under these conditions, 2.3 $\times$ 10^8 counts s^{-1} would be produced at a ratemeter. With an absorptance (the ratio of the radiant energy absorbed by a sample to that incident upon it) of 1.7 $\times$ 10^{-9} and S/N of 2, 4 counts s^{-1} would be produced. If one further assumes a molar absorptivity of 10^5 and a 1-cm path length, this would correspond to a 7.4 $\times$ 10^{-15} M solution. In principle, a tunable laser is more advantageous compared to a xenon arc lamp because a laser linewidth is usually narrower than the absorption linewidth and all the photons incident upon the sample can potentially be absorbed. For example, continuous-wave lasers can deliver 1 W of 555-nm ($\sim$ 2.5 $\times$ 10^{18} photons s^{-1}) radiation to an absorbing sample in solution (104). Using the same conditions discussed previously, the fluorescent solution would be 2.3 $\times$ 10^{-20} M. Such a large improvement in performance compared to a xenon arc lamp is difficult to realize experimentally. Also, it is important to mention that the calculated concentrations for both the xenon arc lamp and laser source are rarely, if at all, achieved experimentally. This occurs mainly because of experimental blank limitations. Omenetto and Winefordner (98) have discussed several aspects of calculated limits of detection with lasers for absorption, fluorescence, and Raman spectrometry. Their discussion should be consulted for more detailed consideration of the topic.

As indicated above, the primary reason the calculated limits of detection are not obtained in luminescence analysis is experimental blank limitations. Most analyses are characterized by a sample matrix that yields an optical blank signal which consists of scattered radiation plus impurity emission. Both Raman scattering and Rayleigh scattering can be reduced by an optical filter or a monochromator. Because of these various interferences, almost all of the analytical laser-based systems have achieved success by studying very well defined chemical systems or by providing additional degrees of selectivity. Additional degrees of selectivity can be achieved by the use of liquid chromatography, fluorescence line narrowing, and temporal separation of the signal from the blank (104). Matthews and Lytle (106) have considered experimentally blank limitations in laser excited solution luminescence. They showed that an improvement in the lower limit of detection can be accomplished by a detailed consideration of the blank luminescence from the cell and the solvent. The laser used in the work was a Phase-R Model N21K nitrogen laser. Some typical limits of detection with laser excitation are 0.52 ng mL^{-1} for an aqueous solution of glucose-6-phosphate using a He:Cd laser and 6 × 10^{-4} ng mL^{-1} riboflavin in water with a N$_2$:dye laser (104). Dovichi et al. (103) obtained a detection limit of 28 attograms using a modified flow cytometer for aqueous rhodamine 6G by laser-induced fluorescence. The detection limit in concentration units was 1.4 × 10^{-13} mol L^{-1}.

It is unlikely that lasers will displace arc lamps as sources for routine analysis in the near future (9). However, the main advantages of lasers relative to mercury and xenon arc lamps are greater intensity, degree of monochromaticity, and spatial coherence. In the pulsed mode, lasers yield very short pulses and are good sources for lifetime measurements. The lasers that have been most widely used for fluorescence excitation are the nitrogen laser (337.1 nm), the argon ion laser (either the 488.0 or 514.5 nm line), and tunable dye lasers. The nitrogen laser is operated in the pulsed mode; thus, to obtain a high signal-to-noise ratio it is necessary to gate the detection system. Because of these factors, in order to use a nitrogen laser in a conventional spectrofluorometer, the detection system has to be modified. Continuous or pulsed operation can be achieved with argon ion lasers. The main limitation of the argon ion laser is its relatively long wavelength of emission, which drastically limits the number of potential applications. It is possible to use frequency doubling, but this results in reduced power and an increase in complexity (9). Dye lasers are more flexible excitation sources because they are tunable (101,102). Dye lasers are normally pumped by a flash lamp or a nitrogen laser. In both situations, the output is pulsed, which requires gated detection. If an argon ion laser is used for pumping, continuous operation is possible; however, this is only possible for long-wavelength excitation. Even though dye lasers are tunable, the dyes employed have a relatively limited range of possible wavelengths. Thus, to vary wavelengths over a wide range of wavelengths, it is necessary to change dyes.

Harrington and Malmstadt (107) described a single modified monochromator that could be used as both a tunable dye laser source of excitation and as a scanning monochromator for fluorescence emission spectra. Richardson and George (108) compared different experimental configurations for pulsed laser-induced molecular fluorescence. A nitrogen-pumped dye laser, a cavity-dumped argon-ion laser, and a externally pulse-picked mode-locked argon-ion laser were compared using rhodamine B. The N$_2$-pumped dye laser was considered the most flexible and generally the most applicable laser system for trace analysis. Omenetto and Winefordner (98) have discussed other laser systems for trace fluorescence analysis. Boutilier and Winefordner (86,109) used a N$_2$ laser and a Chromatix flashlamp pumped dye laser to excite phosphorescence from organic compounds at 77 K using time-resolved measurements. The detection limits with the N$_2$ laser (337 nm) frequently were superior to those with the doubled flashlamp pumped dye laser (270 nm) (98).

For use of lasers as excitation sources, further laser advancement will involve developing better ultraviolet lasers. Excimer lasers will be prominent in these developments. Before an excimer-pumped dye laser can be used to generate tunable radiation throughout the ultraviolet spectrum, further development of dyes is needed (99).

3.3. Wavelength Selection

The optical components used for wavelength selection in luminescence analysis are essentially the same as those employed in ultraviolet–visible analysis work (Chapter 1, Section 1.3.3). A considerable amount has been written on the optical properties and characteristics of various optical components (110,111).

3.3.1. Filters

The use of filters is the simplest way of selecting excitation and emission wavelengths. The filter used in selecting wavelengths for the excitation radiation is called the primary filter, and the one used to select the luminescence wavelengths is called the secondary filter. Figure 8 shows where the wavelength selection components would appear in a typical luminescence instrument. Filters are usually sturdy and can select a very narrow ($\leq$ 1 nm) or very large ($\sim$ 100 nm) bandpass with peak transmissions up to 90%. Filters are frequently used to reduce the stray background from grating monochromators. Both absorption and interference filters can be used in wavelength selection. Guilbault (94) has given a discussion and detailed listing of filters employed in luminescence work. Absorption filters are less expensive than interference filters and are employed for many applications. One of the main advantages of absorption filters is the relatively large fraction of source radiation which is transmitted. This increases the sensitivity of a luminescence analysis. However, the wide bandpass of the absorption filter results in difficulty in choosing a suitable combination of filters so the transmittance of the primary filter does not significantly overlap the transmittance of the secondary filter. The extent to which the transmittances of the primary and secondary filters overlap is an important consideration because the overlap will determine the amount of scattered excitation radiation reaching the detector. The scattered radiation can cause serious error and can be very pronounced even for slightly turbid solutions. In addition, selective excitation of a particular luminescent component is difficult to obtain because of the wide bandpass (9,94).

Interference filters can minimize the above problems with absorption filters because of their narrower bandpass. However, interference filters are usually more expensive than absorption filters. An additional consideration is that the interference secondary filter can transmit scattered excitation radiation if the excitation wavelength corresponds to a higher-order transmission band of the secondary interference filter. This problem can be minimized by using the secondary interference filter in series with a cutoff filter (9).

3.3.2. Monochromators

A monochromator is a device for producing a beam of radiation of narrow bandwidth while allowing the wavelength to be varied. The basic elements of a monochromator are an extrance slit, a dispersing element, and an exit slit. The dispersing element is either a prism or a grating. Prisms are infrequently used in luminescence instruments because they give their greatest dispersion in the ultraviolet region. Most luminescence measurements are made in the visible region. In addition, grating monochromators have a higher light flux throughput than prism monochromators for the same resolution. Gratings are less expensive than prisms and dispersion changes only slightly with wavelength. For any monochromator, there is a tradeoff between throughput and resolving power. For example, decreasing slit widths enhances resolution, but at the expense of source intensity (7,94,95).

In most modern commercial luminescence spectrophotometers, two monochromators are used, one for emission and one for excitation phenomena. Because gratings are the most widely used dispersing elements in luminescence instruments, the properties, advantages, and disadvantages of these devices will be considered.

Figure 10 shows two common monochromator mountings which employ gratings. With these mountings, the grating is rotated to change wavelengths while

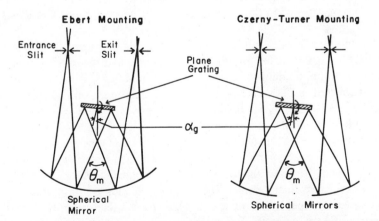

Fig. 10. Grating mountings. The grating is rotated to change wavelengths while the angle between the incident and diffracted rays remains constant. (Reprinted with permission from J. D. Winefordner, S. G. Schulman, and T. C. O'Haver, *Luminescence Spectrometry in Analytical Chemistry*, Wiley, New York, 1972.)

the angle between the incident and diffracted rays remains constant. The fundamental equation for a grating is given by

$$n\lambda = d \sin \theta \tag{34}$$

where n is the order of diffraction, d the distance between adjacent grooves, θ the angle of diffraction, and λ the wavelength of radiation. When source radiation strikes the grating so that the angles of incidence and diffraction are equal but opposite in sign, then $n\lambda = 0$. This condition defines zero order and corresponds to spectral reflection in a reflection grating. When $n = 1$, the diffraction is said to be first order; when $n = 2$, it is second order; etc. Overlapping orders can be a problem with a diffraction grating if the wavelength region to be separated is large. For example, a wavelength at 600 nm in the first order corresponds to 300 nm in the second order, and 200 nm in the third order. All three wavelengths have the same diffraction angle.

The resolving power of a grating is given by

$$R = \frac{\lambda}{\Delta\lambda} = mN \tag{35}$$

where $\Delta\lambda$ is the wavelength difference between two lines that are just distinguishable, λ the average wavelength, N the number of grooves in the grating, and m the length of the grating in centimeters. As seen from Eq. (35), the greater the number of grooves in the grating, the larger the resolution. Gratings with a resolving power of 500,000 have been ruled (94).

Diffraction gratings in luminescence instruments are normally of the blazed reflection type. That is, the grooves are aluminized and have a precise shape so that the maximum amount of radiation is diffracted at a particular angle. If the grating is blazed at 500 nm, for example, its maximum output is at 500 nm. The efficiency of a grating diminishes as the wavelength changes from the wavelength at which it has been blazed.

To summarize, the advantages of gratings are: (1) gratings have uniform resolution and linear dispersion at all wavelengths, (2) up to 50% of the incident radiation can be diffracted into the first order by a blazed grating, and (3) gratings are less expensive than prisms. The major disadvantage of a grating is that several orders of spectra are obtained. This problem can be minimized by use of filters in the optical path. For example, to observe the 600-nm spectral line without interference from the second-order 300-nm spectral line, a filter cutting off radiation below 400 nm should be used (94).

3.4. Slits

Slit width is an important parameter in determining the resolution of a luminescence instrument. The distribution of energy as a function of wavelength for radiation passing through the exit slit of a monochromator can be represented as an isoceles triangle if the entrance and exit slits are of equal width. The wavelength at peak trasmittance is called the nominal wavelength and is the value read on the dial of an instrument. By definition, bandpass is the bandwidth at one-half the peak transmittance and is essentially the width of the exit slit. For grating instruments, the bandpass for a particular slit is constant throughout the spectrum and depends on the number of rulings in the grating (94,110).

There are three types of slits. These are *fixed*, *unilateral*, and *bilateral*. Fixed slits are simply opaque material in which slots have been cut. These can be used in an interchangeable fashion. Unilateral slits are made from two beveled blades or jaws. One jaw is movable and allows continuous variation over a limited range. The main disadvantage of the unilateral slit is that the center of the spectral line is shifted as the slit width is varied. With bilateral slits, two blades move symmetrically to maintain a constant center line (94). The relationship between resolution and slit width has been discussed by several authors (7,94,110). Generally, the smaller the slit width, the better the resolution. Conversely, the larger the slit width, the greater the sensitivity.

3.5. Cell Position and Cells

The most common cell position is one in which there is a 90° angle between the source and detector with the cell at the 90° position. This arrangement provides for minimum background interference from source radiation. The effect of sample geometry on fluorescence intensity has been discussed by Lakowicz (112).

At wavelengths beyond 320 nm, Pyrex cells are suitable for luminescence measurements. It is necessary to use fused silica or quartz cells below 320 nm because Pyrex absorbs strongly below 320 nm. Below 320 nm, it is important to consider the properties of the different kinds of quartz. Various quartz materials possess native fluorescence. Fused silica is preferred over quartz (94). In the 90° configuration, square cells are commonly used in fluorescence measurements because they can be repositioned very reproducibly and scattered and reflected radiation problems are not serious. Round cells can be used in many situations, and they are less expensive.

For measurements at liquid-nitrogen temperature, usually phosphorescence measurements, the sample is normally placed in a quartz tube and immersed in a Dewar flask containing liquid nitrogen. The exciting and emitted radiation pass through the unsilvered part of the Dewar flask. Low temperature measurements involve solid matrices and they are subject to errors because of imperfections in the solid. For example, cracked glasses or snows can form. The errors associated with cracked glasses or snows can be minimized by rotating the sample cell during the measurement (45,113). Lukasiewicz et al. (114) have evaluated open quartz capillary tubes as cells for polar organic molecules in predominantly aqueous solutions at 77 K. McHard and Winefordner (115) developed a windowless, front surface excitation observation cell for low temperature fluorescence and phosphorescence measurements. The sample, which can be a solution, a slurry, or a paste, was cooled by conduction of heat away from the sample through a copper rod. Phosphorimetric limits of detection for several organic compounds were similar to values obtained by conventional spectrophosphorimetric instrumentation. Ward et al. (116,117) have evaluated conduction cooling devices for low-temperature phosphorimetry.

Several microcells and flow cytometers have been developed for biological samples (118–123). With microanalytical techniques, biological tissues can be dispersed into single cell units and analyzed by fluorescence. The cell becomes a microcuvette and has a volume of a few nanoliters. Using a microscope, the microscope objective and eyepiece coupled to a photodetector becomes a spectrofluorometer (122).

Micro flowthrough cells with volumes in the microliter range have been designed for fluorescence detection of compounds separated by high-performance liquid chromatography (94,124–127). Voigtman et al. (90) have compared the fluorescence and photoacoustic characteristics of a windowless flow cell designed for liquid chromatographic applications with the respective characteristics of a static cuvette cell.

Hirschy et al. (128) have described the first reloadable nanoliter cell which is useful for discrete samples. The cell was made from the smallest commercially available Suprasil quartz tubing (0.05 mm i.d.) and had an optical volume of 0.5 nL. It was designed for convenient sample introduction and cleaning.

3.6. Phosphoroscopes

Phosphorescence can be measured in the absence of fluorescence and with minimum scattered radiation by using a phosphoroscope. A phosphoroscope is a device which accomplishes this by allowing the measurement of phosphorescence while the sample is being excited out of phase with the measurement step. O'Haver and Winefordner (80) have discussed the influence of phosphoroscope design in the measured phosphorescence intensity. Mathematical expressions were derived which related the measured and instantaneous phosphorescence intensities to the decay time of the phosphorescent species and to the characteristic

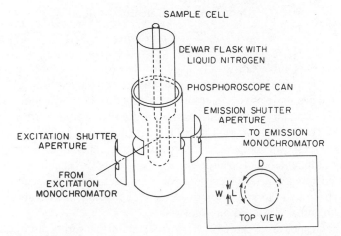

SAMPLE CELL

DEWAR FLASK WITH
LIQUID NITROGEN

PHOSPHOROSCOPE CAN

EMISSION SHUTTER
APERTURE

EXCITATION SHUTTER
APERTURE

TO EMISSION
MONOCHROMATOR

FROM
EXCITATION
MONOCHROMATOR

TOP VIEW

Fig. 11. Schematic diagram of rotating-can phosphoroscope. (Reprinted with permission from T. C. O'Haver and J. D. Winefordner, *Anal. Chem.*, **38**, 602. Copyright 1966 American Chemical Society.)

parameters of the phosphoroscope used. The expressions are important for evaluating essentially any phosphoroscope and for the construction of new phosphoroscopes. Figure 11 shows a diagram of a rotating can phosphoroscope that is widely used in commercial instrumentation. The rotating disk, or Becquerel, phosphoroscope was also considered by O'Haver and Winefordner (80). In addition, Langouet (129) discussed the use of disk phosphorescopes using operational calculus, and Hollifield and Winefordner (130) described a modular phosphorimeter with a single-disk phosphoroscope.

Yen et al. (131) constructed an analog switch phosphoroscope and used it for the measurement of phosphorescence from several organic compounds at 77 K. The apparatus employed a quad analog switch to gate the input of the detector system as in a boxcar integrator. The device was inexpensive and relatively easy to construct. They concluded that the analog switch phosphoroscope was comparable to the rotating can phosphoroscope in rejecting scattered or stray light and fluorescence. In addition, it may be used at higher chopping frequencies than the frequencies used with the commercial phosphoroscope.

In the area of room-temperature phosphorescence, in which the sample is adsorbed on a solid surface, phosphoroscopes have been designed for use with commercial instruments (132,133). Vo-Dinh et al. (132) used a diagonally cut section of an aluminum cylindrical rod for a mirror and reflecting surface. Ford and Hurtubise (133) designed a phosphoroscope and reflection mode assembly for use with a spectrodensitometer.

3.7. Detectors and Detector Systems

3.7.1. Photomultipliers

Photomultiplier tubes are widely used for detection of luminescence in the ultraviolet and visible regions. The most important characteristic of a photomultiplier tube is its relative sensitivity as a function of wavelength. The sensitivity depends mainly on the composition of the photocathode. Materials such as Cs–I, Cs–Te, Sb–Cs, and multialkali have been used in preparing photocathodes. Figure 12 shows anode radiant sensitivity as a function of wavelength for two photomultiplier tubes. The large change in efficiency over certain wavelength regions is an important factor causing recorded luminescence spectra to differ from true spectra and to vary from instrument to instrument. See also Fig. 8 in Chapter 1 for other PMT response curves.

The time response of detectors is important in measuring lifetimes, time-dependent spectra, or in pulse counting. In photomultiplier tubes, the major limiting factor is the width of the output pulse originating from a single photoelectron leaving the photocathode. The width of the pulse depends on the detailed structure of the multiplying stages. The response time of a photomultiplier is about 10^{-8} s, which is sufficiently fast for the measurement of all but the fastest luminescence decay times (7). In complete darkness, there will still be some anode current in a photomultiplier tube. This is called dark current and arises from several sources. Some of these are field emission from sharp points on electrodes, radioactivity, cosmic rays, and, most importantly, thermionic emission from the photocathode. The latter effect has an approximately exponential dependence on cooling; thus, cooling can minimize this contribution greatly. Two important sources of noise in a photomultiplier are the statistical probability of emission of a photoelectron from a photocathode and the fluctuation of secondary emission at the dynodes in a photomultiplier tube. In addition,

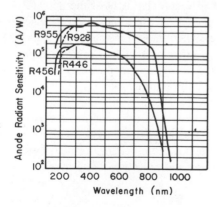

Fig. 12. Typical anode spectral response characteristics. (Reproduced by permission from Hamamatsu Corp., Middlesex, N.J.)

there is a further noise component due to the random arrival of photons at the photocathode. Hamilton et al. (95) and Winefordner et al. (7) have considered these various noise components in some detail.

3.7.2. Photon Counting

Photon counting has been used for the measurement of low light intensities. In photon counting, the photoelectron pulses at the anode of a photomultiplier tube are counted with a high-speed electronic counter. On the average, each photoelectron pulse contains L electrons for a total charge of eL coulombs. Assuming L is approximately 10^6, the resulting charge is large enough to cause an observable voltage pulse across the phototube load resistor R_L. One pulse occurs for each photoelectron ejected at the cathode. Thus, the average pulse count rate is proportional to the light level, assuming that the cathode sensitivity is constant. The photoelectron pulses are often around several millivolts and can be easily amplified by a wide-band AC amplifier and counted by electronic counter circuits. Some of the advantages of photon counting are as follows (7):

1. Digital systems generally drift less than analog systems.
2. At very low light levels, long counting times may be used to accumulate the needed total count.
3. Several types of noise are discriminated against.
4. Digital readouts are easy to read and can be interfaced directly to computers.

A disadvantage of photon-counting detection is that the gain of the photomultiplier tube cannot be varied by changes in applied voltage. In addition, photon counting has a relatively limited range of intensity over which the count rate is linear. Generally, photon-counting detection is inconvenient with high signal levels. To stay within the linear range, slit widths must be adjusted or fluorescence intensity adjusted using neutral density filters. Also, the signal-to-noise ratio becomes unsatisfactory at count rates below 10,000 photons s^{-1} (7).

Franklin et al. (134) have described a relatively simple, inexpensive photon-counting system that can be connected to photomultiplier tubes. Several fundamental instrumental aspects were considered. Jameson et al. (135) discussed the construction and performance of a scanning, photon-counting spectrofluorometer, and Koester and Dowben (136) described their subnanosecond single-photon-counting spectrofluorometer with synchronously pumped tunable dye laser excitation. Cova et al. (137) have given a careful analysis of the requirements for microspectrofluorometric measurements when measuring the fluorescence emitted by single cells. A fully digital instrument using single-photon detection and a multichannel analyzer was described. Darland et al. (138) presented results for the optimum measurement system parameters for several photon-counting systems. They showed that two relatively simple measurements, namely, pulse height distributions and linearity studies, could provide valuable information about the optimum operating parameters for pulse-counting measurement systems under different experimental conditions. Later, Darland et al. (139) presented a method for calculating the relative efficiency of pulse–counting experiments. Meade (140) has reviewed the practical aspects of photon counting. Problems associated with optimizing the detector, amplifier, and discriminator were discussed. In addition, a number of counting configurations found in commercially available equipment were considered. Cline Love and Shaver (141) have discussed the time-correlated single-photon counting technique for lifetime measurements. The time-correlated technique is an instrumental method for measuring changes in light intensity in the nanosecond time region. Upton and Cline Love (142) developed a new method of measuring fluorescence quantum yields by using time-correlated single-photon counting instrumentation. Haugen and Lytle (143) constructed a fluorometer using a single photon counting system and a mode-locked laser excitation source. The fluorometer was used to study various contributions to blank signals.

3.7.3. Optoelectric Imaging Detectors

The use of image or array detectors in luminescence spectrometry is a recent development. Several different types of image detectors are commercially available (144–148). See also Chapter 1, Section 1.3.2. Christian et al. (149) have given an extensive discussion of the properties of array detectors. The major advantage of the imaging detectors is the potentially large multichannel advantage they present. Assuming that equal total observation times are used, the signal-to-noise ratio obtained by using N detectors to simultaneously observe N channels is $N^{1/2}$ greater than that obtained by using a single detector to observe the N channels sequentially. In addition, the time required for N detectors simultaneously observing N channels to attain a given signal-to-noise ratio is a factor of N less than that required for a single detector observing N channels sequentially. In practice, however, the full multichannel advantage is not achieved because of nonideal conditions (149).

The most widely used low-light-level multichannel detector is the silicon intensified target (SIT) vidicon. The main part of this detector is a 16-mm-diameter silicon target. The target consists of a 1000×1000 array of islands of p-type silicon deposited on an n-type silicon wafer. Each of the islands forms a diode whose cathode is addressed by a scanning electron beam. The anode is connected in common with all of the other diodes. While operating, the target is continuously scanned so that each diode is sequentially back-biased and its capacitance is charged.

The spectral response of the most common SIT vidicon (RCA 4804) is defined in the visible and near-IR regions by the composition of the photocathode. Extension into the ultraviolet region is limited by the transmission of the fiber-optic faceplate, which has a sharp cutoff at 380 nm. The response may be extended into the ultraviolet region by special techniques (149). The photometric properties of the SIT vidicon are very good when operated in the continuous scanning mode, and the linearity is comparable to a photomultiplier tube. The dynamic range is limited at the high end by the capacitance of the photodiodes and at the low end by preamplifier noise. It is important to realize that the SIT vidicon is an energy detector rather than a power detector like a photomultiplier tube. Because of this property, it can be operated in an integration mode for enhancing the S/N ratio.

Vo-Dinh et al. (150) replaced a photomultiplier detection system with a commercial SIT–optical multichannel analyzer (OMA) in a commercial spectro-photofluorometer. They considered the SIT–OMA response versus measurement time, resolving power of the SIT system, analytical calibration curves, and limits of detection. They concluded that a SIT image detector tube with a commercial spectrofluorometer was very suitable to analytical applications in molecular spectrometry. Ryan et al. (151) carried out a detailed study of an intensified diode array detector for molecular fluorescence and chemiluminescence measurements. They also compared the intensified diode array detector with a conventional photomultiplier tube. Steady-state fluorescence measurements of quinine sulfate, kinetic-based measurements of thiamine, and chemiluminescence measurements for the lucigenin–H_2O_2 system were investigated. They concluded that the intensified diode array detector was a very useful diagnostic and quantitative tool for low-light-level steady-state and kinetics-based measurements. They found that it was particularly helpful for initial studies of chemical systems to rapidly determine optimum measurement wavelengths and possible spectral overlap problems from potential interfering species. To obtain the lowest possible detection limits a photomultiplier tube was recommended. However, for many analytical situations the intensified diode array detector was sufficiently sensitive for quantitative measurements. In addition, when the intensified diode arrangement is used both quantitative and spectral information are provided simultaneously. Usually, spectral interference problems in samples can be quickly identified, whereas with a fixed-wavelength photomultiplier tube they may go undetected. Also, a variety of simultaneous equilibrium-based, kinetics-based measurements can be implemented with an intensified diode array. For chemiluminescence measurements, Ryan et al. (151) concluded that the intensified diode array would be most useful for initial characterization of new chemical systems and for fundamental mechanistic studies. Its use for quantitative chemiluminescence measurements is more limited because of the very low levels of materials normally measured.

Curtis and Seitz (152) reported the coupling of chemiluminescence thin-layer chromatographic detection to a vidicon rapid scanning detector. Cooney et al. (87) studied the feasibility of using a SIT vidicon as a gas-phase fluorescence detector for gas chromatography. Operation of the SIT in both the real-time and integration modes was discussed. In addition, Cooney et al. (78) reported using the SIT image vidicon as a potential detector for liquid chromatography. A direct comparison was made between a SIT image vidicon and a photomultiplier sequential–linear scanning system for the detection of steady-state fluorescence from molecules in solution. It was concluded that the photomultiplier system was slightly better than the SIT for measuring steady-state fluorescence. However, the multichannel advantage of the SIT made it the better system for measuring transient fluorescence signals. Goeringer and Pardue (153) described the development of a SIT vidicon camera system for phosphorescence studies, and its application to room-temperature phosphorescence of salts of organic acids deposited on filter paper. The instrument permitted time resolved spectra to be recorded with a minimum scan time of 8 ms/scan.

Cooney et al (79) made a comprehensive comparison of image devices versus photomultiplier detectors in atomic and molecular luminescence spectrometry by signal-to-noise ratio calculations. They compaired the signal-to-noise ratios of several image devices (Si-vidicon (V), SIT, intensified SIT (ISIT), secondary electron conduction vidicon (SEC), and image dissector (ID)) with each other and with several sensitive photomultiplier tubes commonly used in optical spectroscopy. Comparisons were made for several hypothetical experimental situations in both atomic and molecular luminescence spectrometry. For the various detectors, the parameters of internal gain, efficiency in the visible and ultraviolet regions, dark count rate, and area per channel (image detectors) were considered. Some of the conclusions that were common to both atomic and molecular luminescence spectrometry are listed below.

1.   For integration times less than about 0.4 s at 32.8 ms per integration, preamplifier noise for image devices dominates over dark-current shot noise at room temperature.

2.   Assuming the photomultiplier tube is shot noise limited, and that signal and noise are measured in one channel of the image device, then the signal-to-noise ratios decrease in the order

$$\text{photomultiplier} > \text{ISIT} > \text{SIT, SEC} > \text{ID} \gg \text{V}$$

3.   The silicon vidicon (V) device is not analytically useful for molecular luminescence spectrometry.

4.   The linear dynamic range of the photomultiplier tube exceeds that of all

the image devices because of its high sensitivity, which results in a lower limit of detection and immunity to target saturation, which in turn results in a higher upper limit compared to the SIT, ISIT, and SEC. The linear dynamic range of the ID is less than that of the photomultiplier, but greater than that of the integrating image devices.

Some of the specific conclusions regarding molecular luminescence spectrometry are listed below.

1. Because the signal-to-noise ratios of image devices (ISIT, SIT, and SEC) are close to the signal-to-noise ratios of photomultiplier tubes for the case of photon noise limitation, the image devices possess a time advantage because they record many spectral components simultaneously.

2. For quantitative analysis, similar detection limits should be obtained with SIT, ISIT, SEC, or the photomultiplier tube.

3. The integrating image devices (ISIT, SIT, SEC, and V) have substantial analytical potential.

Christian et al. (149) have commented on one conclusion made by Cooney et al. (78) concerning the lower detection limit for a photomultiplier tube as a single-channel detector with sequential linear scan compared to the SIT as a parallel detector. Apparently, Cooney et al. (78) did not discuss the sensitivity variation from channel to channel, which is a factor to consider in determining the limit of detection.

Talmi (146) has considered in detail the state of the art of self-scanned photodiode arrays used as a spectrometric multichannel detector for molecular absorption and molecular fluorescence. He compared the fluorescence emission spectra of fluorene measured with a 1024-element silicon photodiode (SPD) linear array, SIT, ISPD, and a photomultiplier tube (R375). He concluded that the ISPD was the multichannel detector of choice because of its photomultiplier-tube-like superior gain and its adequate UV spectral response. Unfortunately, the spectral resolution of the ISPD is substantially worse than that of the SPD. Table 7 gives a comparison between the detection of fluorene obtained with the SPD, SIT, ISPD, and the photomultiplier tube. Recently, Ingle and Ryan (154) have discussed luminescence measurements with an intensified diode array.

In conclusion, image detectors are now widely used in luminescence analysis; however, the photomultiplier tube will continue to be used extensively. Because of the variety of image detectors available, it is important for workers unfamiliar with the properties of these detectors to consult the literature on image detectors about specific properties and possible applications.

**Table 7.   Comparison Between the Detection of Fluorene Obtained with the
SPD, SIT, ISPD, and Photomultiplier Tube[a]**

	Integration Time (s); Readout Mode	Average Signal at Peak (counts)	Average Noise, at Peak, rsd (counts)	S/N at Peak
SPD ($-20°C$)	16; on-target signal integration	78	1.2	65
SIT ($25°C$)	16; 500 consecutive scans, each 0.0325 s	9050	98	92
ISPD ($-15°C$)	1.6; 10 scans, each 0.16 s	43,280	245	177
PMT ($25°C$)	Time constant (system) 0.8 s	56.7 nA	0.8 nA	71

[a]Measurements were performed at peak: 303 nm; λexc, 270 nm. Excitation slits, 10 nm; emission slit, 2 nm. From Talmi (146).

3.8.   Corrected Excitation and Emission Spectra

To determine the luminescence properties of a compound free from instrumental artifacts, it is necessary to correct the observed spectra. Factors that need to be corrected are wavelength-dependent efficiency of the excitation source and detector system plus any wavelength dependency in the monochromators and other optics. Several commercial instruments have accessories available for correcting excitation and emission spectra. Corrected spectra are essential if spectra from different instruments or the same instrument are to be compared. In addition, corrected luminescence spectra are needed in calculating quantum yields and other fundamental luminescence parameters. Parker (4) and Demas and Crosby (10) have reviewed procedures for obtaining corrected spectra. Other authors have discussed various aspects of corrected spectra (155–162). Porro et al. (163) have discussed how corrected excitation spectra can be used in place of absorption spectra to identify small amounts of polycyclic aromatic hydrocarbons. Allen et al. (164) have applied this approach for the characterization of polycyclic aromatic hydrocarbons and hydroaromatics in a distillable coal-derived solvent.

Excitation spectra are distorted mainly by the wavelength-dependent intensity of the exciting radiation. The intensity can be converted to a signal proportional to the number of incident photons by use of a quantum counter. A convenient quantum counter is rhodamine B in ethylene glycol (3 g/L). This solution essentially absorbs all incident radiation from 220 to 600 nm. Generally, the quantum yield and emission maximum ($\sim$ 630 nm) are independent of the excitation wavelength from 200 to 600 nm. Thus, the rhodamine B solution gives a signal of constant wavelength and the signal is proportional to the photon flux of the exciting radiation (112). Quantum counters have been widely used in obtaining corrected excitation spectra and the references cited can be consulted for detailed procedures (155–162).

To obtain corrected emission spectra, one needs to determine the efficiency of the detector system with respect to wavelength. One simple way of obtaining correction factors is to compare the emission spectrum of a standard material with the corrected spectrum for the same substance. Another way to obtain correction factors is by observing the wavelength-dependent output from a calibrated light source. The wavelength distribution of radiation from a tungsten filament lamp can be approximated by a black body of equivalent temperature. Standard lamps are available from the National Bureau of Standards. In addition, one may obtain a corrected emission spectrum by using a quantum counter and a scatterer. The spectral output of the source (Xe lamp) is determined and then the source is used as a calibrated light source. The relative photon output of the source can be obtained with a quantum counter in the sample compartment. Once the intensity distribution of the source is known, the source output is directed onto the detector with a magnesium oxide scatterer. The magnesium oxide is assumed to scatter all the wavelengths with equal efficiency. This procedure is summarized as follows:

1. The excitation wavelengths are scanned with the quantum counter in the sample holder. The results yield the lamp output as a function of wavelength.

2. The scatterer is placed in the sample compartment and the excitation and emission monochromators are scanned simultaneously. This step yields the product of lamp output times the sensitivity of the detector system.

3. Finally the product in step 2 is divided by the function for lamp output and the sensitivity factor for the detector system is obtained (112). Most of the tedium in obtaining data for corrected excitation and emission spectra is minimized with modern computerized instrumentation.

Roberts (165) has considered in detail several aspects of the correction of excitation and emission spectra. Some of his recommendations follow:

1. The excitation source should be corrected by means of a quantum counter.

2. Two compounds that are widely accepted as suitable for quantum counter applications are rhodamine B and 1-dimethylaminonaphthalene-5-sulfate.

3. The primary calibration of the detector system is most conveniently done with the calibrated excitation system and a scatterer, such as $BaSO_4$.
4. There is a need for compounds to serve as secondary emission standards.
5. The usefulness of standard compounds would be improved if they were available as solid blocks. Both heavy-metal-doped inorganic glasses and aromatic fluorophores in plastic deserve further attention.

3.9.   Instrumentation for Lifetime Measurements

Luminescence lifetime measurements are important in many fundamental studies, such as quenching and energy transfer. In analytical work, luminescence lifetimes give an additional factor for characterizing compounds and offer the possibilities of distinguishing various components in a mixture based on lifetime differences.

Phosphorescence lifetimes are normally longer than fluorescence lifetimes and thus are easier to measure. The simplest method is to measure the phosphorescence signal as a function of time after termination of the exciting radiation. For lifetimes longer than approximately 1.0 s, a strip-chart recorder or x–y recorder may be used. For more rapid phosphorescence decay times, a wide-band oscilloscope is used.

Winefordner and co-workers (166,167) have developed pulsed-source phosphorescence instrumentation for time resolved phosphorimetry. Fisher and Winefordner (167) have provided an extensive comparison of pulsed-source and continuous operating mechanical phosphoroscope systems. Recently, Boutilier and Winefordner (86) have described a pulsed-laser (N_2 laser or flash-lamp pumped-dye laser) time-resolved phosphorimeter for phosphorescence lifetime measurements. Mousa and Winefordner (168) developed the new analytical technique of phase-resolved phosphorimetry, which can be used for phosphorescence lifetime measurements. Goeringer and Pardue (153) discussed the development of a time-resolved phosphorescence spectrometer with a silicon-intensified target vidicon camera system and a pulsed source. The instrument permitted time-resolved spectra to be recorded with a minimum scan time of 8 ms/scan. Spectral decay data were processed by a variety of regression methods. Wilson and Miller (169) developed a computer-controlled laser phosphorimeter. The phosphorimeter design allowed time-resolved phosphorimetric studies without the aid of mechanical or electronic chopping devices. Charlton and Henry (170) constructed a very simple apparatus for the determination of relatively long phosphorescence decay times. The main components of the system consisted of a flash gun, a shutter mechanism, and a simple photomultiplier tube. Dyke and Muenter (171) described an inexpensive system for phosphorescent lifetimes constructed of a strobe lamp, optical components, a photomultiplier tube, and an oscilloscope. The system was capable of measuring phosphorescence lifetimes in the millisecond region. In the area of data manipulation, Cline Love and Skrilec (172) have discussed data reduction methods for first-order kinetics of phosphorescence decay.

Several instrumental methods are available for measuring fluorescence lifetimes. Hamilton et al. (95), Rabek (173), and Lakowicz (112) discussed these methods extensively. As Lakowicz (112) has pointed out, there are two widely used methods for the measurement of fluorescence lifetimes. These are the pulse method and the harmonic or phase-modulation method. In the pulse method, the sample is excited with a short pulse of light and the time-dependent decay of fluorescence is measured. For the harmonic method, the sample is excited with sinusoidally modulated light. To calculate the lifetime, the phase shift and demodulation of the emission relative to the incident light are used in the calculation.

For the pulse method, instrumentation is needed to quantify the time-resolved decay of fluorescent intensity. This measurement is experimentally difficult for several reasons. The generally available light sources yield pulses of several nanoseconds in duration and, strictly speaking, simple lifetime theory equations cannot be applied. The observed fluorescence decay must be corrected for the width of the lamp pulse. The correction procedure involves deconvolution (iterative convolution). The problems associated with pulse width can be minimized by pulse lasers with picosecond pulse widths, but pulse lasers are not widely available because of technical and cost considerations. However, if the shape of the lamp pulse remains constant for several pulses, available deconvolution procedures permit measurement of 1-ns lifetimes with pulse widths of 2 ns (112). Another difficulty in lifetime measurements by the pulse method results from measuring the entire time-resolved decay using only a single excitation pulse. For these measurements, a detection system would be needed with a high gain and a subnanosecond response time. To overcome these difficulties, the sample is normally excited with repetitive pulses. It is important that the time between pulses be longer than about 5 decay times to prevent the fluorescence from one pulse from interfering with the observed response from the succeeding pulse. When using repetitive pulses, the time-resolved decay is reconstructed by pulse sampling or photon-counting methods. These methods have been described by Lakowicz (112). Cline Love and Shaver (141,174) have discussed the time-correlated single-photon technique for fluorescence lifetime measurements. Hieftje and Vogelstein (175) have considered a linear response theory approach to time-resolved fluorometry.

The phase and modulation methods of fluorescence lifetime measurements can be briefly described as follows. The sample is excited by light that is modulated in a sinusoidal manner. The emission radiation is, in effect, forced to respond to the excitation, and thus the emission is modulated at the same frequency as the excitation radiation. Because of the finite lifetime of the excited state, the modulated emission is delayed in phase by an angle ϕ relative to the excitation radiation. In addition, the emission radiation is less modulated relative to the excitation radiation. This means the relative amplitude of the variable portion of the emission radiation is smaller for the emission radiation than for the excitation radiation. From these experiments, the phase angle and demodulation factor are obtained and are used to calculate fluorescence lifetime. Weber (176) has discussed in some detail the resolution of the fluorescence lifetimes in a heterogeneous system by phase and modulation methods.

3.10.   Commercial Instruments

Commercial instruments for luminescence measurements are readily available and fall into two general classes, namely, filter instruments and grating instruments. The yearly Labguide issue of *Analytical Chemistry* and yearly Buyer's Guide edition of *American Laboratory* list several companies that manufacture luminescence instrumentation. Guilbault (94) and Lott and Hurtubise (177) have discussed earlier commercial luminescence instrumentation. Many of the design principles of these instruments are still used today. Some of the new features with modern commercial instruments are improved monochromators with holographic gratings and the use of computers for control of instrumental components and data manipulation. Hurtubise (30) has considered commercial instruments that may be employed for solid surface luminescence analysis work. Seitz (92) has discussed commercial instrumentation for chemiluminescence and bioluminescence analysis. In this section, some selected examples of commercial luminescence instruments will be discussed. The discussion is by no means exhaustive.

Farrand Optical Co., Inc. has available filter fluorometers, a spectrodensitometer for measuring fluorescence from thin-layer chromatoplates, spectrofluorometers, and a microscope spectrum analyzer attachment for fluorescence microscopy. An optical schematic diagram for the Farrand MK-2A is shown in Fig. 13. The spectrofluorometer consists essentially of a xenon arc lamp, two grating monochromators, and a detector. Also shown is a thermopile detector for use in obtaining corrected excitation spectra. Wavelength drives and detector signal processing depend on the model. Accessories are available such as cryogenic and phosphorescence attachments, a thin-layer chromatoplate scanner, and flow cells for liquid chromatography experiments. With the Farrand System 3 scanning spectrofluorometer, a programmable data center interface using the Apple II computer can be employed. A variety of data-handling features and accessories are offered with the fluorescence system.

Spex Industries, Inc. has modular luminescence systems available. Many spectral arrangements are possible. Continuous-wave or pulsed sources, single- or double-beam monochromators, single- or dual-beam sample chambers, several

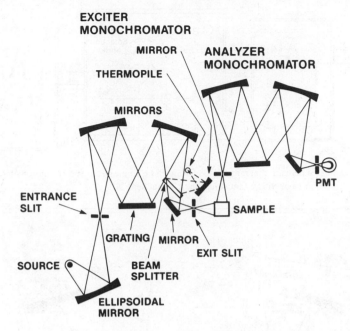

Fig. 13. Optical schematic diagram for the Farrand MK-2A spectrofluorometer. (Reproduced by permission from Farrand Optical Co., Inc., Valhalla, NY.)

detectors for direct current or photon counting, a system control and data manipulation computer, several output storage components, and a variety of accessories can be purchased. Figure 14 shows the optical schematic for the Spex F111 single-beam instrument. One of the several accessories is a phosphorimeter which delivers both pulsed excitation of the sample and selective gating of the signal from the photomultiplier tube. The ability for time-discrimination permits the sorting of competing luminescence signals on the basis of their lifetimes. Because operating parameters are controlled by the Spex computer, data can be acquired and lifetimes calculated automatically through keystroke routines.

The Perkin-Elmer Corporation has a family of fluorescence spectrophotometers which includes both the LS series and the 650 series. Recently, a MPF-66 fluorescence spectrophotometer was introduced that is completely computer controlled. The LS and 650 series instruments can be interfaced to Perkin-Elmer data stations for instrument control, data manipulation, and data storage and retrieval. The Perkin-Elmer LS-5 is a microcomputer-controlled fluorescence spectrophotometer that includes a pulsed source and a gated photomultiplier detector. Several software programs are available for use with the instrument.

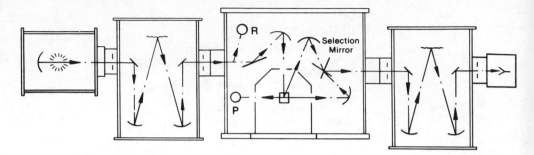

Fig. 14. Optical schematic diagram for the SPEX F111 single beam spectrofluorometer. (Reproduced by permission from SPEX Industries, Inc., Metuchen, NJ.)

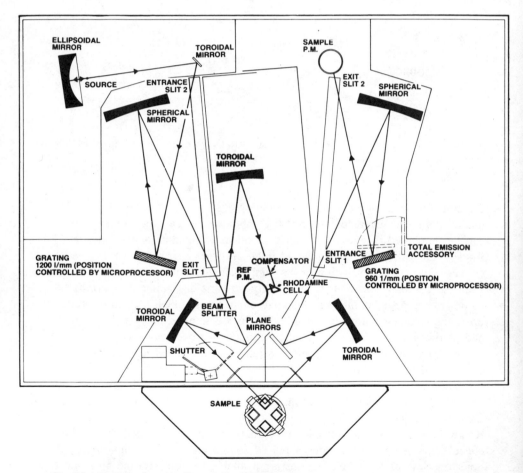

Fig. 15. Optical schematic diagram for the Perkin-Elmer LS-5 spectrofluorometer. (Reproduced by permission from the Perkin-Elmer Corp., Oak Brook, IL.)

116

The LS-5 can be used to measure fluorescence, phosphorescence, chemiluminescence, and bioluminescence. Figure 15 shows the optical path for the LS-5 instrument. Several sampling accessories can be purchased for the LS series of instruments. Some of these are digital printers, microcell holders, high-pressure liquid chromatography flow cells, solid-sample surface accessories, and polarization accessories.

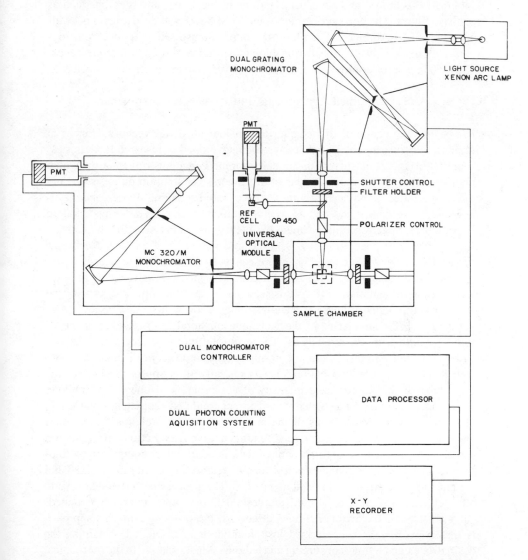

Fig. 16.   Diagram of the SLM 8000 photon-counting spectrofluorometer. (Reproduced by permission from SLM Instruments, Inc./American Instrument Co., Urbana, IL.)

SLM-Aminco sells filter fluorometers and several spectrofluorometers. The SLM 8000 is a photon-counting spectrofluorometer that combines modern electronics and optical design. Microprocessor-based electronics gives spectral data processing as an integral part of the instrument. Features include direct readout of polarization and anisotropy, real-time excitation and emission correction, automatic background subtraction, difference spectra, and derivative spectra. Data can be acquired and saved in any of eight locations, and then operated as required. Figure 16 shows a diagram of the SLM 8000. Other instruments available from SLM-Aminco are a Fluoro-Monitor for use with liquid chromatography, a Chem-Glow for measuring bioluminescence and chemiluminescence, a SPF-500 spectrofluorometer, and an Aminco-Bowman spectrofluorometer.

3.11. Modified Instruments and Research Instruments

It is beyond the scope of this chapter to consider all of the recent advances in luminescence instrumentation that have been developed in research laboratories. Rabek (173) gives a survey of modern luminescence instrumentation which includes comprehensive discussions of instruments for solution fluorescence and phosphorescence, solid-surface luminescence, rapid scanning fluorescence, synchronous fluorescence, and several other instrumental topics. The series "Modern Fluorescence Spectroscopy," edited by Wehry, contains several chapters on fluorescence instrumentation (6). The biennial review in *Analytical Chemistry* on "Molecular Fluorescence, Phosphorescence, and Chemiluminescence Spectrometry" is a very good source of information on recent developments in luminescence instrumentation (74). In this section, selected examples of some of the recent developments in luminescence instrumentation will be discussed.

Giering (178) has discussed the technique of "total luminescence spectroscopy" for multicomponent analysis. The term total luminescence is misleading, however, because in this technique, normally only fluorescence is measured, not both fluorescence and phosphorescence. Generally, this approach involves the acquisition of the fluorescence intensity of a sample as a function of multiple excitation and emission wavelengths. The accumulated dataset is obtained in matrix form with the elements of the matrix representing relative fluorescence intensity, the position of each representing a given excitation and emission wavelength. Because a large amount of data is acquired, a computerized fluorometer is used. By computer control, the wavelength of excitation is fixed and the emission spectrum is obtained and stored. The process is repeated at several other excitation wavelengths until the desired fluorescence matrix is obtained. As Warner and McGown (179) have discussed, the problem with this approach is that even though computerized instrumentation is employed, obtaining the desired data can be time-consuming and tedious. An alternative to the mechanical scanning approach is an instrumental system ("video" fluorometer) developed

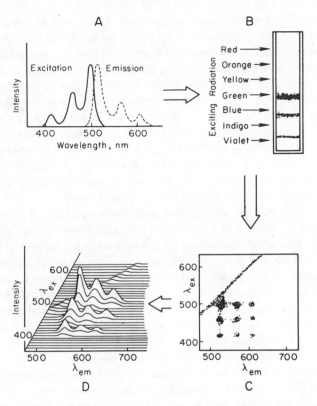

Fig. 17. Production of excitation–emission matrix for hypothetical compound. (*A*) Hypothetical compound's excitation and emission spectra; (*B*) fluorescent cuvet as illuminated by polychromatic beam; (*C*) excitation–emission matrix as viewed by SIT camera; (*D*) isometric projection of excitation–emission matrix as may be obtained on an oscilloscope or graphics terminal. (Reproduced with permission from D. W. Johnson, J. A. Gladden, J. B. Callis, and G. D. Christian, *Rev. Sci. Instrum.*, **50**, 118 (1979).)

by workers at the University of Washington (180–182). The new system uses the novel approach of polychromatic sample excitation to eliminate the necessity of time-consuming mechanical scanning. The fundamental aspects of this approach are illustrated in Fig. 17. In Fig. 17*A*, the solid curve represents an excitation spectrum and the dashed curve represents an emission spectrum of a particular component. Polychromatic illumination can be obtained by exciting the sample cell simultaneously with spatially and spectrally resolved polychromatic radiation (Fig. 17*B*). The spatially resolved polychromatic radiation can be obtained by focusing a continuum source into a monochromator which has been rotated on its side and with the exit slit removed. As an alternative approach, a flat-field polychromator can be used as an excitation source in an analogous fashion (179).

In either situation, the observed sample cell image will be as illustrated in Fig. 17*B*. The bands observed correspond to the relative excitation spectra of the compounds (the fluorescence intensity as a function of the excitation wavelength). The sample cell image is then focused onto the entrance slit of a polychromator or monochromator with the exit slit removed. Each fluorescence band is then dispersed into its component emission wavelengths while the y-axis information is preserved spatially. The resultant image at the exit plane of the analyzing polychromator is shown in Fig. 17*C*. The data can also be displayed as an isometric projection, as given in Fig. 17*D*. The dispersed image is obtained over a specified spectral region without the need for mechanical scanning. Thus, the detection system determines the time limitation in acquiring the data. The acquisition of the two-dimensional spectral data is rapidly obtained with a two-dimensional light transducer such as a multichannel detector. This detector system can yield a 256×256 fluorescence matrix in a time frame of less than 17 ms (179). Warner et al. (183) have discussed a modified version of the "video" fluorometer which was designed to eliminate much unneeded transfer optics. By the combination of holographic polychromators and conventional lenses, spectra as accurate as those obtained on most conventional fluorometers were obtained. In related work, Warner et al. (184) reported a mathematical method for the analysis of a fluorescent sample containing multiple components which uses the experimental emission-excitation matrix (Fig. 17*C*). In addition, Ho et al. (185) discussed simultaneous multicomponent rank annihilation and applications to multicomponent fluorescent data acquired by the "video" fluorometer. Fogarty and Warner (186) described a procedure for the spectral deconvolution of multicomponent fluorescence data by use of a ratio method. Finally, a technique based on the rapid scanning capability of the "video" fluorometer to obtain an emission–excitation matrix has been developed for phosphorimetry (187).

The design and construction of an automated three-dimensional plotter for fluorescence measurements was described by Rho and Stuart (188). The non-computerized approach automatically recorded all the fluorescence spectral parameters of a sample directly on an *X–Y* recorder. Information on the activation and emission wavelengths, emission intensities for each peak, and fluorescence contour plots was obtained. The entire recording was made in about 20 min.

Harrington and Malmstadt (107) developed a new type of spectrofluorometer which employed a single modified monochromator both as a tunable-dye laser excitation source and as a scanning monochromator or polychromator to measure fluorescence emission spectra. The design is unique and the instrument can be assembled with a commercial monochromator and a few attachments, including an easily constructed nitrogen laser. Van Geel and Winefordner (189) compared three laser fluorometers (a pulsed N_2 laser with an emission monochromator, a pulsed N_2 laser-pumped broadband-dye laser with emission monochromator, and a pulsed N_2 laser tunable-dye laser with emission monochromator) with an Aminco-

Table 8. Fluorometric Limits of Detection for Several Polynuclear Aromatic Hydrocarbons by N_2-Laser (No Dye) Fluorometer and a Conventional Spectrofluorometer[a]

	Aminco-Bowman Spectrofluorometer			N_2 Laser—No Dye		
	λexc(nm)	λem(nm)	LOD (μ mL^{-1})	λexc(nm)	λexc(nm)	LOD (μg mL^{-1})
Anthracene	352	399	5×10^{-6}	337	397	1×10^{-5}
Chrysene	260	387	5×10^{-6}	337	388	3×10^{-5}
Fluoranthene	337	464	5×10^{-5}	337	460	1×10^{-5}
Phenanthrene	250	360	3×10^{-5}	337	400	5×10^{-5}
Pyrene	316	380	2×10^{-5}	337	390	2×10^{-5}

[a]Reprinted with permission from Van Geel and Winefordner, *Anal. Chem.* **48**, 337. Copyright 1976 American Chemical Society.

Bowman spectrofluorometer with respect to analytical figures of merit. On the basis of detection limits (concentration of analyte producing a signal-to-noise ratio of 2) the fluorometer with the N_2-laser beam and the Aminco-Bowman spectrofluorometer gave the best results. Table 8 compares the limits of detection obtained for the two fluorometers. Richardson and George (108) compared three laser excitation sources (nitrogen-pumped dye laser, cavity-dumped argon-ion laser, and externally pulse-picked mode-locked argon-ion laser) using rhodamine B as an ideal fluorophor. The complementary characteristics of peak power versus repetition rate were compared. They found that the nitrogen-pumped dye laser offered greater flexibility in potential analytical applications. Tunability was much larger with the nitrogen-pumped dye laser through a wider selection of both dyes and doubling techniques. In addition, the higher peak powers permit the application of nonlinear optical techniques to analytical problems. Haugen and Lytle (143) constructed a fluorometer employing a single photon-counting and mode-locked laser excitation source to test the concept of time-filtered detection. Time-filtered detection will be discussed later in the chapter. A block diagram of the instrument is shown in Fig. 18. This particular instrument design is important for measurements involving temporal resolution. Imasaka et al. (190) reported the design and construction of a fluorometric system with a nitrogen-laser-pumped dye laser excitation source and a gated photon-counting signal processor. The system was applied to the lifetime measurements of nano-second decays. Several researchers have constructed photon-counting spectro-fluorometers. Brief consideration of two of these systems follows. Jameson et al. (135) constructed a photon-counting spectrofluorometer interfaced to a Nuclear Data ND 812 computer to facilitate acquisition of emission spectra at very low signal-to-noise ratios. Photon-counting was combined with repetitive scanning

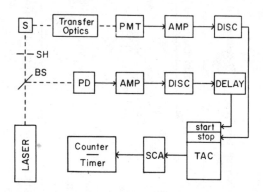

Fig. 18. Instrument block diagram for time-filtered detection: BS, beam splitter; S, sample; SH, shutter; PD, photodiode; PMT, photomultiplier; AMP, amplifier; DISC, discriminator; TAC, time-to-amplitude converter; and SCA, single-channel analyzer. (Reprinted with permission from G. R. Haugen and F. E. Lytle, *Anal. Chem.*, **53**, 1554. Copyright 1981 American Chemical Society.)

and allowed the detection and quantitation of emission from strong fluorophors near the picomolar level. Koester and Dowben (136) described a synchronously pumped tunable-dye laser which was interfaced with a modified Ortec 9200 photon-counting system for the purpose of measuring subnanosecond relaxation phenomena. The shortest lifetime measured was 68 ps.

Wehry and Mamantov (191) have developed the technique of analytical matrix isolation fluorescence spectroscopy. In matrix isolation, the sample constituents are vaporized and then mixed thoroughly with a large excess of a diluent gas. The resulting gaseous mixture is deposited at cryogenic temperatures on an acceptable optical window for spectroscopic examination as a solid. Figure 19 shows a schematic diagram of a cryostat head design used for matrix isolation experiments. Dickinson and Wehry (192) discussed time-resolved matrix-isolation fluorescence spectrometry of mixtures of polycyclic aromatic compounds. In this work, they described a time-resolution fluorometer for matrix isolation studies. In the instrumental system, an argon-ion laser was used as an excitation source. Fluorescence from the matrix isolated sample was then analyzed with a grating spectrometer and photomultiplier system.

Khalil et al. (193) developed a new optical arrangement for the Abbott VP bichromatic analyzer. They converted it into a fluorometer by the insertion of a

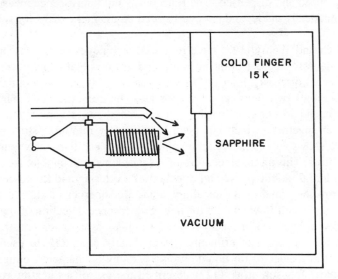

Fig. 19. Schematic diagram of cryostat head design used for matrix isolation of polycyclic aromatic compounds. Samples placed in Knudsen cell (wrapped with heating wire maintained at voltage V), mixed with N_2 and deposited on a window (in this case, sapphire for fluorescence spectrometry). Head maintained at low temperature by closed-cycle (helium) refrigerator and evacuated to pressure of $\sim 10^{-5}$ Torr. (Reprinted with permission from E. L. Wehry and G. Mamantov, *Anal. Chem.*, **51**, 643A. Copyright 1979 American Chemical Society.)

specially designed filter carriage. Fluorescence intensity was measured by using straight-through excitation geometry and was corrected for light-source intensity fluctuations by using the attentuated transmitted beam through the solution as a reference. Ritter et al. (194) described procedures and an electronic interface for a microcomputer-controlled spectrofluorometer. The instrument was used to obtain corrected fluorescence spectra which were then integrated for use in the calculation of quantum yields. Corrections were made for photomultiplier wavelength response profile, light source emission profile, and monochromator/optics transmission properties. Possible absorbance of excitation and emission radiation by the sample was not considered. The instrument was used to calculate and output derivative spectra and differential spectra to correct for background interferences. Thompson and Pardue (195) evaluated a silicon-intensified target vidicon as a detector for synchronous fluorescence spectroscopy. Data were acquired in the form of 180 emission spectra at each of 100 excitation wavelengths and stored in computer memory. Synchronous spectra were obtained as diagonals through the matrix that corresponded to fixed wavelength differences between excitation and emission wavelengths. Howard et al. (196) constructed an automated instrument for measurement of fluorescence and reflectance from reagent strips. The reagent strips could be automatically loaded, and samples automatically sequenced and dispensed. The reflectance and fluorescence measurements were automatically made, and data was stored on magnetic disks for subsequent analyses.

Froehlich and Wehry (197) have reviewed techniques, examples, and prospects for fluorescence detection in liquid and gas chromatography. Yeung and Sepaniak (125) discussed laser fluorometric detection in liquid chromatography. Figure 20 shows two flow cells used for laser fluorometric detection in high-performance liquid chromatography. Figure 20a shows a capillary tube and a solid rod arrangement which has been employed to minimize stray radiation. Figure 20b shows an alternative arrangement using a capillary tube coupled with an optical fiber. This arrangement miminizes problems with bubbles from degassing of the liquid mobile phase and rejection of scattering and fluorescence originating from the walls of the capillary tube. Hershberger et al. constructed a submicroliter sheath-flow cuvette for laser fluorescence detection of liquid chromatography effluents (126), and utilized a similar system for real-time video fluorometric monitoring of effluents (197a). Shelly et al. (197b) used a video fluorometer as a detector for liquid chromatography to characterize highly complex samples. Johnson et al. (127) described the use of a filter fluorometer in liquid chromatography. The detector was useful at the picogram level. Sepaniak and Yeung (198) characterized and described a laser two-photon excited fluorometric detector for high-performance liquid chromatography. Excitation was provided by the absorption of two photons of radiation at 514.5 Å from an argon-

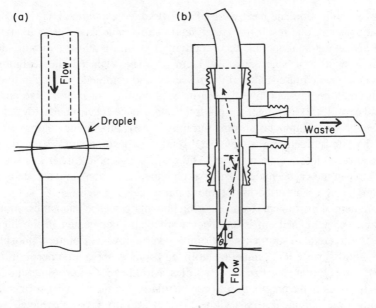

Fig. 20. Two flow cells for laser fluorometric detection in HPLC. (*a*) A flowing droplet suspended between a capillary tube and a solid rod. (*b*) A capillary tube-optical fiber combination. (Reprinted with permission from E. S. Yeung and J. M. Sepaniak, *Anal. Chem.*, **52**, 1465A. Copyright 1980 American Chemical Society.)

ion laser. Boutilier et al. (199) reported a computer-controlled liquid chromatograph. The absorption, fluorescence, and fluorescence-excitation detection system was centered around a SIT vidicon detector. The system allowed fluorescence and absorption spectra to be recorded and stored every 3 s without stopping flow. Folestad et al. (200) studied the effluent from different kinds of columns by a laser-induced fluorescence detection technique with respect to postcolumn band broadening, sensitivity, and utility. The effluent from conventional high-performance columns was arranged as a free-falling thin jet. To avoid drop formation, microcolumn effluents were passed through a quartz capillary. For excitation of the sample, radiation from a krypton-ion laser in the ultraviolet mode was used for excitation.

Cooney and Winefordner (88) compared several different optical systems and excitation sources for the detection of gas-phase fluorescence from gas chromatographic effluents. It was concluded for the systems studied that improved performance can be attained by the use of sources with intense output in the ultraviolet, or by use of conventional sources with excitation monochromators which are very efficient in the ultraviolet. Hayes and Small (201) commented on the simplification and narrowing of laser-excited excitation and fluorescence emission spectra that result from analytes in supersonic jets. The dramatic rotational and vibrational cooling accompanying expansion allows vibronic absorption line widths of approximately 1 cm $^{-1}$. They labeled the technique as rotationally cooled laser-induced fluorescence and combined this technique with gas chromatography and investigated its applicability for quantitative analysis. Conrad et al. (202) used matrix isolation fluorescence spectrometry for detection in open tubular column gas chromatography. Chromatographic column effluents were deposited directly on a movable 12-sided deposition surface positioned in the head of a closed-cycle cryostat. The chromatographic carrier gas served as the matrix in the cryogenic deposit. An undispersed xenon lamp source and SIT vidicon detector were used in obtaining fluorescence spectra.

Commercial instruments for measuring phosphorescence intensities, lifetimes, excitation spectra, and emission spectra are well documented in the literature (94). Commercial instruments are employed in low-temperature phosphorescence, solid-surface RTP, micelle-stabilized room-temperature phosphorescence (RTP), and solution-sensitized RTP studies. Modifications to commercial instruments for work in the previous areas can be relatively minor or very sophisticated, depending on individual needs. Cline Love et al. (38) have constructed a phosphorimeter from commercially available components and have used it in micelle-stablized RTP. Ford and Hurtubise (203) constructed a phosphoroscope and

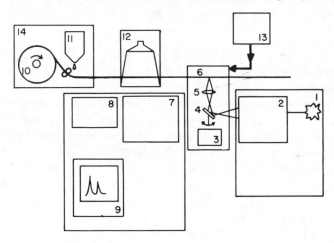

Fig. 21. Schematic diagram of an AutoAnalyzer continuous filter with the room-temperature phosphorescence detection system. (1) Light source; (2) excitation monochromator; (3) rotation motor–phosphoroscope; (4) reflecting surface; (5) optics; (6) filter paper; (7) emission monochromator; (8) detection unit; (9) recorder; (10) filter paper roll; (11) spotting syringe; (12) drying IR lamp; (13) dry air supply; (14) AutoAnalyzer continuous filter. (Reprinted with permission from T. Vo-Dinh, G. L. Walden, and J. D. Winefordner, *Anal. Chem.*, **49**, 1126. Copyright 1977 American Chemical Society.)

modified a reflection mode assembly for a spectrodensitometer to measure RTP from solid surfaces. Vo-Dinh et al. (132) designed an automatic phosphorimetric instrument for solid-surface RTP with a continuous filter paper device, as shown in Fig. 21. Later, Yen-Bower and Winefordner (204) reported a modified version of the above system with a filter paper guide that allowed continuous sampling of organic phosphors adsorbed on filter paper. Walden and Winefordner (205) emphasized that one problem with most commercial luminescence instruments is that only a small fraction of the total emitted luminescence is collected and measured. They made a comparison of ellipsoidal and parabolic mirror systems that partly compensated for this problem and employed the systems in fluorometry and solid-surface RTP work. Hurtubise (30) has considered other aspects of the instrumentation developed for solid-surface RTP.

Fisher and Winefordner (167) first showed the experimental importance of pulsed-source time-resolved phosphorimetry. This approach is useful for analyzing mixtures of fast-decaying phosphors. After an initial pulse of source energy, with a duration t_f, the phosphorescence intensity climbs to a maximum peak value and then decays exponentially. At a delay time t_d after the source flash has decayed significantly, the multiplier phototube is turned on and the phosphorescence signal is monitored. The multiplier phototube is then turned off and the sequence is repeated. The integrated luminescence intensity is measured during the "on" time t_p of the phototube. Three variations of pulsed-source phosphorimetry that can be used to determine phosphors in binary and multicomponent mixtures have been discussed (167). The application of pulsed-source phosphorimetry to the quantitative and qualitative analysis of drugs has been reported (206). In this application, a xenon flashtube source was used instead of the standard continuously operated xenon-arc source and phosphoroscope. In addition, either a boxcar averager for temporal information at one value of t_d or a signal averager for information at several values of t_d was used. Recently, Boutilier and Winefordner (86,109) used both a pulsed N_2 laser or a flashtube-pumped dye laser in place of the pulsed xenon source. The main reasons for using a laser were the clean temporal characteristics of the laser pulse and the ability to perform time resolution to enhance the sensitivity and selectivity of the measurement (28). Figure 22 shows a block diagram of the pulsed-laser time-resolved phosphorimeter.

Wilson and Miller (169) developed a computer-controlled laser phosphorimeter in which the phosphorescence spectra were recorded on magnetic tape as signal-averaged families of decay curves. With the phosphorescence emission data in this form, time-resolved phosphorimetric studies were possible without the aid of mechanical or electronic chopping devices. Generally the data were acquired once and then could be displayed as desired using almost any time window of interest. This allowed one to conduct detailed kinetic analyses of the decay processes influencing emission. Wilson and Miller emphasized that further

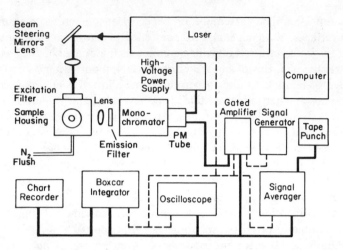

Fig. 22. Block diagram of pulsed-laser time-resolved phosphorimeter. (Reprinted with permission from G. D. Boutilier and J. D. Winefordner, *Anal. Chem.*, **51**, 1384. (Copyright 1979 American Chemical Society.)

refinement of this technique is needed, such as adjustment of the time lapse between the data acquisition and the display of the reconstructed spectrum.

Goeringer and Pardue (153) discussed the development of a time-resolved phosphorescence spectrometer with a SIT vidicon camera system and a pulsed source. The instrument permitted time-resolved spectra to be recorded with a minimum scan time of 8 ms/scan. Spectral decay data were processed by a variety of regression methods to obtain rate constants, lifetimes, and initial intensities. Both temporal and spectral data were used to analyze data from the room-temperature phosphorescence of compounds adsorbed on filter paper for single-, two-, and three-component mixtures.

Ho and Warner (187,207) discussed a new approach using phosphorimetry for multicomponent samples by the rapid acquisition of multiparametric data. The phosphorescence data were acquired in the form of emission–excitation matrices. As discussed earlier, workers at the University of Washington (182) developed a new fluorescence instrument, called a "video" fluorometer. Mathematical expressions were used in representing the fluorescence emission–excitation matrix, and linear algebra and computer algorithms were used for both qualitative and quantitative analysis of fluorescence data (184,186). Similar mathematical techniques were used by Ho and Warner (187,207) for the phosphorescence emission–excitation matrix. The "video" fluorometer system was modified slightly to obtain the phosphorescence emission–excitation matrix. For fluorescence measurements, the excitation source was continuous, and the observed fluorescence signal reached steady-state conditions before the data were acquired.

With the phosphorescence experiments, the excitation source was cut off completely so only the phosphorescence signal was observed. The multidimensional phosphorimetric approach developed by Ho and Warner (187,207) illustrates the need to couple instrumental methods with mathematical algorithms for simultaneous multiparametric luminescence measurements.

As part of the overall microprocessor revolution, there has been an increased use of computers with luminescence instrumentation. Wampler (208) has considered methods and instrumentation associated with computers in fluorescence instrumentation. Weiner and Goldberg (209) discussed optimizing the selectivity of standard spectrofluorometers with computer control. Lyons et al. (210) considered spectral interpretation of fluorescence spectra by automated file searching. In other areas, Fitzgerald (211) described digital and analog measurements in fluorescence spectroscopy. O'Haver (212) reported on modulation and derivative techniques in luminescence spectroscopy of increased analytical selectivity. Warner and McGown (179) have given a review of instrumentation for multicomponent fluorescence analysis.

3.12. Instrumental Limitations and Improvements

The chemist interested in luminescence measurements has a wide range of instruments and instrumental systems available, from simple filter fluorometers to highly sophisticated computer-controlled instruments. However, there are still some limitations associated with modern luminescence instrumentation. An ideal luminescence instrument should have a light source that yields a constant photon output at all wavelengths, a monochromator that passes all wavelengths with equal efficiency, a monochromator that is independent of polarization, and a detector that detects photons at all wavelenths with equal efficiency (112). It is well known that light sources, monochromators, and detectors with such ideal characteristics are not available. Thus, it is necessary to compromise on the selection of components and to correct for the nonideal responses of the instrumental components. Many of the corrections can be carried out with modern electronics and computers, and one should see many more computer-controlled instruments in the future.

As Weiner and Goldberg (209) have pointed out, most commercial spectrofluorometers are designed to use sensitivity advantages. Only the newer instruments have the capability of taking advantage of the selectivity advantages of luminescence. One way of improving selectivity is to use the total excitation–emission matrix described by Christian and co-workers (182). Weiner and Goldberg (209) have described a modification to a commercial spectrofluorometer which employs computer control of the spectrofluorometer to obtain the excitation–emission matrix automatically.

The widespread use of lasers in commercial instruments has not occurred.

This is partly due to cost and the technical complexity of some laser systems. Lytle (104) has commented that unless additional forms of selectivity (chemical or instrumental) are combined with steady-state fluorometry, it is not likely that chemically complicated samples will benefit much from laser excitation. Lytle further states that major instrumental concerns in molecular solution fluorometry have recently been directed toward selectivity rather than sensitivity. This trend acknowledges the fact that such methodology has been blank limited. In the future, combination instruments using several laser properties should allow the handling of very complicated samples. Much remains to be done to transfer laser technology to the analytical laboratory. For example, modern microprocessor controlled pulsed dye lasers are a step in the right direction (104).

Christian et al. (149) have emphasized that they have yet to be held back by array detectors in any application attempted. The final replacement of the photomultiplier tube for various applications must await the availability of the large-area backside-illuminated charge-coupled device array with its excellent quantum efficiency and near-photon-counting sensitivity. Also, it should be emphasized that the development of new strategies for data reduction have to keep up with the increasing amount of data obtained per sample, as with array detectors.

For solid-surface luminescence analysis, commercial and research instruments are adequate for obtaining good luminescence data. However, several improvements can be made. Positioning of the adsorbed component in the beam of source radiation does cause some difficulty and can cause errors that exceed 2% (213). More extensive use of computers and digital electronics will minimize this source of error. Ebel and Hocke (213–215) and Ebel et al. (216) have discussed some of these aspects. In one application, they used a programmable desk calculator with a spectrodensitometer to maximize the position of each adsorbed component with respect to the source beam (213). Because highly scattering media are usually employed as solid surfaces, new detection systems can be employed to improve signal-to-noise ratio. Television-type multichannel detectors (image devices) should find use in this area.

4. APPLICATIONS

In this section, relatively new applications that have appeared in the literature will be discussed. It is obvious that complete coverage of all analytical applications in luminescence analysis is beyond the scope of this chapter.

4.1. Derivative Spectroscopy

O'Haver (212) has discussed modulation and derivative techniques in luminescence spectroscopy. To obtain derivative spectra, two general experimental approaches have been used, those that operate on the output signal of the spectrometer, such as electronic differentiation, and those which operate on the light beam in the optical part of the spectrometer, such as wavelength modulation. Electronic differentiation is a relatively simple approach, and with modern computers and recorders it has become relatively easy to obtain derivative spectra. O'Haver (212) has considered both general approaches for obtaining derivative spectra in terms of signal-to-noise ratio and also has discussed qualitative and quantitative applications. Several authors have detailed the advantages and disadvantages of derivative spectroscopy (217–222). Two important advantages are the resolution effect and discrimination effect (220). Generally, the resolution effect allows the resolution of bands that cannot be resolved in the zeroth derivative spectrum. The discrimination effect permits the enhancement of the sharper features of the zeroth-order spectrum.

Green and O'Haver (217) first showed the qualitative and quantitative applicability of derivative spectrometry in luminescence spectrometry. Figure 23 shows the zeroth and first derivative fluorescence excitation spectra of 1 ppm benzo[a]pyrene in a methanol–water solvent mixture. Corresponding spectral features are numbered for comparison. Minor features, such as those numbered 3, 7, and 12, are much more evident in the derivative spectrum. Green and O'Haver illustrated several potential applications of the derivative technique to fluorescence crude oil fingerprinting. In addition, they showed the quantitative application of fluorescence derivative spectroscopy for a mixture of pyrene and excess anthracene.

Fox and Staley (223) investigated the determination of polycyclic aromatic

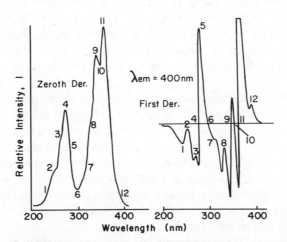

Fig. 23.   Normal and first derivative fluorescence excitation spectra of 1 ppm benzo[a]pyrene in a methyl alcohol–water solvent mixture. (Reprinted with permission from G. L. Green and T. C. O'Haver, *Anal.Chem.*, **46**, 2191. (Copyright 1974 American Chemical Society.)

hydrocarbons in atmospheric particulate matter by high-performance liquid chromatography and fluorescence techniques. Both derivative fluorescence spectroscopy and selective modulation of the fluorescence spectra were employed for difficult identifications. Vo-Dinh and Gammage (224) applied the second derivative technique to room-temperature phosphorescence analysis. They found some reduction in signal-to-noise values, but the second derivative technique was particularly useful in improving the selectivity of an assay. Christenson and McGlothlin (225) reported the elucidation of metanephrine to normetanephrine and epinephrine to norepinephrine ratios by fluorescence derivative spectrometry. First and second derivative fluorescence spectroscopy were performed over the 325–380 nm excitation region on the fluorescent products of the metanephrines and catecholamines. A fixed 505-nm emission wavelength was employed in the work. The ratios of methanephrine to normetanephrine and epinephrine to norpinephrine were obtained by calculations using first and second derivative fluorometric spectra.

4.2. Synchronous Luminescence Spectroscopy

Synchronous luminescence spectroscopy involving scanning both excitation and emission monochromators at the same time at a fixed wavelength difference ($\Delta\lambda$). Vo-Dinh (226) has considered the details of synchronous luminescence spectrometry and discussed conditions for choosing $\Delta\lambda$. The main advantages of the synchronous luminescence technique are its simplicity and the enhanced selectivity achieved for relatively complex mixtures. Vo-Dinh recently reviewed synchronous excitation spectroscopy (227) and the overall methodology and applicability of synchronous luminescence spectroscopy (228). Lloyd and Evett (229) discussed the prediction of peak wavelengths and intensities in synchronously excited fluorescence emission spectra. Latz et al. (230) pointed out some of the limitations of synchronous luminescence spectrometry in multicomponent analysis.

Lloyd (231) introduced synchronous excitation spectrofluorometry. Later he showed the applicability of the approach in forensic science (232,233). For example, automobile engine oils, bitumen and coal tar pitch, and soots from automobile exhausts were characterized by synchronous fluorescence spectroscopy. Various complex mixtures were characterized by partly quenched, synchronously excited fluorescence emission spectra (234). It was shown that specificity of synchronously excited fluorescence spectra was enhanced by the use of a variety of quenching conditions. Used oils, soots, gear oils, and mineral oil were employed in the experimental work.

John and Soutar (235) have considered several factors influencing synchronous excitation fluorometry. They were mainly interested in the identification of crude oils, and they investigated the effects of solvent, wavelength increment, concentration, temperature, and frequency bandpass. Several synchronous excitation fluorescence spectra were presented for various crude oils. Wakeham (236) compared conventional fluorescence emission spectra and spectra produced by the synchronous technique on a series of standard oils to demonstrate the increased resolution obtained by synchronous scanning. In addition, the indigenous and petroleum-derived aromatic hydrocarbons were characterized in sediments from Lake Washington.

Vo-Dinh and Gammage (237) proposed the use of the singlet–triplet energy difference as a new parameter for improved selectivity in room-temperature phosphorescence. The approach was based on the synchronus excitation technique and took advantage of the specificity of energy gaps (Δ_{st}) between the phosphorescence emission band and absorption bands. Table 9 shows the results for various polynuclear aromatic compounds. These data show the specificity of the $\Delta\lambda$ values and the general applicability of the approach. The $\Delta\lambda$ values were determined mainly by the energy gaps Δ_{st}. Vo-Dinh et al. (238) reported the application of the synchronous technique to the monitoring of trace organic pollutants, mainly polycyclic aromatic hydrocarbons, originating from coal conversion processes. In another application, room-temperature phosphorimetry was used to identify and quantify polynuclear aromatic compounds in a synthetic fuel sample (239). Several trace and major components, benzo[a]pyrene, chrysene, fluorene, phenanthrene, and pyrene, were identified and quantitated at concentrations ranging from tens to thousands of parts per million. Selectivity was improved by employing selective heavy-atom perturbation and synchronous excitation scanning. Vo-Dinh et al. (240) reported the results of the analysis of 13 polynuclear aromatic compounds from a workplace air particulate sample by synchronous luminescence and room-temperature phosphorescence. Also, Vo-Dinh and Martinez (241) developed methods for the direct determination of selected polynuclear aromatic hydrocarbons in a coal liquefaction product by synchronous fluorescence and room-temperature phosphorescence methods. Standard deviations ranged from 10–30% for a concentration range of 0.1–6 mg g^{-1}. As mentioned in Section 3.11, Thompson and Pardue (195) investigated synchronous fluorescence spectroscopy with a silicon-intensified target vidicon. Single- and two-component mixtures of anthracene, 9,10-diphenylanthracene, perylene, and tetracene were studied. Detection limits at the 95% confidence level were approximately between 0.005 mg L^{-1} (perylene) and 0.17 mg L^{-1} (anthracene).

Inman and Winefordner (242,243) developed the technique of constant-energy synchronous luminescence spectroscopy. Polycyclic aromatic hydrocarbons at low temperature (242) and room temperature (243) were investigated. In constant-energy synchronous luminescence spectroscopy, the excitation and emission

Table 9.   RTP Excitation and Emission Bands and Experimental Δλ Values for
            Several Polynuclear Aromatic Compounds[a]

Compounds	Excitation Peaks (nm)	Emission Peaks (nm)	Experimental Optimal Δλ Values (nm)
Acridine	360	640	280
Chrysene	325	515	190
Fluorene	305	430	125
Naphthalene	275	472	197
Phenanthrene	250	460	210
Pyrene	350	600	250
Quinoline	315	465	150
Benzo[e]pyrene	335	543	208
Benzo[a]pyrene	392	690	298
1,2,5,6-Dibenzanthracene	301	555	254
1,2,3,4-Dibenzanthracene	296	570	274

[a]Reprinted with permission from Vo-Dinh and Gammage, *Anal. Chem.*, **50,** 2058. Copyright 1978
American Chemical Society.

monochromators are scanned simultaneously, and are synchronized so that a
constant energy difference $\Delta \overline{V}$ is maintained between the monochromators.
Improved selectivity over conventional fluorometry was obtained. When applied
at low temperature, the combination of band narrowing and improved selectivity
with the constant-energy-difference technique proved to be very useful for analysis of relatively complex samples (242,243).

4.3. Matrix-Isolation and Site-Selection Fluorescence Spectroscopy

As discussed earlier in Section 3.11, Wehry and Mamantov (191) developed the
analytical aspects of matrix-isolation fluorescence spectroscopy. Several quantitative and qualitative applications of this technique have appeared for polycyclic
aromatic hydrocarbons (PAH) (244). Stroupe et al. (245) investigated matrix-
isolated samples of PAH which were prepared by mixing the PAH vapor effusing
from a Knudsen cell with a large excess of nitrogen gas. The gaseous mixture
was then deposited onto an optical window at approximately 15 K. The detection
limit for PAH in nitrogen matrices was about 10^{-11} g. Calibration curves for
the compounds were linear over five decades or more in concentration. Stroupe
et al. (245) demonstrated the use of matrix-isolation fluorometry for identification
and quantitation of PAH in coal-derived samples. Tokousbalides et al. (246)

combined matrix-isolation fluorescence and Fourier transform infrared spectrometry for the analysis of isomeric methylchrysenes. Each one of the six methylchrysene isomers could be identified in mixtures containing all six compounds by matrix-isolation fluorescence spectrometry. The positions and relative intensities of the principal matrix-isolation fluorescence peaks for the six isomers are given in Table 10. Maple et al. (247) explored the analytical utility of laser-induced fluorescence spectrometry of PAH matrix isolated in vapor-deposited, annealed n-heptane deposits. Greater spectral resolution was obtained in heptane compared to the conventional nitrogen or argon matrices. The linear dynamic range for fluorescence of PAH in the vapor-deposited alkane matrices was comparable to and usually exceeded that observed for N_2 media. Maple and Wehry (66) investigated the usefulness of site-selection techniques (fluorescence line narrowing) for the characterization of multicomponent, matrix-isolated mixtures of polar PAH. In the site-selection technique, guest molecules occupy several different microenvironments or "sites" in a low-temperature matrix. Thus, the purely electronic energy levels of different molecules of the same solute are shifted to different extents. Maple and Wehry (66) found that by laser selective excitation of the polar molecules residing in different "sites," much greater spectral resolution was achieved compared to conventional lamp sources. The linearity of the fluorescence for model compounds was from approximately 5 ng to greater than 5 μg. Typical relative standard deviation values were 7% for 2,3-dihydroxynaphthalene and 4% for 2,7-dihydroxynaphthalene. In later work, Maple and Wehry (248) reported on the analytical importance of fluorescence photoselection for distinguishing overlapping spectral bands in mixtures of matrix-isolated PAH. After excitation by a polarized laser source, they used polarizers

Table 10. Positions and Relative Intensities of Main Bands in Matrix-Isolated Fluorescence Spectra of Methylchrysenes[a,b]

Position of Substitution	Relative Intensity				
	100	55–60	20–25	10–20	5–10
1	359.3	378.6	363.6	399.9	382.6
2	358.5	377.1	362.0	398.6	381.3
3	359.7	379.4	364.3	400.6	383.4
4	363.8	383.3	368.1	405.5	387.5
5	365.7	386.2	370.0	408.0	390.0
6	360.0	379.4	364.3	400.5	383.7

[a] All entries are band maxima in nanometers.
[b] Reprinted with permission from Tokosbalides et al., Anal. Chem., 50, 1190. Copyright 1978 American Chemical Society.

to selectively pass the fluorescence of one component from a mixture of fluorescent compounds. Significant improvement in spectral discrimination was found. Perry et al. (249) used an untreated solvent-refined coal sample to show that matrix-isolation fluorescence spectra were suitable for identification and determination of PAH. Selective excitation of fluorescence in Shpol'skii matrices and background suppression with the time-resolution capabilities of a dye-laser spectrometer permitted identification of PAH. Four PAH (benzo[a]pyrene, perylene, benz[a]anthracene, and benzo[b]fluorene) were determined at the ppm level in a solvent-refined coal sample. Conrad and Wehry (250) investigated laser-induced matrix-isolation fluorescence spectrometry of methyl and methoxyl derivatives of benzo[a]pyrene. The analytical utility of matrix-isolation Shpol'skii fluorometry for the resolution of multicomponent PAH samples was investigated.

4.4. Shpol'skii and Fluorescence Line Narrowing Spectrometry

When polycyclic aromatic hydrocarbons are dissolved in n-alkanes, such as n-heptane, and the resulting liquid is frozen at 77 K, the solid solution formed emits sharp-line luminescence under ultraviolet excitation. The sharp-line spectra are a result of the Shpol'skii effect (251,252). In the previous section, the work by Maple et al. (247) with annealed n-heptane deposits, and the work by Conrad and Wehry (250), were discussed as examples of the use of the Shpol'skii effect. Causey et al. (251) compared excitation sources, sample cells, and detection systems for the detection and determination of PAH with the Shpol'skii effect. One of their conclusions was that the photon-counting technique with their instrumental system did not result in a large gain in sensitivity. Fassel and co-workers (252–254) have considered X-ray excited optical luminescence of polynuclear aromatic compounds in n-heptane at 77 K. One advantage of X-ray excitation over ultraviolet excitation is the freedom from optical cross-talk between the exciting and luminescence radiation frequently encountered with ultraviolet radiation. Another possible advantage may result in populating electronic levels in molecules not available to ultraviolet excitation. Colmsjo and Ostman (255) demonstrated the selectivity of Shpol'skii fluorescence spectrometry for PAH in a sample from the pyrolysis of a petroleum product. Fassel and co-workers investigated laser-excited Shpol'skii spectrometry (256–259). Good selectivity could be obtained with the Shpol'skii effect because of the inherent sharp absorption bandwidths obtained. Sharp-line laser radiation can be used to advantage for exciting high-resolution, quasi-linear luminescent spectra from either frozen solutions (256) or with the matrix-isolation technique (248). In addition, site-specific excitation of PAH in complex mixtures can be obtained with narrow-bandwidth tunable-dye laser excitation (257). Yang et al. (257) have commented that for laser-excited Shpol'skii spectra, the preparation of frozen solutions is a simpler process and is more readily adaptable to quantitative transfer than the matrix-isolation approach. Yang et al. (258) used deuterated analogues as internal reference compounds for the direct determination of benzo[a]pyrene and perylene in liquid fuels by laser-excited Shpol'skii spectrometry. For one coal liquid sample from an SRC II process, they found 145 ppm benzo[a]pyrene by a previous standard addition method (256), and 163 ppm by the deuterated analogue internal reference method (257). Renkes et al. (259) determined aromatic hydrocarbons in particulate samples by high-temperature extraction and Shpol'skii spectrometry. Direct quantitation was accomplished by laser excitation after an aliquot of the extract was diluted with n-octane.

Lai et al. (91) assembled a simple fluorometer from commercial components and used it for the study and application of the Shpol'skii effect. They were interested in the correlation between dimensions and geometries of PAH and their corresponding Shpol'skii solvents. In addition, they emphasized that with commercial spectrofluorometers, conventional Shpol'skii luminescence could easily be developed into a routine analytical technique. Rima et al. (260) applied the Shpol'skii effect to the quantitative analysis of monomethylphenanthrene isomers. The method was used to obtain the distribution of these isomers in a petroleum fraction. Ewald et al. (261) identified geochemical PAH from terpenes by Shpol'skii fluorescence spectrometry. In addition, Garrigues and Ewald (262) identified monomethylated PAH in crude oils by liquid chromatography and high-resolution Shpol'skii fluorescence spectrometry. Bykovskaya et al. (263) discussed methods of obtaining the resolved line spectra by selective laser excitation of solutions of complex molecules at low temperature (4.2 K). The determination of benzo[a]pyrene in gasoline without preliminary separation was given as an example of the applicability of their approach.

Small and co-workers used fluorescence line-narrowing spectrometry (site-selection spectrometry) in organic glasses for the characterization of PAH (264,265) and amino polycyclic aromatic hydrocarbons. Brown et al. (264) commented that the multiplet structure characteristic of Shpol'skii matrices posed a problem, since the multiplet fluorescence pattern depends on the sample cooling rate. This problem is not insurmountable, but increases sample analysis time. Although low-temperature (~4 K) electronic molecular absorption spectra in glasses are broadened because of site inhomogeneity, narrow laser line excitation near the absorption origin of the fluorescent state gives sharp-lined fluorescence spectra. Fluorescence line-narrowing spectroscopy is a manifestation of the fact that only impurity sites whose excitation profiles overlap with the laser frequency profile are able to fluoresce. This approach is essentially the same as site-selection spectrometry, discussed in Section 4.3. The effect is very pronounced at low temperatures, where impurity site interconversion rates are not competitive with fluorescence (264). After surveying several different glasses, it was found that 1:1 glycerol:H_2O was almost ideal. Two particular advantages of line-narrowing spectroscopy are the ability to resolve certain structural isomers and to analyze contaminated water samples directly. Brown et al. (265) employed fluorescence line-narrowing spectrometry in glasses for the direct determination of PAH in solvent-refined coal. A tunable pulsed N_2 pumped-dye laser was used in the work. It was shown that fluorescence line-narrowing spectrometry in glasses gave detection limits of ~1 part per trillion for strongly fluorescent species and could be used for the analysis of complex mixtures of PAH. These authors gave an interesting comparison of their approach with other solid-state fluorescence-based techniques. In particular, they stated that the Shpol'skii matrix offers only one advantage over the glassy matrix; namely, their reduced site inhomogeneously broadened absorption line widths for the S_1 state of PAH allows for a greater degree of selective excitation. Chiang et al. (266) also used fluorescence line-narrowing spectrometry for amino polycyclic aromatic hydrocarbons in an acidified organic glass. The compounds investigated were 1-aminopyrene, 2-aminoanthracene, and 1-aminoanthracene. Heisig et al. (267) investigated fluorescence-line-narrowed spectra of polycyclic aromatic carcinogen–DNA adducts. They demonstrated that fluorescence-line-narrowed spectra for substituted polycyclic aromatic hydrocarbon derivatives bound to polymeric materials could be obtained and that fluorescence line narrowing in polar glasses was a promising approach for the analysis of DNA.

4.5. Fluorescence and Phosphorescence Detection in Gas and Liquid Chromatography

In Section 3.11, several instruments and instrumental techniques were discussed for fluorescence detection in gas and liquid chromatography (88,125,127,197–200). In this section, several examples will be discussed which show the analytical applicability of luminescence detection in gas and liquid chromatography. Froehlich and Wehry (197) have given a rather extensive review of fluorescence detection in liquid and gas chromatography. By far most applications have appeared for liquid chromatography. Fluorescence detection has been used in liquid chromatography to improve both sensitivity and selectivity. As Froehlich and Wehry (197) have emphasized, to optimize an assay that combines chromatography and fluorescence detection, the entire analytical process has to be considered. For example, a change in one part of the analysis, such as the chromatographic step, might have an important impact upon some other part of the analysis, such as detection.

Johnson et al. (127) have given several examples of fluorescence detection in high-performance liquid chromatography (HPLC). Examples were given for polynuclear aromatic compounds, aflatoxins, and N,N-diethyl-D-lysergamide (LSD). In addition, both pre- and post-column derivatization were discussed. Frei et al. (268) have considered some important parameters such as the pump system, mixing devices, and detector design in instrumentation for post-column derivatization in HPLC. Various nona-peptides were used as model compounds. Both adsorption and reversed-phase chromatographic systems were employed to separate the peptides prior to the post-column reaction. They concluded that reversed-phase chromatography had the advantage of simpler sample preparation, better reaction control, and optimization of solvent conditions. Detection limits of between 5 and 10 ng per injection could be obtained and a relative standard deviation of better than ±2% was obtained. Table 11 lists the structure of the nona-peptides investigated. Lloyd (269) considered ultraviolet absorbance and corrected fluorescence spectral data and fluorescence quantum yields from some representative fatty acid esters of 4-hydroxymethyl-7-methoxycoumarin. The fluorescence quantum yields varied extensively between different solvents. In methanol, values less than 0.1 were typical. With the addition of water, the yield from the lower-molecular-weight derivatives rose to values of about 0.4. In nonhydrogen-bonding solvents, the yields were less than 0.02. Lloyd's results showed the considerable restrictions on the choice of HPLC conditions. Sepaniak and Yeung (270) reported the determination of the antitumor drugs adriamycin and daunorubicin in urine by HPLC with laser fluorometric detection. Reversed-phase HPLC was used and the drugs were detected down to the low picogram level by laser-excited fluorescence with a unique fiber optic based flow-cell (Figure 20a). Excellent detector selectivity and linearity were reported. Naka-mura and Tamura (271) investigated the simultaneous fluorometric determination of biogenic thiols and disulfides by liquid chromatography. The compounds were separated by anion-exchange chromatography with gradient elution, reacted with sodium sulfite at pH 6–7 and o-phthalaldehyde at pH 9.4–10.5 in the presence of taurine to produce the fluorescent isoindole fluorophores. The relative standard deviations for five determinations were 1.35, 4.09, and 6.87% for the analyses of 5 nmol of L-cystine and oxidized glutathione and 8.5 nmol of the oxidized form of coenzyme A, respectively.

Su et al. (124) developed a new fluorescence detector with good sensitivity for both fluorescent and nonfluorescent species. With the device, a fluorescent probe (aniline) was used in the liquid chromatographic mobile phase. Fluorescent species gave positive peaks and nonfluorescent or weakly fluorescent species gave negative peaks. Detection limits were in the part per billion range for fluorescent species and in the part per million range for nonfluorescent species. The compounds investigated were o-bromobenzoic acid, 2,4-dinitrophenol, car-bazole, indole, Tegretol, and Dianabol. Voigtman et al. (90) compared laser-excited fluorescence and photoacoustic limits of detection for static and flow cells. In addition, a photoionization mode of operation for the flow cell was discussed which used the ionization products of two-photon excitation of poly-nuclear aromatic hydrocarbons. The main advantage of their detector was that photoacoustic detection is complementary to fluorescence detection, and two-photon photoionization detection may be useful in quantifying substances which

Table 11. Structures of the Nona-Peptides Studied[a]

Nona-Peptide	Structure
Oxytocin	H—Cys—Tyr—Ile—Gln—Asn—Cys—Pro—Leu—Gly—NH$_2$
Lysine-vasopressin	H—Cys—Tyr—Phe—Gln—Asn—Cys—Pro—Lys—Gly—NH$_2$
Ornipressin	H—Cys—Tyr—Phe—Gln—Asn—Cys—Pro—Orn—Gly—NH$_2$

Leucine

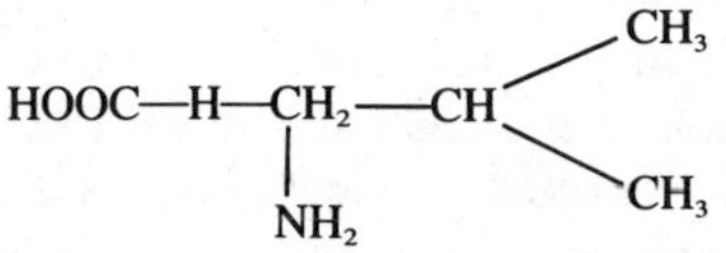

$$HOOC-H-CH_2-CH\begin{matrix}CH_3\\\\CH_3\end{matrix}$$
$$\underset{NH_2}{|}$$

Isoleucine

$$\overset{CH_3}{\overset{|}{HOOC-CH-CH-CH_2-CH_3}}$$
$$\underset{NH_2}{|}$$

Lysine

$$HOOC-CH-CH_2-CH_2-CH_2-CH_2-NH_2$$
$$\underset{NH_2}{|}$$

Ornithine

$$HOOC-CH-CH_2-CH_2-CH_2-NH_2$$
$$\underset{NH_2}{|}$$

Tyrosine

$$HOOC-CH-CH_2-\bigcirc-OH$$
$$\underset{NH_2}{|}$$

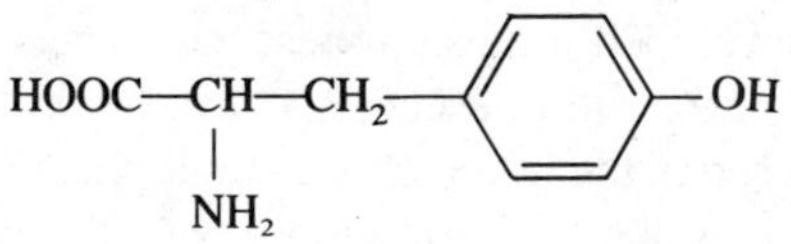

Phenylalanine

$$HOOC-CH-CH_2-C_6H_5$$
$$\underset{NH_2}{|}$$

Cystine

$$HOOC-CH-CH_2SH$$
$$\underset{NH_2}{|}$$

[a]From Frei et al. (268).

do not appreciably luminesce and are relatively easy to ionize. Furuta and Otsuki (272) used time-resolved fluorometry in the detection of ultratrace polycyclic aromatic hydrocarbons in lake waters by liquid chromatography. They replaced the Xe lamp of a conventional fluorometer with a N_2 laser-pumped dye laser as an excitation source and improved the detection capability for polycyclic aromatic hydrocarbons by 1 or 2 orders of magnitude. A detection limit $(S/N=2)$ of 180 fg was obtained for benzo[a]pyrene, and the precision was 3.9%. The authors demonstrated that time-resolved fluorometry based on a pulsed laser along with HPLC provided both high sensitivity and high selectivity for PAH.

Hershberger et al. (197a) investigated real-time fluorescence monitoring at multiple wavelengths of excitation and emission by interfacing a high-performance liquid chromatograph to the "video" fluorometer using a laminar flow cell to minimize scattered light and dead volume. A total fluorescence chromatogram and two selected excitation–emission wavelength chromatograms were displayed in real time. Perylene gave a limit of detection of 1 ng, and selected fluorescence monitoring was capable of spectrally separating and quantifying benzo[a]pyrene and benzo[e]pyrene, whose chromatographic retention profiles overlapped extensively. Gianelli et al. (273) evaluated a multichannel detector for in situ analysis of fluorescent materials on thin-layer chromatoplates. One optical system was designed to obtain a fluorescence spectrum from each position along the elution axis of a one-dimensional plate without the need for mechanical scanning. Another optical system was designed to obtain a quantitative two-dimensional fluorescence image of a single, two-dimensional chromatogram or of multiple, one-dimensional chromatograms. A fluorescence intensity dynamic range of over 500:1 could be measured on a single plate, and detection limits extended to 3 pg for tetraphenylporphine.

Weinberger et al. (274) described liquid chromatographic room-temperature phosphorescence detection with micellar chromatography and postcolumn reaction modes. A micellar mobile phase consisting of 0.15 M sodium/thallium lauryl sulfate (70/30) was used for both the separation (micellar chromatography) step and detection step. This could be done because all reagents necessary for micellization and spin–orbit coupling enhancement were present. For reversed-phase separation, the micellar reagents were introduced postcolumn. The general approach was less sensitive than classical fluorescence detection, but dramatic improvement in selectivity was achieved for the analysis of a mixture of β-naphthol, biphenyl, and phenanthrene. Limits of detection as low as 5 ng were obtained for these compounds. Armstrong et al. (275) studied the effect of a micellar mobile phase on the fluorescence and room-temperature liquid phosphorescence detection of several PAHs separated by liquid chromatography. They found that the fluorescence of the separated PAH was enhanced up to ten times by sodium dodecylsulfate micellar mobile phase. Donkerbroek et al. (276–278) have considered room-temperature phosphorescence in liquid solutions as applied to chromatographic analysis. They showed that sensitized room-temperature phosphorescence in liquid solutions could be successfully applied as a detection method in continuous-flow and chromatographic systems (276). In later work, this general approach was used for the analysis of mixtures of polychloronaphthalenes and polychlorobiphenyls (277). In addition, the potential of quenched room-temperature phosphorescence in liquids was demonstrated as a detection technique for flow injection analysis and liquid chromatography (278).

The feasibility of using a SIT vidicon as a gas-phase fluorescence detector for gas chromatography was demonstrated by Cooney et al. (87). They found the detection sensitivity was comparable to that of a fast-scanning spectrometer with a photomultiplier detector. In addition, several different optical systems and excitation sources were compared for the detection of gas-phase fluorescence from gas chromatographic effluents (88). Conrad et al. (202) investigated the use of matrix-isolation fluorescence spectrometry for detection in the open tubular column gas chromatography of PAH. Chromatographic column eluents were deposited directly on a movable 12-sided deposition surface positioned in the head of a closed-cycle cryostat.

4.6. Multicomponent Fluorescence and Phosphorescence Analysis

Warner and McGown (179) have extensively reviewed the area of multicomponent fluorescence analysis. Some of the multicomponent analysis techniques, such as derivative spectroscopy, synchronous spectroscopy, and matrix isolation spectroscopy, have been discussed in earlier sections. Probably the most powerful instrument available for multicomponent fluorescence analysis is the "video" fluorometer and the resulting excitation–emission matrix (EEM) obtained. Several aspects of the approach were discussed earlier (126,149,180,187,273). Ho et al. (279,280) considered in detail a scheme for quantitative analysis of a multicomponent fluorescent mixture using the EEM acquired by the "video" fluorometer. The method is termed rank annihilation. Warner and McGown stated that (179) the method is more useful than either the least squares or linear programming method when all the components in the mixture are not known. Warner et al. (281) have discussed the details of the least squares and linear programming approaches for use with EEM. Essentially, the method of rank annihilation allows one to obtain quantitative information for known compounds without having to be concerned about other species present in the sample. Ho et al. (280) demonstrated the usefulness of the approach by using ten different samples of a six-component PAH solution. The compounds had a wide range of fluorescence quantum efficiencies and spectral overlaps.

Ho and Warner (187) investigated multicomponent mixture analysis by multidimensional phosphorimetry. The technique was based on the rapid scanning capability of the "video" fluorometer to obtain an EEM. To illustrate the applicability of the approach, three sets of synthetic data were obtained and analyzed. Three PAH were used in the study, namely, coronene ($10^{-8}\,M$), phenanthrene ($10^{-7}\,M$), and triphenylene ($10^{-8}\,M$). The manner in which the phosphorescence EEM was obtained allowed for time resolution. One particular advantage of the time-resolution approach employed was that short-lived phosphors did not cause problems in the acquisition of the data.

As Warner and McGown (179) have commented, multicomponent fluorescence analysis is a new field in chemical analysis experiencing relatively rapid growth. Multicomponent phosphorescence analysis is an even newer field in trace analysis. Both fields should see continued developments in the future.

4.7. Solid-Surface Luminescence Analysis

Solid-surface luminescence analysis involves the measurement of fluorescence or phosphorescence of components adsorbed on solid materials. Hurtubise (30) has discussed the theory, instrumentation, and analytical applications associated with this technique. The instrumentation needed for measurement of the luminescent signals from compounds adsorbed on solid surfaces was considered in Sections 3.10 and 3.11. In addition, some of the phosphorescence applications were mentioned in Section 4.2. Also, the theoretical aspects of solid-surface luminescence were discussed in Section 2.9 and in Section 2.11. Additional solid-surface room-temperature luminescence applications will be discussed in Section 4.8. In this section, solid-surface fluorescence applications will be emphasized. Applications have appeared in such areas as environmental research, forensic science, pesticide analysis, food analysis, pharmaceutical process control, biochemistry, medicine, and clinical chemistry. Solid-surface luminescence analysis has some distinct advantages over solution luminescence analysis. Generally, the sample needed for analysis is much smaller. This is important with toxic materials, and with materials that are not available in large quantities, such as certain biological compounds. In addition, the fluorescence of a series of compounds on a thin-layer chromatoplate or on filter paper can be measured very rapidly. Also, as mentioned earlier, room-temperature phosphorescence can be obtained from a variety of compounds adsorbed on solid surfaces. This means it is possible to measure both fluorescence and phosphorescence from compounds adsorbed on solid surfaces at room temperature. In the remaining part of this section, a few representative examples of solid-surface fluorescence analysis will be discussed.

Sawicki and co-workers pioneered in the development and application of solid-surface fluorescence analysis in air pollution research. Sawicki (20) has reviewed fluorescence analysis in air pollution research, and Sawicki and Sawicki

(282) have reviewed the role of thin-layer chromatography in air pollution research, with an emphasis on fluorescence characterization, identification, and quantitation.

Gibson (283) has reviewed the applications of luminescence in forensic science. One important area in forensic science in which solid-surface fluorescence analysis has been used is questioned-document work. Watermarks can be faked by waxing the document to make it more translucent; thus, it will show a different fluorescence from the original paper. Also, most papers possess their own fluorescence that would be altered by chemical erasure. Many fluorescent powders have been developed to show mechanical erasure alterations, and much work has gone into characterizing the luminescence of inks. Those inks that luminesce do so mainly in the infrared region. Apparently, there is a high probability that any additions made to a document using inks of similar appearance may be detected by a difference in infrared luminescence.

Numerous applications have appeared for pesticide analysis in which the fluorescence of pesticides or the fluorescence derivatives of pesticides are measured from thin-layer chromatoplates (30,284). One such application has been described by Zakrevsky and Mallet (285). They based their approach on the reaction of fluorescamine with primary amines to yield highly fluorescent derivatives on silica gel chromatoplates. Table 12 shows the fluorescence properties of the fluorescence derivatives. It can be seen that the detection limits are in the nanogram range. Also, it was found that spraying the chromatoplates with a

Table 12. Fluorescence Properties of the Fluorescamine Derivatives on Thin Layers of Silica Gel[a]

Compound	Excitation Maximum (nm)	Emission Maximum (nm)	Limit of Detection (μg)	Upper Limit of Linear Range (μg)
Aminofenitrothion	382	500	0.01	0.5
4-Amino-3-methyl-phenol	382	492	0.02	0.5
Aminotriazole	382	488	0.01	0.5
Fenitrooxon	383	490	0.01	1.0
Fenitrothion	383	490	0.01	1.0
Parathion-methyl	390	510	0.01	1.0
3-Methyl-4-nitrophenol	385	490	0.08 (0.2)[b]	1.0
Parathion	390	510	0.01	1.0

[a]From Zakrevsky and Mallet (285).
[b]Limit of detection after spraying with 10% triethanolamine in ethanol.

10% solution of ethanolamine solution enhanced the fluorescence. The reproducibility of the approach was evaluted for aminofenitrothion and fenitrothion using 0.4-μg spots. The average relative standard deviation was 9.2% for fenitrothion and 7.2% for aminofenitrothion. The authors commented that the diffusion of spots and greater background irregularities on the chromatoplates, caused by spraying with the inorganic reagents, explained the lower precision for fenitrothion compared to aminofenitrothion. It should be mentioned that Kaiser (286) discussed the accuracy and precision in modern instrumental TLC in quantitative and qualitative analysis. He stated that thin-layer chromatoplates of highly improved quality are now available, and that their chromatographic characteristics have come close to those of high-performance liquid chromatography columns. Reproducibility of repetitive scanning of a single spot can be obtained at the 0.1% level. The reproducibility with the same sample and repeated separation in the nanogram range was at the 1% level. However, Kaiser did not mention fluorescence specifically, and his comments refer to ideal analytical conditions. Nevertheless, they do suggest the best possible analytical precision for fluorescence measurements from thin-layer chromatoplates.

Guilbault and co-workers (287–291) have developed methods for the assay of enzymes, substrates, activators, and inhibitors using solid-surface fluorescence analysis. Generally, a cell holder for an Aminco filter instrument was adapted to accept a metal slide on which a silicone rubber pad was placed that contained the necessary reagents for an assay. The change in fluorescence with time was measured and related to the concentration of the substance to be determined. Guilbault and Zimmerman (291) reported the first successful measurement of the rate of an enzyme reaction on a solid surface using fluorescence detection. Cholinesterase was determined in horse serum. For the determination of cholinesterase, pads were prepared by placing 10 μL of a 10^{-2}–M solution of N-methylindoxyl acetate on a silicone rubber pad and evaporating to dryness. To start the determination, 20 μL of a sample enzyme solution was applied to the pad. When the pad was placed in the light beam of a fluorometer, a recorder was started immediately and the rate of change of fluorescence measured. From a calibration curve of the change in fluorescence units per minute versus enzyme concentration, the activity of cholinesterase in the sample solution was obtained. Calibration curves were obtained from 10^{-6} to 10^{-2} units mL^{-1}, and the precision and accuracy of the method were 2 and 2.2%, respectively. Studies indicated that pads were stable for at least 30 days if kept in a cold, dark place.

Bicking et al. (292) reported the determination of aflatoxins in air samples of refuse-derived fuel by thin-layer chromatography with laser-induced fluorescence spectrometric detection. A detection limit of 10 pg for four aflatoxins was obtained. Aflatoxin B$_1$ was found at levels up to 17 ppb in solid samples collected from the air at a plant which produced refuse-derived fuel. Huff and Sepaniak

(293) used different modes of laser-fluorescence detection in high-performance thin-layer chromatography. The low limits of detection (picogram range) and the unique focusing properties of the laser were the two main advantages of the approach developed.

4.8. Phosphorimetry

The theoretical and instrumental aspects of phosphorescence analysis were considered in earlier sections. In addition, some applications of solid-surface room-temperature phosphorescence were discussed earlier. In this section, both the present status of phosphorimetry and selected applications will be elaborated upon.

As Winefordner and co-workers (28) have discussed, little has been done since 1975 in the area of solution low-temperature phosphorimetry. The lack of widespread use probably is related to the need for cryogenic equipment and the problems involved with introducing the sample into the phosphorimetric system. For example, generally a long quartz capillary cell is slowly lowered into a quartz Dewar flask filled with liquid nitrogen. Depending on the rate of cooling of the sample cell and the chemical nature and composition of the solvent system, the cooled matrix can be a clear glass, cracked glass, or a snow. A skilled technician or chemist, however, can obtain very good data with low-temperature phosphorimetry. The major advantages of solution low-temperature phosphorimetry are low detection limits (nanomolar range), linear analytical calibration curves of wide range, and very good selectivity. Wehry (74) has reviewed some recent low-temperature phosphorescence applications. Gifford et al. (294) reported the phosphorimetric analysis of phenothiazine derivatives. Other low-temperature phosphorescence applications have appeared for polymers (295), pharmaceuticals (296), porphyrins (297), drugs (298,299), fuel oil identification (300,301), and nitrated polynuclear aromatic hydrocarbons (72). For the nitrated polynuclear aromatic hydrocarbons, detection limits lower than 10 ng mL^{-1} were obtained in several cases. Gifford et al. (302) constructed and evaluated a phosphorimeter for the direct measurement of low-temperature phosphorescence from separated components on thin-layer chromatoplates. Miller et al. (303) evaluated solvent enhancement effects for phosphorescence in thin-layer chromatography. Aaron et al. (68) reported a comparative analytical study of substituted quinolines by a fluorimetric and phosphorimetric method. McCall and Winefordner (304) discussed the enhancement of phosphorescence signals of several analytes on paper substrates measured at 90 K compared to room-temperature filter-paper phosphorescence.

As discussed in Section 2.9, Cline Love and co-workers (38,39,305–308) have developed the new analytical approach of solution micelle-stabilized RTP. Analytical figures of merit were reported for naphthalene, pyrene, and biphenyl in Tl/Na and Ag/Na mixed counterion lauryl sulfate micelles (38). The average precision of the measurements was about 6%, and the sensitivities obtained were competitive with other phosphorescence techniques. Skrilec and Cline Love (39) investigated micelle stabilized room temperature phosphorescence characteristics of functionally substituted aromatic compounds in aqueous Tl/Na lauryl sulfate micellar solution. Compound types investigated were ketones, aldehydes, alcohols, carboxylic acids, phenols, amines, and some compounds of pharmaceutical interest. Table 13 shows the phosphorescence lifetimes and micelle-stabilized RTP sensitivities for a series of pyrene compounds. As indicated in Table 13, the fluorescence of pyrene, 1-pyrenebutyric acid, and 1-aminopyrene was quenched 95% or more by addition of Tl^+. However, with the quenching of the fluorescence, there was a corresponding increase in the phosphorescence due to the addition of Tl^+. Cline Love et al. (305) also considered the analytical utility of micelle-stabilized RTP lifetimes. Single-component systems and a two-component system in mixed-heavy-atom micelles were investigated. A background correction method for interfering radiation present in micelle-stabilized RTP was developed by Cline Love and Skrilec (307). The method corrected for fluorescence and scatter and was based on oxygen quenching of the RTP. In the presence of oxygen, phosphorescence was quenched in fluid solutions at room temperature, and it was assumed that the oxygen had no significant effect on the fluorescence intensity. For the method developed, the luminescence intensities of the deaerated and aerated solutions were measured, and the difference in the two emission intensities was calculated. By this calculation, the signal and spectra due to the phosphorescence alone were obtained.

Donkerbroek et al. (40,41,276–278,309) investigated several aspects of sensitized solution RTP. In one application, they considered solution sensitized RTP with 1,4-dibromonaphthalene and biacetyl as acceptors. After excitation, the analyte molecule transferred its triplet-state energy to an acceptor molecule, which subsequently emitted phosphorescence. The limits of detection for a number of substituted benzophenones and biphenols in acetonitrile/water (1:1) were about 10^{-8} M(40). Donkerbroek et al. (309) considered sensitized liquid phase RTP applied to the detection of polychloronaphthalenes. The approach involved the sensitized RTP of biacetyl in azeotropic acetonitrile/water. The detection limits were on the order of 10^{-8} to 10^{-9} M. For some of the polychloronaphthalenes, the influence of the reversed energy transfer on the sensitized room-temperature phosphorescence in liquids was also discussed.

By far most of the recent papers in phosphorimetry have appeared in solid-surface RTP. Hurtubise (30) has reviewed some earlier applications in this area. Vo-Dinh (310) has reported an overview of his work in solid-surface RTP for trace organic analysis. Parker et al. (27) have given a comprehensive review of the area of solid-surface RTP. Winefordner and co-workers have discussed several aspects of solid-surface RTP (26,28,29,35,204,311–313). Recently, McCall

Table 13. Phosphorescence Lifetimes and Micelle-Stabilized-RTP Sensitivities for a Pyrene Series[a]

| Compound | Concentration for MS–RTP Lifetime ($M \times 10^4$) | Wavelength of Excitation for MS–RTP (nm) | Lifetime[b] | | Fluorescence Quenched[d] (%) | Limit of Detection by MS–RTP ($M \times 10^8$)[e] |
			293 K (ms)[c]	77K (s)		
Pyrene	0.25	336	0.93	0.47 (0.5)	99	0.50
1-Pyrenebutyric acid	0.19	345	1.1	0.47	99	0.45
1-Aminopyrene	0.18	345	0.12	0.27	95	1.0
1-Pyrenesulfonic acid	1.0	342	f	g	g	f

[a]Reprinted with permission from Skrilec and Cline Love, *Anal. Chem.* **52**, 1559. Copyright 1980 American Chemical Society.
[b]Correlation coefficient > 0.995 in all cases; concentrations for 77-K lifetime determinations about 0.5×10^{-4} M in ethanol; literature value for 77 K in EPA in parentheses.
[c]Sample temperature ranged between 291 and 295 K.
[d]Reflects decrease in intensity going from NaLS micellar solution.
[e]Based on signal-to-noise ratio of 3.
[f]MS-RTP not observed.
[g]Not measured.

and Winefordner (304) investigated low-temperature filter paper phosphorescence.

Experimentally, in solid-surface RTP, a solution of the solute is applied to a surface, the adsorbed solute is dried for 5–30 min, and then the RTP signal is measured. Some of the surfaces that have been used to induce RTP are sodium acetate, several varieties of filter paper, silica gel chromatoplates with polyacrylic acid binder, polyacrylic acid–sodium chloride mixtures, diethylenetriaminepentaacetic acid treated filter paper, and polyacrylic acid treated filter paper. For 100 ng of benzo[f]quinoline adsorbed on a silica gel chromatoplate which contained polyacrylic acid binder, the relative standard deviation was 3.2% (133). For diethylenetriaminepentaacetic acid treated filter papers, the relative standard deviation ranged from 2.2 to 4.7% for the RTP of p-aminobenzoic acid (313). In addition, for polyacrylic acid treated filter paper, the relative standard deviation ranged from 1.7 to 6.1% for various compounds (89). Limits of detection have been reported in the nanogram and subnanogram ranges.

Some RTP applications were already considered in Sections 4.1 and 4.2 (224,237–239). One of the first applications of solid-surface RTP was the determination of p-aminobenzoic acid in multicomponent vitamin tablets (314). The limit of detection was about 0.5 ng and the method was shown to be more sensitive than the USP XVIII method. The selectivity of RTP is illustrated by the previous example because p-aminobenzoic acid was determined without prior separation of the p-aminobenzoic acid. There were ten other tablet ingredients present. Ford and Hurtubise (315) used RTP excitation and emission spectra and fluorescence emission spectra to identify benzo[f]quinoline and phenanthridine after high-performance liquid chromatographic separation from shale oil. Vo-Dinh and Hooyman (316) used selective heavy-atom perturbation for the analysis of mixtures of polynuclear aromatic compounds in coal-derived products. A unique feature offered by this approach was the ability to pick out specific target polynuclear aromatic compounds of interest in an environmental sample using appropriate heavy atoms. Warren et al. (317) investigated RTP with internal

Table 14. Selected Compounds and Compound Types Yielding
Analytical Solid-Surface RTP

Compounds or Compound Types	Reference
Imidazoles	318
Several pharmaceuticals	70
Aromatic carboxylic acids	34
Cinoxacin (a cinnoline derivative)	319
1,8-Naphthyridine derivatives	320
Selected pteridines	321
Tryptophan and tyrosine	322

standard and standard addition techniques. Precision was improved to 1–3%. The approach developed allowed the measurement of analytes in samples without the use of complicated standards or a previous knowledge of the spectral distribution or intensity. Table 14 lists selected examples of some of the compounds that have given useful analytical solid-surface RTP.

4.9. Lasers in Luminescence Analysis

In this section, several examples of the use of lasers in luminescence analysis will be given. Some of the properties of lasers were discussed in Section 3.2., and several applications in which lasers were used in luminescence analysis were considered in earlier sections. Wehry (74) has given a reasonably thorough coverage of the applications of lasers in luminescence. In addition, the applications of lasers to fluorescence analysis have been considered by Richardson (99), Demtroder (100), Lytle (104), and Wright (105). Omenetto and Winefordner (98) have discussed the application of lasers in both fluorimetry and phosphorimetry. Table 15 gives representative examples of limits of detection of several luminescent compounds excited by various lasers. The appropriate reference should be consulted for experimental details.

Sacchi et al. (327) have discussed the use of a tunable-dye laser to study the fluorescence of a signal cell (frog erythrocytes). Dolbeare (122) detailed the fluorometric quantification of specific chemical species in single cells and discussed dual-laser flow cytometers. When two fluorophores are employed to stain a cell, a single excitation wavelength may not be sufficient to excite both fluorophores. In addition, when using single-laser sources, spectral matching may be poor. Thus, two different lasers which provide two different excitation wavelengths allow analyses of two fluorophores simultaneously. Dolbeare emphasized

Table 15.   Fluorescence and Phosphorescence Limits of Detection by
Laser Excitation

Compound	Laser	Limit of Detection	Reference
Morphine	N_2	1 μg/mL^{-1} (low-temperature phosphorescence)	86
Benzo[a]pyrene	N_2	0.004 ng/mL^{-1} (fluorescence)	323
Rhodamine 6G	Argon-ion	8.9×10^{-14} M (fluorescence)	324
Benzene	N_2 dye	19 ng mL^{-1} (fluorescence)	325
Gluscose-6-phosphate	He:C$_d$	0.52 ng mL^{-1} (fluorescence)	326
Procaine	N_2	3 ng mL^{-1} (low-temperature phosphorescence)	98

that there are several dual-laser instruments described in the literature, and he discussed one such instrument. Berman and Zare (328) reported the laser fluorescence analysis of aflatoxins on thin-layer chromatoplates at the subnanogram level. Dalrymple et al. (329) used lasers in fingerprint detection. Their approach consisted of laser illumination of the exhibit under investigation, and then either direct viewing or photographing of the laser-induced fingerprint luminescence. The exhibit under investigation was illuminated with the 514.5-nm line of a continuous-wave argon-ion laser. The laser beam illuminated an area of about 65 cm^2. Richardson and Ando (325) reported the less than parts-per-trillion detection of polycyclic aromatic hydrocarbons in aqueous solutions by laser-induced molecular fluorescence. The limits of detection of benzene, naphthalene, anthracene, fluoranthene, and pyrene were found to be 19 ppb, 1.3 ppt (parts per trillion), < 4.4 ppt, 1.0 ppt, and 0.5 ppt, respectively. The polycyclic aromatic hydrocarbons fluorescence exhibited a linear dependence on concentration that extended in some cases over six orders in magnitude.

Allegrini and Omenetto (330) investigated laser-induced fluorescence and Raman scattering for real-time measurement of suspended particulate matter. They stressed that both Raman and fluorescence spectra were obtained in their experiments, but they could not claim any definitive conclusions on the analytical feasibility of both techniques for *in situ* measurements of particulates. However, preliminary results were encouraging. Rodgers et al. (331) discussed photofragmentation-laser induced fluorescence. The approach was used for detecting atmospheric trace gases. In the method discussed, the species to be detected was first laser photolyzed at a wavelength λ_1, producing one or more vibrationally excited photofragments. Before vibrational relaxation occurred, one of the photofragments was pumped into a bonding excited state by a second laser pulse centered at wavelength λ_2. Fluorescence was sampled at a wavelength λ_3, where $\lambda_3 < \lambda_2$ and λ_1. Specific sampling schemes for detecting NO_2, NO_3, and HNO_2 were proposed. Gehlhaar et al. (332) explored the potential of a compact and highly sensitive helicopter-born fluorescence lidar for oceanographic measurements. From an altitude of 70 m, selective detection was made of less than 10^{-10} g cm^{-3} of the tracer dye rhodamine B in natural waters. An airborne laser fluorosensor for the detection of oil spills was described by O'Neil et al. (333). The fluorosensor was designed to detect and identify targets by the characteristic fluorescence emission spectrum. Information on field trials of the sensor over marine oil and dye spills was reported. A correlation technique was developed that differentiated among dye, crude oils, and the general fluorescence background of ocean water.

Lidofsky et al. (334) investigated a new technique for competitive binding assay. Bound and free fluorescent-labeled antigens were separated by HPLC and detected by laser fluorometry. The potential of the method was illustrated by the detection of insulin in well characterized buffer solutions. The 448-nm line of a CW argon-ion laser was used for excitation, and the limit of detection of insulin was 0.4 ng mL^{-1}. Detection limits of 0.02 and 0.6 ppt were obtained for fluorescein and riboflavin, respectively, by Ishibashi et al. (335). A nitrogen-laser-pumped dye laser and a pulse-gated photon counter were used in the investigation. Strojny and de Silva (336) investigated the laser induction of molecular fluorescence for the analysis of intrinsically fluorescent drugs and fluorescent derivatives of nonfluorescent drugs. They concluded that the sensitivity limit (2 × blank reading) of reliable quantitation using laser-induced fluorescence was equal to or better than that obtained by a conventional spectrofluorometer.

Yamada et al. (337) reported a highly sensitive fluorometric system consisting of a nitrogen laser and a pulse-gated counter. They used the system for the fluorometric determination of tris(1,1,1-trifluoro-4-(2-thienyl)-2,4-butane-diono)europium(III). The detection limit was 2 pg L^{-1}. With a conventional Xe lamp, a detection limit of 10 ng L^{-1} was obtained. Dovichi et al. (324) investigated laser-induced fluorescence detection of flowing samples. A flow cytometer system was used to detect aqueous Rhodamine 6G by laser-induced fluorescence. At the detection limit (2 × the statistical fluctuation of the blank), 18 ag or 22,000 molecules of Rhodamine 6G flowed through the probed volume during the signal integration period. Their results, compared with other workers', showed that the limits of detection were nearly two orders of magnitude superior to previous results. They further discussed the possibility of single-molecule detection. In addition, other authors have also considered fluorescence detection of single atoms and molecules (338).

4.10. Other Applications

Luminescence techniques have been applied extensively to organic, biological, clinical, environmental, and energy-related samples. In addition, fluorescence techniques have been used for metal analysis via metal chelate fluorescence. It is beyond the confines of this chapter to discuss more than a small fraction of the luminescence applications that have appeared in the literature. Several applications have been detailed in earlier sections. Guilbault (94) has given an extensive survey of luminescence applications. In addition, Schulman (15) has discussed luminescence applications. A portion of the recent book by Lakowicz (112) is slated toward biochemical fluorescence applications. For earlier applications in biology and medicine, the volumes by Udenfriend (339) can be consulted. Pesez and Bartos (340) have provided an extensive discussion of procedures for the colorimetric and fluorimetric analysis of organic compounds and drugs. Rost (119) reviewed microspectrofluorometry in histochemistry. Froehlich (341) has discussed the use of luminescence spectroscopy in clinical and biological analysis. Sawicki (20) has surveyed earlier work for fluorescence analysis in air pollution research. Lawrence and Frei (284) reviewed fluorimetric derivatization for pesticide residue analysis. Gibson (283) considered the applications of luminescence in forensic science, and Morawetz (342) the applications of fluorimetry to synthetic polymer studies. Hurtubise (30) surveyed applications in solid-surface luminescence analysis. Four volumes have appeared on modern fluorescence spectroscopy which have been edited by Wehry (6). These volumes contain several discussions on the uses of fluorescence in analysis. In the remaining part of this section, a cross section of luminescence analysis examples which could not be readily categorized into the earlier sections will be given. By these examples, the reader should get a flavor for the diversity of luminescence analysis.

De Silva et al. (343) reported the spectrofluorometric determination of substituted tetrahydrocarbazoles by a methylene blue sensitized photolytic reaction. The use of 4-bromomethyl-7-methoxycoumarin as a fluorescence label for fatty acids was considered by Dunges (344). Pillai and Patel (118) investigated microcells for fluorescence and spectrophotometric studies of biologically active compounds. Arakawa et al. (345) discussed the extraction and fluorometric determination of organotin compounds with morin. Quantification of neuronal and extraneuronal fluorescence with the aid of histogram analysis and microfluorometric scanning of sympathetic nerve fibers was developed by Schipper et al. (346). Kiang et al. (347) reported the enzymatic determination of nitrate. They used fluorometric detection after reduction with nitrate reductase. Lavallee et al. (348) investigated the fluorometric properties of N-methyltetraphenylporphine and several derivatives.

Jadamec et al. (349) discussed an optical multichannel analyzer for characterization of fluorescent liquid chromatographic petroleum fractions. Schuresko (350) developed a portable fluorometric monitor for detection of surface contamination by polynuclear aromatic compounds. A spectrofluorometer was described by Christmann et al. (351) in which a dedicated microcomputer was used to shift the fluorescence sample cell so that the effective pathlengths for excitation and emission were changed. From the fluorescence intensities measured at three cell positions, the absorption-free fluorescence intensity of the sample was calculated. Kelly and Christian (352) discussed a fluorometer for flow injection analysis with application to oxidase enzyme dependent reactions. Holland et al. (353) considered the effect of excitation beam absorption on measured values of fluorescence with a computer-centered spectrofluorometer capable of measuring fluorescence and absorbance simultaneously. Tromberg et al. (354) investigated optical fiber fluoroprobes for biological measurements.

Killeen et al. (355) considered a digitized fluorescence spectrum containing n points as a vector in n-dimensional space. This idea was used to develop a vector model using angular distance in n-space as a parameter for comparison of fluorescence spectra of weathered oils. Gold et al. (356) analyzed a series of binary and tertiary mixtures of PAH by using principal component and decomposition analysis of molecular fluorescence spectra. The results demonstrated the ability to determine the number and identity of components that were present. Knorr and Harris (357) found that the resolution of overlapped fluorescence emission spectra from mixtures could be accomplished by increasing the dimensionality of the measurement. The emission spectra were gathered as a function of fluorescence decay time to form a data matrix D. The matrix could then be decomposed into its two factors: A, which contained the spectra of the individual components in its columns, and C, which contained the time behavior of the components in its rows.

Rosenberg et al. (358) used fluorometry for the determination of the dissociation constants of sparingly soluble heterocyclic bases. Lampert et al. (359) carried out an investigation of standards for nanosecond fluorescence decay time measurements. A critical evaluation of literature fluorescence decay parameters for standard compounds was made and new standards were proposed for lifetime measurements. Lochmuller et al. (360–362) examined chemically modified silica surfaces using fluorescence spectroscopy. Aminated microparticulate silica gel was derivatized with the fluorescent tag dansyl chloride to yield a chemically modified silica surface that was examined by fluorescence spectroscopy (360). It was also shown that the fluorescence maxima at room temperature of dansylamide groups covalently attached to the surface of silica particles in dry form and slurried in acetonitrile showed varying degrees of dependence upon excitation wavelength (361). Lochmuller et al. (362) studied the fluorescence of pyrene silane molecules chemically bonded to microparticulate gel at several surface concentrations to assess the proximity and distribution of chemically bound molecules on the native silica gel.

Selective fluorescence quenching with 80% ethanol/20% chloroform was used to analyze phenolic antioxidant mixtures by Hurtubise (60). Ryan and Weber (363) employed fluorescence quenching titration for the determination of complexing capacities and stability constants of fulvic acid. Wolfbeis and Urbano (364) investigated a fluorescence quenching method for the determination of two or three components in solution. The Stern–Volmer equation was extended for cases of two or more dynamic quenchers present in solution. It was shown that the determination of n quenches requires n indicators whose Stern–Volmer constants have to be different.

Halfman and Schneider (365) demonstrated that the polarization or anisotropy of a solution containing fluorescent components could be determined from the ratio V/I_δ, where V is the vertical component of the emission and I_δ is the intensity with the transmission angle of the polarizer at δ. Roemelt and Seitz (366) studied perylene–fulvic acid binding by fluorescence polarization. They determined an equilibrium constant of 1.5×10^6 for the binding of perylene to fulvic acid in 75% glycerol. The conformation of soil fulvic acid was studied by fluorescence polarization as a function of pH, concentration, and ionic strength by Lapen and

Seitz (367). Seitz (368) has given a detailed evaluation of analytical derivatization reactions based on changes in fluorescence polarization. He has discussed such aspects as the possibility of distinguishing analyte fluorescence from excess reagent fluorescence on the basis of polarization even if the analytical reaction does not cause a change in fluorescence behavior.

4.11. Chemiluminescence and Bioluminescence

Several surveys have appeared on the fundamental and applied aspects of chemiluminescence and bioluminescence (42,43,74,92,369–371). As with other sections in this chapter, only selected aspects and applications will be considered because of the numerous publications in chemiluminescence and bioluminescence analysis. Seitz (92) has stated that analytical methods based on chemiluminescence and bioluminescence are not widely used presently in clinical laboratories. However, they are receiving close attention as potential methods in the future. There is particular focus on four systems: (1) the firefly reaction which is widely used to determine adenosine triphosphate, (2) bacterial bioluminescence for measuring the reduced form of nicotinamide adenine dinucleotide, (3) the luminol reaction, (4) chemiluminescence accompanying phagocytosis (92). The above approaches offer a wide range of analytical possibilities. For example, they can be coupled to other analyses by enzymes that generate or consume materials such as hydrogen peroxide. Additionally, they can be employed for immunological assays by labeling antigen or antibody with one of the reactants required for light production.

Seitz (92) has emphasized that there are several reasons for the recent surge of interest in chemiluminescence and bioluminescence. These are: (1) a need for alternatives to the widespread use of radioisotopes as labels in immunological assays, (2) the possibility of increasing sample throughput for high volume analyses in clinical laboratories and to minimize required sample volumes for some assays, and (3) the recent commercial availability of improved instrumentation and reagents.

The most widespread application of bioluminescence is the determination of adenosine triphosphate (ATP) using the firefly reaction. Other important methods are the determination of pyridine–adenine dinucleotide employing bacterial bioluminescence and the determination of hydrogen peroxide (9). Gas-phase chemiluminescence has attained importance in the determination of oxides of nitrogen and ozone (9,369,370). In addition, chemiluminescence analysis has been used in water pollution studies (372).

Of all the bioluminescing systems, that of the firefly has been the most studied. The chemistry of the firefly reaction is summarized below (92):

$$LH_2 + E + ATP \xrightarrow{Mg^+} E{:}LH_2HAMP + PP$$

$$E{:}LH_2{:}AMP + O_2 \rightarrow E + L{=}O + CO_2 + AMP + light\ (\lambda_{max} = 562\ nm)$$

where LH_2 is firefly luciferin, $L{=}O$ is oxyluciferin, E is firefly luciferase, PP is pyrophosphate, and AMP is adenosine monophosphate. The detection limit for ATP varies with the luciferase preparation. Typical values range from 10^{-11} to 10^{-14} moles of ATP. The limits of detection will be lower depending on the purity of the luciferase. With crude luciferase, there is background emission without added ATP which establishes the detection limit. The luminescence response is linear with the ATP concentration over several orders of magnitude.

Several strains of marine bacteria have been used in bioluminescence analytical work. The reaction producing the bioluminescence has been studied extensively, and while some differences exist between strains, the fundamental reaction is the same for all of them. The reaction producing the light emissions is (92)

$$FMNH_2 + O_2 + R{-}CHO \xrightarrow{\text{luciferase}} FMN + H_2O + light$$

where $FMNH_2$ is the reduced form of flavin mononucleotide and RCHO is a long-chain aldehyde. The bacterial bioluminescence system gives a minimum level of detection for the determination of FMN which is better than other popular methods. For example, the minimum detectable concentration by paper chromatography is 1000 μg/100 μL, by fluorometry 10^{-4} μg/100 μL. It appears that fluorometry might compete for the determination of FMN; however, compounds normally found in samples of biosystems interfere with the fluorometric determination of FMN (370).

Barry et al. (373) evaluated the optical emissions of *in vivo* bioluminescence pulses by time-resolved spectrometry. The firefly *Photinus pyralis* was used as the test species. The light pulses were found to have several characteristics: (1) a duration of 0.3–0.4 sec, (2) a nonsymmetrical wavelength envelope, (3) an emission peak near 585 nm, (4) continuous emission over the interval 500–720 nm, (5) a power emission of the order of 1×10^{-6} W per pulse, (6) an intrapulse peak wavelength shift to shorter wavelengths during the pulse period. The authors recommended time-resolved spectrometry for intrapulse evaluation of other bioluminescence systems. Schroeder and Yeager (374) evaluated several chemiluminescent compounds as potential nonisotopic labels for monitoring competitive protein-binding reactions. The chemiluminescence yields and detection limits of some isoluminol derivatives in various oxidative systems were presented. Auses et al. (375) developed a chemiluminescent enzyme method for the determination of glucose. In the method, the oxidation of glucose in the presence of glucose oxidase at pH 7 provided a source for H_2O_2 which reacted with $Fe(CN)_6^{3-}$ in the presence of luminol to yield chemiluminescence proportional to the initial glucose concentration.

Matsumoto et al. (376) reported a method for the determination of phosphorus by gas-phase chemiluminescence after hydride generation. The detection limit obtained for phosphorus was 8 ng mL^{-1}. Honda et al. (377) investigated bis(2,4-dinitrophenyl)oxalate as a chemiluminescence reagent in the determination of fluorescent compounds by flow injection analysis. Factors affecting the chemiluminescence intensities were studied, and a limit of detection of 5 fmol was reported for dansylalanine. Keszthelyi et al. (378) discussed the efficiency of electrogenerated chemiluminescence of several systems using both potassium ferrioxalate actinometry and calibrated-photodiode measurements. Experimental methods and necessary corrections in the efficiency determinations were discussed, and electrogenerated chemiluminescence in mixed solvent systems was described. Rubinstein et al. (379) investigated the reaction between electrogenerated Ru(2,2′-bipyridine) and oxalate and found that electrogenerated chemiluminescence was produced in aqueous solution. The effect of oxalate concentration on the intensity of electrogenerated chemiluminescence was studied. The intensity was linearly related to oxalate concentration over the region 10^{-6} to 10^{-4} M. The results suggested the approach was selective for the determination of oxalate in urine. Ratzlaff and Crouch (380) reported a mathematical model of secondary inner-filter effects which allowed the derivation of a correction factor based on a determination of the ratio of chemiluminescence intensities from two pathlengths. A microprocessor-controlled instrument equipped with a unique dual-path cell which obtained intensity measurements from a collimated, monochromatic optical system was used in the work. Sigvardson and Birks (381) applied peroxyoxalate chemiluminescence to the HPLC detection of polycyclic aromatic hydrocarbons. Polycyclic aromatic hydrocarbons were excited by energy transfer from the decomposition products of the reaction between hydrogen peroxide and bis(2,4,6-trichlorophenyl)oxalate. The reagents were introduced by post-column mixing, and the emission was observed with a conventional fluorescence detector with the source turned off. The method yielded linear responses over three orders of magnitude. Detection limits were in the picogram range. Nelson et al. (382) developed a new gas chromatography detector with selectivity for reduced sulfur compounds. The detector monitored the chemiluminescence resulting from the reaction of the GC effluent with molecular fluorine at reduced pressures. The reaction of sulfur compounds with fluorine produced vibrationally excited HF whose emission was detected with a red-sensitive photomultiplier tube. The detection limits for a variety of sulfides, disulfides, and mercaptans were in the 24–260 pg range.

In the area of environmental pollution, chemiluminescence has been applied rather extensively to air pollution measurements (369,383). Chemiluminescence has also been used in water pollution studies (374). The development of automated instrumentation for air pollution analysis using chemiluminescent reactions has resulted in low-cost, sensitive and specific methods for ozone, sulfur compounds, and oxides of nitrogen. Commercial instruments are readily available (369). It should be mentioned that generally certain concentration ranges are of interest for gaseous pollutants. For example, for NO, NO_2, O_3, SO_2, and nonmethane hydrocarbons in ambient air monitoring, 1 ppb represents a lower limit and 1–10 ppm an upper limit (383). The concentration ranges of the air pollutants are one factor in deciding on what analytical techniques to use for air pollution monitoring. Other important factors are cost and specificity. There are essentially two basic types of monitors based on homogeneous gas-phase chemiluminescence. These are the compound-specific detector and the element-specific detector, which is a flame chemiluminescence detector. The latter detector is used for routine air pollution monitoring of sulfur compounds (383).

An important homogeneous gas-phase chemiluminescence technique for the detection of ozone is the chemiluminescent reaction between ozone and ethylene. This reaction yields chemiluminescent emission in the 300–600 rm range, and the intensity of the emission is directly proportional to ozone concentration. The detection system responds linearly to ozone concentrations between 0.003 and 30 ppm. No interferences have been observed. Flame chemiluminescence is routinely used for monitoring atmospheric sulfur compounds. The emission results from the recombination of sulfur atoms formed in the reducing flame environment. This is represented by the following reaction sequence:

$$S + S \rightarrow S_2^*$$
$$S_2^* \rightarrow S_2 + h\nu \ (350\text{--}450 \text{ nm})$$

Because the recombination involves two sulfur atoms, the intensity of the S_2^* is proportional to the square of the concentration of sulfur compound in the flame. The main reaction which has been employed for the detection of oxides of nitrogen is illustrated by the chemiluminescent reaction between nitric oxide and ozone as indicated below.

$$NO + O_3 \rightarrow NO_2^* + O_2$$
$$NO_2^* \rightarrow No_2 + h\nu$$

A typical linear range that has been obtained with a NO detection system is 0.001 to 10,000 ppm NO or a dynamic range of 10^7 (369). Several other researchers have considered detection of oxides of nitrogen (384–387). Mehrabzadeh et al. (388) have given an extensive discussion of optimization of response of chemiluminescence analyzers. They emphasized that the large number of variables operative in a chemiluminescent analyzer is difficult to optimize by an empirical approach. They developed equations which illustrated how these variables interact, and which ones most influenced the response of a chemiluminescence analyzer under different operating conditions.

5.   LIMITS OF DETECTION AND SELECTIVITY

Throughout this chapter, the limits of detection that can be obtained by several luminescence techniques have been discussed. Also, in Tables 8, 12, 13, and 15, the limits of detection were given for a variety of components. In Section 3.2.2, general comments were presented about the potential limits of detection with lasers and conventional sources. One of the important characteristics of luminescence analysis is the low limits of detection that can be obtained. Indeed, the low limits of detection by laser excitation for some samples are the lowest obtained in the history of luminescence analysis (324). However, as considered in Section 3.2.2, blank limitations are a very important factor for both laser and conventionally excited luminescence in discussing limits of detection (104,106). For a solute dissolved in a purified solvent, it would be possible to obtain a limit of detection in the 10^{-3}–10^{-5} ppb range by direct laser excitation. For components in complex samples, detection limits greater than 1 ppb would not be

Table 16. Detection Limits of Rhodamine B[a]

Exciting Source	Detection Apparatus	λ_{ex} (nm)	$\Delta\lambda_{ex}$ (nm)	λ_{em} (nm)	$\Delta\lambda_{em}$ (nm)	Detection Limit (ng L^{-1})	Source of Noise
Dye laser	Pulse-grated photon counter	560.0	0.66	592	1.3	0.04	Blank
Dye laser	d.c. amplifier	560.0	0.66	592	5.0	1	Dark current
Dye laser	Pulse-gated photon counter	514.5	0.33	585	1.0	0.21	Blank
Argon laser	d.c. amplifier	514.5	0.01	585	1.3	0.16	Blank
Xe lamp	d.c. amplifier	560.0	16.5	585	15.5	2.5	Dark current

[a]From Miyaishi et al (390).

unusual. Lytle (104) has concluded that unless additional forms of selectivity, such as chemical or instrumental, are combined with those normally associated with steady-state fluorometry, chemically complex samples will not benefit greatly from laser excitation. Several researchers have compared limits of fluorescence detection between laser sources and conventional sources (98,336,389,390). Table 16 shows the results from one such study. Experimental conditions, sources of noise, and detection limits are given, and rhodamine.B was used as a standard. The detection limit of 0.04 ng L^{-1} in Table 16 was achieved by combining a dye laser system with a pulse-gated photon counter. It is apparent from the literature that laser methods are very sensitive, capable of excellent precision, long linear dynamic ranges, and microsampling (98). Omenetto and Winefordner (98) have given an excellent discussion of calculated limits of detection based on signal-to-noise expressions.

Selectivity is another important advantage of luminescence analysis. However, the broad emission bands obtained with normal luminescence techniques prevent optimal selectivity from being achieved in many analytical situations. Selectivity can be substantially improved by using several new approaches. The chemist has available such techniques as derivative, synchronous, and fluorescence line narrowing spectroscopy, and, in addition, temporal resolution, the matrix isolation technique, and the Shpol'skii effect. Also, laser excitation can improve selectivity, quenching techniques, and a variety of chromatographic techniques employed in conjunction with luminescence analysis can enhance selectivity. Because of blank limitations there has been a trend by researchers away from sensitivity enhancement in luminescence analysis work and towards improved selectivity (104).

6.   FUTURE IMPROVEMENTS AND NEEDS

A very impressive array of instruments, techniques, and methods is available in luminescence analysis. However, luminescence analysis is still in the evolving stages and several improvements are still needed. In the area of theory, a better understanding is needed of the chemical and physical processes that cause luminescence. For example, one can use ethanol as a solvent and strong fluorescence is obtained from a component; however, if the same component is put in chloroform, the fluorescence is completely quenched. Similarly, in room-temperature phosphorescence, a compound on one surface will give room-temperature phosphorescence, whereas on another surface it will give no room-temperature phosphorescence. These are just two examples of hundreds of situations where improved luminescence theory would be important in explaining luminescent and nonluminescent phenomena. If explanations were in hand, conditions could be adjusted to improve sensitivity and/or selectivity. Great strides have been made in deriving signal-to-nose expressions. However, with new developments in instrumentation there will be a continuing need for developments in this area. Hirschfeld (391) and Ingle (392) have given an interesting exchange of ideas on signal-to-noise ratio in reference to the choice between absorption and fluorescence techniques.

Some very impressive luminescence instruments are in use today in analytical work. However, we should continue to see greater use of lasers and computers with luminescence instrumentation in the future. Lytle (104) has predicted that combination instruments using several laser properties could be used to handle very complicated mixtures. In addition, it is important to assemble laser-centered instruments which can be kept operative by personnel not necessarily well versed with laser techniques. Modern microprocessor-controlled pulsed-dye lasers would be a useful addition to luminescence instrumentation. In solid-surface luminescence analysis, the position of the absorbed luminescent component in the beam of source radiation can cause some difficulty. More extensive use of computers and digital electronics would minimize this error. With the development of instruments like the "video" fluorometer, a very large amount of luminescence data is generated in a short period of time. This necessarily means that new data-collection and data-handling techniques will be needed.

Luminescence analysis will continue to play a very important role in organic trace analysis. The demands for greater sensitivity and selectivity in biological, environmental, industrial, energy-related, forensic, and other areas will continue to appear. To take added advantage of luminescence analysis, it is necessary to find ways to minimize blank luminescence. One of the significant future challenges in luminescence analysis is the selective identification and quantification of components in highly complex samples (twenty and more components).

REFERENCES

1. T. C. O'Haver, *J. Chem. Educ.*, **55**, 423 (1978).
2. M. Kasha, *Radiat. Res. Suppl.*, **2**, 243 (1960).
3. R. S. Becker, *Theory and Interpretation of Fluorescence and Phosphorescence*, Wiley, New York, 1969.
4. C. A. Parker, *Photoluminescence of Solutions*, Elsevier, New York, 1968.
5. D. M. Hercules, Ed., *Fluorescence and Phosphorescence Analysis*, Wiley, New York, 1966.
6. E. L. Wehry, Ed., *Modern Fluorescence Spectroscopy*, Vol. 1, 1976; Vol. 2, 1976; Vol. 3, 1981; Vol. 4, 1981, Plenum, New York.
7. J. D. Winefordner, S. G. Schulman, and T. C. O'Haver, *Luminescence Spectrometry in Analytical Chemistry*, Wiley, New York, 1972.
8. R. J. Hurtubise, *Anal. Chem.*, **55**, 669A (1983).
9. W. R. Seitz, "Luminescence Spectrometry," in P. J. Elving, E. J. Meehan, and

I. M. Kolthoff, Eds., *Treatise on Analytical Chemistry*, 2nd ed., Part I, Vol. 7, Sec. H, Wiley, New York, 1981, Chap. 4.

10. J. N. Demas and G. A. Crosby, *J. Phys. Chem.*, **75**, 991 (1971).

11. J. W. Bridges, "The Determination of Quantum Yields," in J. N. Miller, Ed., *Standards in Fluorescence Spectrometry*, Chapman and Hall, New York, 1981, Chap. 8.

12. W. J. McCarthy and J. D. Winefordner, *J. Chem. Educ.*, **44**, 136 (1967).

13. R. A. Badley, "Fluorescence Probing of Dynamic and Molecular Organization of Biological Membranes," in E. L. Wehry, Ed., *Modern Fluorescence Spectroscopy*, Vol. 2, Plenum, New York, 1976, pp. 101–112.

14. S. G. Schulman, "Acid–Base Chemistry of Excited Singlet States," in E. L. Wehry, Ed., *Modern Fluorescence Spectroscopy*, Vol. 2, Plenum, New York, 1976, Chap. 6.

15. S. G. Schulman, *Fluorescence and Phosphorescence Spectroscopy: Physiochemical Principles and Practice*, Pergamon, Elmsford, NY, 1977.

16. G. Jackson and G. Porter, *Proc. Roy. Soc., Ser. A*, **260**, 13 (1961).

17. H. H. Richtol and B. R. Fitch, *Anal. Chem.*, **46**, 1860 (1974).

18. E. Vander Donckt, *Prog. React. Kinet.*, **5**, 273 (1970).

19. C. A. Parker, "Phosphorescence and Delayed Fluorescence in Solution," in W. A. Noyes, G. S. Hammond, and J. N. Pitts, Eds., *Advances in Photochemistry*, Vol. 2, Interscience, New York, 1964, pp. 305–383.

20. E. Sawicki, *Talanta*, **16**, 1231 (1969).

21. P. Froehlich and E. L. Wehry, "The Study of Excited-State Complexes ("Exciplexes") by Fluorescence Spectroscopy," in E. L. Wehry, Ed., *Modern Fluorescence Spectroscopy*, Vol. 2, Plenum, New York, 1976, Chap. 8.

22. N. J. Turro, *Modern Molecular Photochemistry*, Benjamin-Cummings, Menlo Park, CA, 1978.

23. M. Roth, *J. Chromatogr.*, **30**, 276 (1967).

24. E. M. Schulman and C. Walling, *Science*, **178**, 53 (1972).

25. E. M. Schulman and C. Walling, *J. Phys. Chem.*, **77**, 902 (1973).

26. R. A. Paynter, S. L. Wellons, and J. D. Winefordner, *Anal. Chem.*, **46**, 736 (1974).

27. R. T. Parker, R. S. Freedlander, and R. B. Dunlap, *Anal. Chim. Acta*, **119**, 189 (1980); **120**, 1 (1980).

28. J. L. Ward, G. L. Walden, and J. D. Winefordner, *Talanta*, **28**, 201 (1981).

29. E. Lue Yen-Bower, J. L. Ward, G. Walden, and J. D. Winefordner, *Talanta*, **27**, 380 (1980).

30. R. J. Hurtubise, *Solid Surface Luminescence Analysis: Theory, Instrumentation, Applications*, Dekker, New York, 1981.

31. E. M. Schulman and R. J. Parker, *J. Phys. Chem.*, **81**, 1932 (1977).

32. R. M. A. Von Wandruszka and R. J. Hurtubise, *Anal. Chem.*, **49**, 2164 (1977).

33. G. J. Niday and P. G. Seybold, *Anal. Chem.*, **49**, 1577 (1978).

34. R. J. Hurtubise and G. A. Smith, *Anal. Chim. Acta*, **139**, 315 (1982).

35. E. Leu Yen-Bower and J. D. Winefordner, *Anal. Chim. Acta*, **102**, 1 (1978).

36. S. M. Ramasamy and R. J. Hurtubise, *Anal. Chim. Acta*, **152**, 83 (1983).

37. R. A. Dalterio and R. J. Hurtubise, *Anal. Chem.*, **56,** 336 (1984).
38. L. J. Cline Love, M. Skrilec, and J. G. Habarta, *Anal. Chem.*, **52,** 754 (1980).
39. M. Skrilec and L. J. Cline Love, *Anal. Chem.*, **52,** 1559 (1980).
40. J. J. Donkerbroek, C. Gooijer, N. H. Velthorst, and R. W. Frei, *Anal. Chem.*, **54,** 891 (1982).
41. J. J. Donkerbroek, J. J. Elzas, C. Gooijer, R. W. Frei, and N. H. Velthorst, *Talanta,* **28,** 717 (1981).
42. W. R. Seitz and M. P. Neary, "Recent Advances in Bioluminescence and Chemiluminescence Assay," in D. Glick, Ed., *Methods of Biochemical Analysis,* Vol. 23, Wiley, New York, 1976, pp. 161–168.
43. M. A. DeLuca and W. D. McElroy, Eds., *Bioluminescence and Chemiluminescence: Basic Chemistry and Analytical Applications*, Academic, New York, 1981.
44. D. M. Hercules, *Anal. Chem.*, **38,** 29A (1966).
45. R. Zweidinger and J. D. Winefordner, *Anal. Chem.*, **42,** 639 (1970).
46. G. Kortum, *Reflectance Spectroscopy,* Springer-Verlag, New York, 1969.
47. V. Pollak and A. A. Boulton, *J. Chromatogr.,* **72,** 231 (1972).
48. V. Pollak, *Opt. Acta,* **21,** 51 (1974).
49. V. Pollak, *J. Chromatogr.,* **133,** 49 (1977).
50. J. Goldman, *J. Chromatogr.,* **78,** 7 (1973).
51. R. J. Hurtubise, *Anal. Chem.*, **49,** 2160 (1977).
52. T. Vo-Dinh and J. D. Winefordner, *Appl. Spectrosc. Rev.,* **13,** 261 (1977).
53. E. L. Wehry, "Effects of Molecular Structure and Molecular Environment on Fluorescence," in G. G. Guilbault, Ed., *Practical Fluorescence: Theory, Methods, and Techniques,* Dekker, New York, 1973, Chap. 3.
54. N. Mataga and T. Kubota, *Molecular Interactions and Electronic Spectra,* Dekker, New York, 1970.
55. H. Suzuki, *Electronic Absorption Spectra and Geometry of Organic Molecules,* Academic, New York, 1967.
56. T. C. Werner, "Use of Fluorescence Spectroscopy to Study Structural Changes and Solvation Phenomena in Electronically Excited Molecules," in E. L. Wehry, Ed., *Modern Fluorescence Spectroscopy,* Vol. 2, Plenum, New York, 1976, Chap. 7.
57. M. Goldman and E. L. Wehry, *Anal. Chem.*, **42,** 1178.
58. B. L. Van Duuren, *Chem. Rev.,* **63,** 325 (1963).
59. R. F. Chen, *Fluorescence News,* **9,** 9 (1975).
60. R. J. Hurtubise, *Anal. Chem.*, **48,** 2092 (1976).
61. J. N. Demas, *J. Chem. Educ.,* **53,** 657 (1976).
62. M. Zander, *Phosphorimetry,* Academic, New York, 1968.
63. E. L. Wehry and L. B. Rogers, "Fluorescence and Phosphorescence of Organic Molecules," in D. M. Hercules, Ed., *Fluorescence and Phosphorescence Analysis,* Interscience, New York, 1966.
64. R. T. Williams and J. W. Bridges, *J. Clin. Path.,* **17,** 371 (1964).
65. I. Chiang, J. M. Hayes, and G. J. Small, *Anal. Chem.*, **54,** 315 (1982).
66. J. R. Maple and E. L. Wehry, *Anal. Chem.*, **53,** 266 (1981).
67. D. Alkalay, L. Khemani, and M. F. Bartlett, *J. Pharm. Sci.,* **61,** 1746 (1972).

68. J. J. Aaron, J. L. Ward, and J. D. Winefordner, *Analusis,* **10,** 98 (1982).
69. R. J. Hurtbubise, *Anal. Chem.,* **47,** 2457 (1975).
70. E. Lue Yen-Bower and J. D. Winefordner, *Anal. Chim. Acta,* **101,** 319 (1978).
71. R. J. Hurtubise, G. T. Skar, and R. E. Poulson, *Anal. Chim. Acta,* **97,** 13 (1978).
72. O. S. Wolfbeis, W. Posch, G. Gubitz, and P. Tritthart, *Anal. Chim. Acta,* **147,** 405 (1983).
73. H. Nakamura and Z. Tamura, *Anal. Chem.,* **50,** 2047 (1978).
74. E. L. Wehry, *Anal. Chem.,* **52,** 75R (1980); **54,** 131R (1982).
75. P. A. St. John, W. J. McCarthy, and J. D. Winefordner, *Anal. Chem.,* **38,** 1828 (1966).
76. J. D. Winefordner, W. J. McCarthy, and P. A. St. John, *J. Chem. Educ.,* **44,** 80 (1967).
77. C. G. Enke and T. A. Nieman, *Anal. Chem.,* **48,** 705A (1976).
78. R. P.Cooney, T. Vo-Dinh, G. Walden, and J. D. Winefordner, *Anal. Chem.,* **49,** 939 (1977).
79. R. P. Cooney, G. D. Boutilier, and J. D. Winefordner, *Anal. Chem.,* **49,** 1048 (1977).
80. T. C. O'Haver and J. D. Winefordner, *Anal. Chem.,* **38,** 602 (1966).
81. V. Pollak, *J. Chromatogr.,* **123,** 11 (1976).
82. G. L. Long and J. D. Winefordner, *Anal. Chem.,* **55,** 712A (1983).
83. "Guidelines for Data Acquisition and Data Quality Evaluation in Environmental Chemistry," *Anal. Chem.,* **52,** 2242 (1980).
84. P. A. St. John, W. J. McCarthy, and J. D. Winefordner, *Anal. Chem.,* **39,** 1495 (1967).
85. J. J. Cetorelli, W. J. McCarthy, and J. D. Winefordner, *J. Chem. Educ.,* **45,** 98 (1968).
86. G. D. Boutilier and J. D. Winefordner, *Anal. Chem.,* **51,** 1384 (1979).
87. R. P. Cooney, T. Vo-Dinh, and J. D. Winefordner, *Anal. Chim. Acta,* **89,** 9 (1977).
88. R. P. Cooney and J. D. Winefordner, *Anal. Chem.,* **49,** 1057 (1977).
89. V. P. Senthilnathan, S. M. Ramasamy, and R. J. Hurtubise, *Anal. Chim. Acta,* **157,** 203 (1984).
90. E. Voigtman, A. Jurgensen, and J. D. Winefordner, *Anal. Chem.,* **53,** 1921 (1981).
91. E. P. Lai, E. L. Inman, and J. D. Winefordner, *Talanta,* **29,** 601 (1982).
92. W. R. Seitz, *Crit. Rev. Anal. Chem.,* **13,** 1 (1981).
93. P. F. Lott, J. R. Dias, and R. J. Hurtubise, *J. Chromatogr. Sci.,* **14,** 488 (1976).
94. G. G. Guilbault, *Practical Fluorescence: Theory, Methods, and Techniques,* Dekker, New York, 1973.
95. T. D. S. Hamilton, I. H. Munro, and G. Walker, "Luminescence Instrumentation," in M. D. Lumb, Ed., *Luminescence Spectroscopy,* Academic, New York, 1978, pp. 149–238.
96. G. M. Hieftje, J. C. Travis, and F. E. Lytle, *Lasers in Chemical Analysis,* Humana, Clifton, NJ, 1981.
97. J. D. Winefordner, "Laser-Excited Luminescence Spectrometry," in G. M. Hieftje, Ed., *New Applications of Lasers to Chemistry* (ACS Symposium Series, No. 85), American Chemical Society, Washington, D.C., 1978, pp. 50–79.

98. N. Omenetto and J. D. Winefordner, *Crit. Rev. Anal. Chem.*, **13**, 59 (1981).

99. J. H. Richardson, "Applications of Lasers in Analytical Molecular Fluorescence Spectroscopy," in E. L. Wehry, Ed., *Modern Fluorescence Spectroscopy*, Vol. 4, Plenum, New York, 1981, Chap. 1.

100. W. Demtroder, "Molecular Absorption and Fluorescence Spectroscopy with Lasers," in N. Omenetto, Ed., *Analytical Laser Spectroscopy*, Wiley, New York, 1979, Chap. 5.

101. H. W. Latz, "Dye Lasers: Fundamentals and Analytical Applications," in E. L. Wehry, Ed., *Modern Fluorescence Spectroscopy*, Vol. 1, Plenum, New York, 1976, Chap. 4.

102. R. B. Green, *J. Chem. Educ.*, **54**, A365, A407 (1977).

103. N. J. Dovichi, J. C. Martin, J. H. Jett, and R. A. Keller, *Science*, **219**, 845 (1983).

104. F. E. Lytle, *J. Chem. Educ.*, **59**, 915 (1982).

105. J. C. Wright, "Laser-Excited Fluorescence Spectroscopy," in G. M. Hieftje, J. C. Travis, and F. E. Lytle, Eds., *Lasers in Chemical Analysis*, Humana, Clifton, NJ, 1981, Chap. 9.

106. T. G. Matthews and F. E. Lytle, *Anal. Chem.*, **51**, 583 (1979).

107. D. C. Harrington and H. V. Malmstadt, *Anal. Chem.*, **47**, 271 (1975).

108. J. H. Richardson and S. M. George, *Anal. Chem.*, **50**, 616 (1978).

109. G. D. Boutilier and J. D. Winefordner, *Anal. Chem.*, **51**, 1391 (1979).

110. E. D. Olsen, *Modern Optical Methods of Analysis*, McGraw-Hill, New York, 1975, Chaps. 1 and 8.

111. E. J. Meehan, "Spectroscopic Apparatus and Measurements," in P. J. Elving, E. J. Meehan, and I. M. Kolthoff, Eds. *Treatise on Analytical Chemistry*, Part I, Vol. 7, Sec. H, Wiley, New York, 1981, Chap. 3.

112. J. R. Lakowicz, *Principles of Fluorescence Spectroscopy*, Plenum, New York, 1983.

113. H. C. Hollifield and J. D. Winefordner, *Anal. Chem.*, **40**, 1759 (1968).

114. R. J. Lukasiewicz, P. A. Rozynes, L. B. Sanders, and J. D. Winefordner, *Anal. Chem.*, **44**, 237 (1972).

115. J. A. McHard and J. D. Winefordner, *Can. J. Spectrosc.*, **18**, 31 (1973).

116. J. L. Ward, R. P. Bateh, and J. D. Winefordner, *Appl. Spectrosc.*, **34**, 15 (1980).

117. J. L. Ward, G. L. Walden, R. P. Bateh, and J. D. Winefordner, *Appl. Spectrosc.*, **34**, 348 (1980).

118. G. C. Pillai and R. C. Patel, *Anal. Biochem.*, **106**, 506 (1980).

119. F. W. D. Rost, *Med. Biol.*, **52**, 73 (1974).

120. G. G. Vurek and R. L. Bowman, "Assay Methods Using Capillary Tubes," in M. Werner, Ed., *Microtechniques for the Clinical Laboratory: Concepts and Applications*, Wiley, New York, 1976, Chap. 4.

121. F. Ruch and U. Leeman, "Cytofluorometry," in V. Neuhoff, Ed., *Micromethods in Molecular Biology*, Springer-Verlag, New York, 1973, Chap. 10.

122. F. A. Dolbeare, "Fluorometric Quantification of Specific Chemical Species in Single Cells," in E. L. Wehry, Ed., *Modern Fluorescence Spectroscopy*, Vol. 3, Plenum, New York, 1981, Chap. 6.

123. E. Kohen, B. Thorell, J. G. Hirschberg, A. W. Weuters, C. Kohen, P. Bartick,

J. M. Salmon, P. Viallet, D. O. Schachtschabel, A. Rabinovitch, D. Mintz, P. Meda, H. Westerhoff, J. Nestor, and J. S. Ploem, "Microspectrofluorometric Procedures and Their Applications in Biological Systems," E. L. Wehry, Ed., *Modern Fluorescence Spectroscopy,* Vol. 3, Plenum, New York, 1981, Chap. 7.

124. S. Y. Su, A. Jurgensen, D. Bolton, and J. D. Winefordner, *Anal. Lett.,* **14,** 1 (1981).

125. E. S. Yeung and M. J. Sepaniak, *Anal. Chem.,* **52,** 1465A (1980).

126. L. W. Hershberger, J. B. Callis, and G. D. Christian, *Anal. Chem.,* **51,** 1444 (1979).

127. E. Johnson, A. Abu-Shumays, and S. R. Abbot, *J. Chromatogr.,* **134,** 107 (1977).

128. L. Hirschy, B. Smith, E. Voigtman, and J. D. Winefordner, *Anal. Chem.,* **54,** 2387 (1982).

129. L. Langouet, *Appl. Opt.,* **11,** 2358 (1972).

130. H. C. Hollifield and J. D. Winefordner, *Chem. Instru.,* **1,** 341 (1969).

131. E. L. Yen, G. D. Boutilier, and J. D. Wincfordner, *Can. J. Spectrosc.,* **22,** 120 (1977).

132. T. Vo-Dinh, G. L. Walden, and J. D. Winefordner, *Anal. Chem.,* **49,** 1126 (1977).

133. C. D. Ford and R. J. Hurtubise, *Anal. Chem.,* **51,** 659 (1979).

134. M. L. Franklin, G. Horlick, and H. V. Malmstadt, *Anal. Chem.,* **41,** 2 (1969).

135. D. M. Jameson, R. D. Spencer, and G. Weber, *Rev. Sci. Instrum.,* **47,** 1034 (1976).

136. V. J. Koester, and R. M. Dowben, *Rev. Sci. Instrum.,* **49,** 1186 (1978).

137. S. Cova, G. Prenna, and G. Mazzini, *Histochem. J.,* **6,** 279 (1974).

138. E. J. Darland, G. E. Leroi, and C. G. Enke, *Anal. Chem.,* **51,** 240 (1979).

139. E. J. Darland, G. E. Leroi, and C. G. Enke, *Anal. Chem.,* **52,** 714 (1980).

140. M. L. Meade, *J. Phys. E.,* **14,** 909 (1981).

141. L. J. Cline Love and L. A. Shaver, *Anal. Chem.,* **48,** 364A (1976).

142. L. M. Upton and L. J. Cline Love, *Anal. Chem.,* **51,** 1941 (1979).

143. G. R. Haugen and F. E. Lytle, *Anal. Chem.,* **53,** 1554 (1981).

144. Y. Talmi, *Anal. Chem.,* **47,** 658A, 697A (1975).

145. Y. Talmi, D. C. Baker, J. R. Jadamec, and W. A. Saner, *Anal. Chem.,* **50,** 936A (1978).

146. Y. Talmi, *Appl. Spectrosc.,* **36,** 1 (1982).

147. Y. Talmi, Ed., *Multichannel Image Detectors* (ACS Symposium Series, Vol. 102), American Chemical Society, Washington, D.C., 1979.

148. Y. Talmi, Ed., *Multichannel Image Detectors, Volume 2* (ACS Symposium Series, Vol. 236), American Chemical Society, Washington, D.C., 1983.

149. G. D. Christian, J. B. Callis, and E. R. Davidson, "Array Detectors and Excitation-Emission Matrices in Multicomponent Analysis," in E. L. Wehry, Ed., *Modern Fluorescence Spectroscopy,* Vol. 4, Plenum, New York, 1981, Chap. 4.

150. T. Vo-Dinh, D. J. Johnson, and J. D. Winefordner, *Spectrochim. Acta,* **33A,** 341 (1977).

151. M. A. Ryan, R. J. Miller, and J. D. Ingle, Jr., *Anal. Chem.,* **50,** 1772 (1978).

152. T. G. Curtis and W. R. Seitz, *J. Chromatogr.,* **134,** 513 (1977).

153. D. E. Goeringer and H. L. Pardue, *Anal. Chem.,* **51,** 1054, (1979).

154. J. D. Ingle and M. A. Ryan, "Luminescence Measurements with an Intensified

Diode Array," in Y. Talmi, Ed., *Multichannel Image Detectors, Volume 2* (ACS Symposium Series, Vol. 236), American Chemical Society, Washington, D.C., 1983, Chap. 7.

155. C. A. Parker and W. T. Rees, *Analyst,* **85,** 587 (1960).
156. R. J. Argauer and C. E. White, *Anal. Chem.,* **36,** 368 (1964).
157. W. H. Melhuish, *J. Opt. Soc. Am.,* **52,** 1265 (1962).
158. H. V. Drushel, A. L. Sommers, and R. C. Cox, *Anal. Chem.,* **35,** 2166 (1963).
159. I. Landag and J. C. Kremen, *Anal. Chem.,* **46,** 1694 (1974).
160. R. F. Chen, *Anal. Biochem.,* **20,** 339 (1967).
161. D. A. Corliss, S. S. West, and J. F. Golden, *Appl. Opt.,* **19,** 3290 (1980).
162. K. D. Mielenz, Ed. *Optical Radiation Measurements: Measurement of Photoluminescence,* Vol. 3, Academic, New York, 1982.
163. T. J. Porro, R. E. Anacreon, P. S. Flandreau, and I. S. Fagerson, *J. Assoc. Off. Anal. Chem.,* **56,** 607 (1973).
164. T. W. Allen, R. J. Hurtubise, and H. F. Silver, *Anal. Chim. Acta,* **141,** 411 (1982).
165. G. C. K. Roberts, "Correction of Excitation and Emission Spectra," in J. N. Miller, Ed., *Standards in Fluorescence Spectrometry,* Chapman and Hall, New York, 1981, Chap. 7.
166. T. C. O'Haver and J. D. Winefordner, *Anal. Chem.,* **38,** 1258 (1966).
167. R. P. Fisher and J. D. Winefordner, *Anal. Chem.,* **44,** 948 (1972).
168. J. J. Mousa and J. D. Winefordner, *Anal. Chem.,* **46,** 1195 (1974).
169. R. M. Wilson and T. L. Miller, *Anal. Chem.,* **47,** 256 (1975).
170. J. L. Charlton and B. R. Henry, *J. Chem. Educ.,* **51,** 753 (1974).
171. T. R. Dyke and J. S. Muenter, *J. Chem. Educ.,* **52,** 251 (1975).
172. L. J. Cline Love and M. Skrilec, *Anal. Chem.,* **53,** 2103 (1981).
173. J. F. Rabek, *Experimental Methods in Photochemistry and Photophysics,* Part 2, Wiley, New York, 1982, Chaps. 21, 22, and 23.
174. L. J. Cline Love and L. A. Shaver, *Anal. Chem.,* **52,** 154 (1980).
175. G. M. Hieftje and E. E. Vogelstein, "A Linear Response Theory Approach to Time-Resolved Fluorometry," in E. L. Wehry, Ed., *Modern Fluorescence Spectroscopy,* Vol. 4, Plenum, New York, 1981, Chap. 2.
176. G. Weber, *J. Phys. Chem.,* **85,** 949 (1981).
177. P. F. Lott and R. J. Hurtubise, *J. Chem. Educ.,* **51,** A315, A357 (1974).
178. L. P. Giering, *Ind. Res.,* **20,** 134 (1978).
179. I. M. Warner and L. B. McGown, *Crit. Rev. Anal. Chem.,* **13,** 155 (1982).
180. D. W. Johnson, J. A. Gladden, J. B. Callis, and G. D. Christian, *Rev. Sci. Instrum.,* **50,** 118 (1979).
181. I. M. Warner, J. B. Callis, E. R. Davidson, and G. D. Christian, *Clin. Chem.,* **22,** 1483 (1976).
182. D. W. Johnson, J. B. Callis, and G. D. Christian, *Anal. Chem.,* **49,** 747A (1977).
183. I. M. Warner, M. P. Fogarty, and D. C. Shelly, *Anal. Chim. Acta.,* **109,** 361 (1979).
184. I. M. Warner, G. D. Christian, E. R. Davidson, and J. B. Callis, *Anal. Chem.,* **49,** 564 (1977).
185. C. N. Ho, G. D. Christian, and E. R. Davidson, *Anal. Chem.,* **53,** 92 (1981).

186. M. P. Fogarty and I. M. Warner, *Anal. Chem.*, **53**, 259 (1981).

187. C. N. Ho and I. M. Warner, *Anal. Chem.*, **54**, 2486 (1982).

188. J. H. Rho and J. L. Stuart, *Anal. Chem.*, **50**, 620 (1978).

189. T. F. Van Geel and J. D. Winefordner, *Anal. Chem.*, **48**, 335 (1976).

190. T. Imasaka, I. Ogawa, and N. Ishibashi, *Anal. Chem.*, **51**, 502 (1979).

191. E. L. Wehry and G. Mamantov, *Anal. Chem.*, **51**, 643A (1979).

192. R. B. Dickinson and E. L. Wehry, *Anal. Chem.*, **51**, 778 (1979).

193. O. S. Khalil, W. S. Routh, K. Lingenfelter, D. B. Carr, and P. Ladouceur, *Clin. Chem.*, **27**, 1586 (1981).

194. A. W. Ritter, P. C. Tway, L. J. Cline Love, and H. A. Ashworth, *Anal. Chem.*, **53**, 280 (1981).

195. J. E. Thompson and H. L. Pardue, *Anal. Chim. Acta.*, **152**, 73 (1983).

196. W. E. Howard, A. Greenquist, B. Walter, and F. Wogoman, *Anal. Chem.*, **55**, 878 (1983).

197. P. Froehlich and E. L. Wehry, "Fluorescence Detection in Liquid and Gas Chromatography," in E. L. Wehry, Ed., *Modern Fluorescence Spectroscopy*, Vol. 3, Plenum, New York, 1981, Chap. 2.

197a. L. W. Hershberger, J. B. Callis, and G. D. Christian, *Anal. Chem.*, **53**, 971 (1981).

197b. I. C. Shelly, M. P. Fogarty, and I. M. Warner, *HRC CC, J. High Resolut. Chromatogr. Chromatogr. Commun.*, **4**, 616 (1981).

198. M. J. Sepaniak and E. S. Yeung, *Anal. Chem.*, **49**, 1554 (1977).

199. G. D. Boutilier, R. M. Irwin, R. R. Antcliff, L. B. Rogers, and L. A. Carreira, *Appl. Spectrosc.*, **35**, 576 (1981).

200. S. Folestad, L. Johnson, B. Josefsson, and B. Galle, *Anal. Chem.*, **54**, 925 (1982).

201. J. M. Hayes and G. J. Small, *Anal. Chem.*, **54**, 1202 (1982).

202. V. B. Conrad, W. J. Carter, E. L. Wehry, and G. Mamantov, *Anal. Chem.*, **55**, 1340 (1983).

203. C. D. Ford and R. J. Hurtubise, *Anal. Chem.*, **52**, 656 (1980).

204. E. Leu Yen-Bower and J. D. Winefordner, *Appl. Spectrosc.*, **33**, 9 (1979).

205. G. L. Walden and J. D. Winefordner, *Appl. Spectrosc.*, **33**, 166 (1979).

206. K. F. Harbaugh, C. M. O'Donnell, and J. D. Winefordner, *Anal. Chem.*, **46**, 1206 (1974).

207. C. N. Ho and I. M. Warner, *Trends. Anal. Chem.*, **1**, 159 (1982).

208. J. E. Wampler, "Fluorescence Spectroscopy with On-Line Computers," in E. L. Wehry, Ed., *Modern Fluorescence Spectroscopy*, Vol. 1, Plenum, New York, 1976, Chap. 1.

209. E. R. Weiner and M. C. Goldberg, *Am. Lab.* **14(9)**, 91 (1982).

210. J. W. Lyons, P. T. Hardesty, C. S. Baer, and L. R. Faulkner, "Structural Interpretation of Fluorescence Spectra by Automated File Searching," in E. L. Wehry, Ed., *Modern Fluorescence Spectroscopy*, Vol. 3, Plenum, New York, 1981, Chap. 1.

211. J. M. Fitzgerald, "Digital and Analog Measurements in Fluorescence Spectroscopy," in E. L. Wehry, Ed., *Modern Fluorescence Spectroscopy*, Vol. 1, Plenum, New York, 1976, Chap. 2.

212. T. C. O'Haver, "Modulation and Derivative Techniques in Luminescence Spec-

troscopy: Approaches to Increased Analytical Selectivity," in E. L. Wehry, Ed., *Modern Fluorescence Spectroscopy*, Vol. 1, Plenum, New York, 1976, Chap. 3.

213. S. Ebel and J. Hocke, *J. Chromatogr.*, **126**, 449 (1976).

214. S. Ebel and J. Hocke, *Chromatographia*, **9**, 78 (1976).

215. S. Ebel and J. Hocke, *Chromatographia*, **10**, 123 (1977).

216. S. Ebel, G. Herold, and J. Hocke, *Chromatographia*, **8**, 573 (1975).

217. G. L. Green and T. C. O'Haver, *Anal. Chem.*, **46**, 2191 (1974).

218. T. C. O'Haver and W. M. Parks, *Anal. Chem.*, **46**, 1886 (1974).

219. G. Talsky, L. Mayring, and H. Kreuzer, *Angew. Chem. Int. Ed. Engl.*, **17**, 785 (1978).

220. J. E. Cahill, *Am. Lab.* **11**(11), 79 (1979).

221. T. C. O'Haver, *Anal. Chem.*, **51**, 91A (1979).

222. T. R. Griffiths, K. King, H. V. St. A. Hubbard, M. J. Schwing-Weill, and J. Meullemeestre, *Anal. Chim. Acta.*, **143**, 163 (1982).

223. M. A. Fox and S. W. Staley, *Anal. Chem.*, **48**, 992 (1976).

224. T. Vo-Dinh and R. B. Gammage, *Anal. Chim. Acta.*, **107**, 261 (1979).

225. R. H. Christenson, and C. D. McGlothlin, *Anal. Chem.*, **54**, 2015 (1982).

226. T. Vo-Dinh, *Anal. Chem.*, **50**, 396 (1978).

227. T. Vo-Dinh, *Appl. Spectrosc.*, **36**, 576 (1982).

228. T. Vo-Dinh, "Synchronous Excitation Spectroscopy," in E. L. Wehry, Ed., *Modern Fluorescence Spectroscopy*, Vol. 4, Plenum, New York, 1981, Chap. 5.

229. J. B. F. Lloyd and I. W. Evett, *Anal. Chem.*, **49**, 1710 (1977).

230. H. W. Latz, A. H. Ullman, and J. D. Winefordner, *Anal. Chem.*, **50**, 2148 (1978).

231. J. B. F. Lloyd, Nature (London), *Phys. Sci.*, **231**, 64 (1971).

232. J. B. F. Lloyd, *J. Forensic Sci. Soc.*, **11**, 83 (1971).

233. J. B. F. Lloyd, *J. Forensic Sci. Soc.*, **11**, 153 (1971).

234. J. B. F. Lloyd, *Analyst*, **99**, 729 (1974).

235. P. John and I. Soutar, *Anal. Chem.*, **48**, 520 (1976).

236. S. G. Wakeham, *Environ. Sci. Technol.*, **11**, 272 (1977).

237. T. Vo-Dinh and R. B. Gammage, *Anal. Chem.*, **50**, 2054 (1978).

238. T. Vo-Dinh, R. B. Gammage, A. R. Hawthorne, and J. H. Thorngate, *Environ. Sci. Technol.*, **12**, 1297 (1978).

239. T. Vo-Dinh, R. B. Gammage, and P. R. Martinez, *Anal. Chim. Acta*, **118**, 313 (1980).

240. T. Vo-Dinh, R. B. Gammage, and P. R. Martinez, *Anal. Chem.*, **53**, 253 (1981).

241. T. Vo-Dinh and R. B. Martinez, *Anal. Chim. Acta*, **125**, 13 (1981).

242. E. L. Inman and J. D. Winefordner, *Anal. Chim. Acta*, **141**, 241 (1982).

243. E. L. Inman and J. D. Winefordner, *Anal. Chem.*, **54**, 2018 (1982).

244. E. L. Wehry and G. Mamantov, "Low-Temperature Fluorometric Techniques and Their Application to Analytical Chemistry," in E. L. Wehry, Ed., *Modern Fluorescence Spectroscopy*, Vol. 4, Plenum, New York, 1981, pp. 211–231.

245. R. C. Stroupe, P. Tokousbalides, R. B. Dickinson, E. L. Wehry, and G. Mamantov, *Anal. Chem.*, **49**, 701 (1977).

246. P. Tokousbalides, E. Ray Hinton, R. B. Dickinson, P. V. Bilotta, E. L. Wehry, and G. Mamantov, *Anal. Chem.*, **50**, 1189 (1978).

247. J. R. Maple, E. L. Wehry, and G. Mamantov, *Anal. Chem.*, **52**, 920 (1980).

248. J. R. Maple and E. L. Wehry, *Anal. Chem.*, **53**, 1244 (1981).
249. M. B. Perry, E. L. Wehry, and G. Mamantov, *Anal. Chem.*, **55**, 1893 (1983).
250. V. B. Conrad and E. L. Wehry, *Appl. Spectrosc.*, **37**, 46 (1983).
251. B. S. Causey, G. F. Kirkbright, and C. G. deLima, *Analyst*, **101**, 367 (1976).
252. A. P. D'Silva, G. J. Oestreich, and V. A. Fassel, *Anal. Chem.*, **48**, 915 (1976).
253. C. S. Woo, A. P. D'Silva, V. A. Fassel, and G. J. Oestreich, *Environ. Sci. Tech.*, **12**, 173 (1978).
254. C. S. Woo, A. P. D'Silva, and V. A. Fassel, *Anal. Chem.*, **52**, 159 (1980).
255. A. L. Colmsjo and C. E. Ostman, *Anal. Chem.*, **52**, 2093 (1980).
256. Y. Yang, A. P. D'Silva, V. A. Fassel, and M. Iles, *Anal. Chem.*, **52**, 1350 (1980).
257. Y. Yang, A. P. D'Silva, and V. A. Fassel, *Anal. Chem.*, **53**, 894 (1981).
258. Y. Yang, A. P. D'Silva, and V. A. Fassel, *Anal. Chem.*, **53**, 2107 (1981).
259. G. D. Renkes, S. N. Walters, C. S. Woo, M. K. Iles, A. P. D'Silva, and V. A. Fassel, *Anal. Chem.*, **55**, 2229 (1983).
260. J. Rima, M. Lamotte, and J. Joussot-Dubien, *Anal. Chem.*, **54**, 1059 (1982).
261. M. Ewald, A. Moinet, A. Saliot, and P. Albrecht, *Anal. Chem.*, **55**, 959 (1983).
262. P. Garrigues and M. Ewald, *Anal. Chem.*, **55**, 2155 (1983).
263. L. A. Bykovskaya, R. I. Personov, and Yu. V. Romanovskii, *Anal. Chim. Acta.*, **125**, 1 (1981).
264. J. C. Brown, M. C. Edelson, and G. J. Small, *Anal. Chem.*, **50**, 1394 (1978).
265. J. C. Brown, J. A. Ducanson, and G. J. Small, *Anal. Chem.*, **52**, 1711 (1980).
266. I. Chiang, J. M. Hays, and G. J. Small, *Anal. Chem.*, **54**, 315 (1982).
267. V. Heisig, A. M. Jeffrey, M. J. McGlade, and G. J. Small, *Science*, **223**, 289 (1984).
268. R. W. Frei, L. Michel, and W. Santi, *J. Chromatogr.*, **126**, 665 (1976).
269. J. B. F. Lloyd, *J. Chromatogr.*, **178**, 249 (1979).
270. M. J. Sepaniak and E. S. Yeung, *J. Chromatogr.*, **190**, 377 (1980).
271. H. Nakamura and Z. Tamura, *Anal. Chem.*, **54**, 1951 (1982).
272. N. Furuta and A. Otsuki, *Anal. Chem.*, **55**, 2407 (1983).
273. M. L. Gianelli, J. B. Callis, N. H. Andersen, and G. D. Christian, *Anal. Chem.*, **53**, 1357 (1981).
274. R. Weinberger, P. Yarmchuk, and L. J. Cline Love, *Anal. Chem.*, **54**, 1552 (1982).
275. D. W. Armstrong, W. L. Hinze, K. H. Bui, and H. N. Singh, *Anal. Lett.*, **14**, (**A19**), 1659 (1981).
276. J. J. Donkerbroek, N. J. R. van Eikema Hommes, C. Gooijer, N. H. Velthorst, and R. W. Frei, *Chromatographia*, **15**, 218 (1982).
277. J. J. Donkerbroek, N. J. R. van Eikema Hommes, C. Gooijer, N. H. Velthorst, and R. W. Frei, *J. Chromatogr.*, **255**, 581 (1983).
278. J. J. Donkerbroek, A. C. Veltkamp, C. Gooijer, N. H. Velthorst, and R. W. Frei, *Anal. Chem.*, **55**, 1886 (1983).
279. C. N. Ho, G. D. Christian, and E. R. Davidson, *Anal. Chem.*, **50**, 1108 (1978).
280. C. N. Ho, G. D. Christian, and E. R. Davidson, *Anal. Chem.*, **52**, 1071 (1980).
281. I. M. Warner, E. R. Davidson, and G. D. Christian, *Anal. Chem.*, **49**, 2155, (1977).

282. C. R. Sawicki and E. Sawicki, "Thin-Layer Chromatography in Air Pollution Research," in A. Niederwieser and G. Pataki, Eds., *Progress in Thin-Layer Chromatography and Related Methods,* Vol. III, Ann Arbor Science Publishers, Inc., Ann Arbor, MI, 1972, Chap. 6.

283. E. P. Gibson, *J. Forensic Sci., 22,* 680 (1977).

284. J. F. Lawrence and R. W. Frei, *J. Chromatogr., 98,* 253 (1974).

285. J. G. Zakrevsky and V. N. Mallet, *J. Chromatogr., 132,* 315 (1977).

286. R. E. Kaiser, *Abstracts of Papers,* 177th National Meeting of the American Chemical Society, Honolulu, Hawaii, April, 1979, Abstract Anal. 15.

287. H. K. Y. Lau and G. G. Guilbault, *Enzym. Tech. Dig., 3,* 164 (1974).

288. G. G. Guilbault, "Analytical Uses of Immobilized Enzymes," in M. Salomona, C. Saronio, and S. Garattini, Eds., *Immobilized Enzymes,* Raven, New York, 1974.

289. G. G. Guilbault, "Fluorescence Analysis on Solid Surfaces," in E. Wanninen, Ed., *Analytical Chemistry: Essay in Memory of Anders Ringbom,* Pergamon, New York, 1977.

290. G. G. Guilbault, *Photochem. Photobiol., 25,* 403 (1977).

291. G. G. Guilbault and R. J. Zimmerman, *Anal. Lett., 3,* 133 (1970).

292. M. K. L. Bicking, R. M. Kniseley, and H. J. Svec, *Anal. Chem., 55,* 200 (1983).

293. P. B. Huff and M. J. Sepaniak, *Anal. Chem., 55,* 1992 (1983).

294. L. A. Gifford, J. N. Miller, D. L. Phillipps, D. Thorburn Burns, and J. W. Bridges, *Anal. Chem., 47,* 1699 (1975).

295. N. S. Allen, J. Homer, and J. F. McKellar, *Analyst, 101,* 260 (1976).

296. J. A. F. DeSilva, N. Strojny, and K. Stika, *Anal. Chem., 48,* 144 (1976).

297. D. K. Lavallee and J. Andrew, *Anal. Chem., 49,* 1482 (1977).

298. L. A. Gifford, J. N. Miller, J. W. Bridges, and D. Thorburn Burns, *Talanta, 24,* 273 (1977).

299. J. N. Miller, D. L. Phillipps, D. Thorburn Burns, and J. W. Bridges, *Talanta, 25,* 46 (1978).

300. S. H. Fortier and D. Eastwood, *Anal. Chem., 50,* 334 (1978).

301. M. M. Corfield, H. L. Hawkins, P. John, and I. Soutar, *Analyst, 106,* 188 (1981).

302. L. A. Gifford, J. N. Miller, D. Thorburn Burns, and J. W. Bridges, *J. Chromatogr., 103,* 15 (1975).

303. J. N. Miller, D. L. Phillipps, D. Thorburn Burns, and J. W. Bridges, *Anal. Chem., 50,* 613 (1978).

304. S. L. McCall and J. D. Winefordner, *Anal. Chem., 55,* 391 (1983).

305. L. J. Cline Love, J. G. Habarta, and M. Skrilec, *Anal. Chem., 53,* 437 (1981).

306. L. J. Cline Love and M. Skrilec, *Am. Lab., 13*(3), 103 (1981).

307. L. J. Cline Love and M. Skrilec, *Anal. Chem., 53,* 1872 (1981).

308. M. Skrilec and L. J. Cline Love, *J. Phys. Chem., 85,* 2047 (1981).

309. J. J. Donkerbroek, A. C. Veltkamp, A. J. J. Praat, C. Gooijer, R. W. Frei, and N. H. Velthorst, *Appl. Spectrosc., 37,* 188 (1983).

310. T. Vo-Dinh, "Rapid Screening Luminescence Techniques for Trace Organic Analysis," in D. Eastwood, Ed., *New Directions in Molecular Luminescence,* ASTM, Philadelphia, 1983, pp. 5–16.

311. T. Vo-Dinh, E. L. Yen, and J. D. Winefordner, *Anal. Chem.*, **48**, 1186 (1976).
312. J. J. Aaron, E. M. Kaleel, and J. D. Winefordner, *J. Agric. Food Chem.*, **27**, 1233 (1979).
313. R. P. Bateh and J. D. Winefordner, *Talanta*, **29**, 713 (1982).
314. R. M. A. Von Wandruszka and R. J. Hurtubise, *Anal. Chem.*, **48**, 1784 (1976).
315. C. D. Ford and R. J. Hurtubise, *Anal. Lett.*, **13**(A6), 485 (1980).
316. T. Vo-Dinh and J. R. Hooyman, *Anal. Chem.*, **51**, 1915 (1979).
317. M. W. Warren, J. P. Avery, and H. V. Malmstadt, *Anal. Chem.*, **54**, 1853 (1982).
318. F. Abdel Fattah, W. Baeyens, and P. De Moerloose, "Room Temperature Phosphorescence of Some Pharmaceutical Important Imidazoles," in M. A. DeLuca and W. D. McElroy, Eds., *Bioluminescence and Chemiluminescence: Basic Chemistry and Analytical Applications*, Academic, New York, 1981, pp. 335–346.
319. I. M. Jakovljevic, *Anal. Chem.*, **49**, 2048 (1977).
320. C. G. DeLima and E. M. de M. Nicola, *Anal. Chem.*, **50**, 1658 (1978).
321. R. T. Parker, R. S. Freedlander, E. M. Schulman, and R. B. Dunlap, *Anal. Chem.*, **51**, 1921 (1979).
322. M. L. Meyers and P. G. Seybold, *Anal. Chem.*, **51**, 1609 (1979).
323. J. D. Winefordner and E. Voightman, "Laser-Excited Fluorescence, Photoacoustic, and Photoionization Detection of Polyaromatic Hydrocarbons and Drugs," in D. Eastwood, Ed., *New Directions in Molecular Luminescence*, ASTM, Philadelphia, 1983, pp. 17–31.
324. N. J. Dovichi, J. C. Martin, J. H. Jett, M. Trkula, and R. A. Keller, *Anal. Chem.*, **56**, 348 (1984).
325. J. H. Richardson and M. E. Ando, *Anal. Chem.*, **49**, 955 (1977).
326. T. Imasaka and R. N. Zare, *Anal. Chem.*, **51**, 2082 (1979).
327. C. A. Sacchi, O. Svelto, and G. Prenna, *Histochem. J.*, **6**, 251 (1974).
328. M. R. Berman and R. N. Zare, *Anal. Chem.*, **47**, 1200 (1975).
329. B. E. Dalrymple, J. M. Duff, and E. R. Menzel, *J. Forensic Sci.*, **22**, 106 (1977).
330. I. Allegrini and N. Omenetto, *Environ. Sci. Tech.*, **13**, 349 (1979).
331. M. O. Rodgers, K. Asai, and D. D. Davis, *Appl. Opt.*, **19**, 3597 (1980).
332. U. Gehlhaar, K. P. Gunther, and J. Luther, *Appl. Opt.*, **20**, 3318 (1981).
333. R. A. O'Neil, L. Buja-Bijunas, and D. M. Rayner, *Appl. Opt.*, **19**, 863 (1980).
334. S. D. Lidofsky, T. Imasaka, and R. N. Zare, *Anal. Chem.*, **51**, 1602 (1979).
335. N. Ishibashi, T. Ogawa, T. Imasaka, and M. Kunitake, *Anal. Chem.*, **51**, 2096 (1979).
336. N. Strojny and J. A. F. de Silva, *Anal. Chem.*, **52**, 1554 (1980).
337. S. Yamada, F. Miyoshi, K. Kano, and T. Ogawa, *Anal. Chim. Acta*, **127**, 195 (1981).
338. Proceedings of the Society of Photo-Optical Instrumentation Engineers (The International Society for Optical Engineering), "Laser-based Ultrasensitive Spectroscopy and Detection V," R. A. Keller, Ed., Vol. 426, Aug. 23–24, 1983, San Diego, California.
339. S. Udenfriend, "Fluorescence Assay in Biology and Medicine," Academic, New York, Vol. I, 1962, Vol. II, 1969.
340. M. Pesez and J. Bartos, *Colorimetric and Fluorimetric Analysis of Organic Compounds and Drugs*, Dekker, New York, 1974.

341. P. Froehlich, *Appl. Spectrosc. Rev.*, **12**, 83 (1976).
342. H. Morawetz, *Science*, **203**, 405 (1979).
343. J. A. F. de Silva, N. Strojny, F. Rubio, J. C. Meyer, and B. A. Koechlin, *J. Pharm. Sci.*, **66**, 353 (1977).
344. W. Dunges, *Anal. Chem.*, **49**, 442 (1977).
345. Y. Arakawa, O. Wada, and M. Manabe, *Anal. Chem.*, **55**, 1901 (1983).
346. J. Schipper, F. J. H. Tilders, R. G. Wassink, H. F. Boleij, and J. S. Ploem, *J. Histochem. Cytochem.*, **28**, 124 (1980).
347. C. H. Kiang, S. S. Kuan, and G. G. Guilbault, *Anal. Chem.*, **50**, 1323 (1978).
348. D.K. Lavallee, T. J. McDonough, and L. Cioffi, *Appl. Spectrosc.*, **36**, 430 (1982).
349. J. R. Jadamec, W. A. Saner, and Y. Talmi, *Anal. Chem.*, **49**, 1316 (1977).
350. D. D. Schuresko, *Anal. Chem.*, **52**, 371 (1980).
351. D. R. Christmann, S. R. Crouch, and A. Timnick, *Anal. Chem.*, **53**, 276 (1981).
352. T. A. Kelly and G. D. Christian, *Anal. Chem.*, **53**, 2110 (1981).
353. J. F. Holland, R. E. Teets, P. M. Kelly, and A. Timnick, *Anal. Chem.*, **49**, 706 (1977).
354. B. J. Tromberg, J. F. Eastham, and M. J. Sepaniak, *Appl. Spectrosc.*, **38**, 38 (1984).
355. T. J. Killeen, D. Eastwood, and M. S. Hendrick, *Talanta*, **28**, 1 (1981).
356. H. S. Gold, G. T. Rasmussen, J. A. Mercer-Smith, D. G. Whitten, and R. P. Buck, *Anal. Chim. Acta*, **122**, 171 (1980).
357. F. J. Knorr and J. M. Harris, *Anal. Chem.*, **53**, 272 (1981).
358. L. S. Rosenberg, J. Simons, and S. G. Schulman, *Talanta*, **26**, 867 (1979).
359. R. A. Lampert, L. A. Chewter, D. Phillips, D. V. O'Connor, A. J. Roberts, and S. R. Meech, *Anal. Chem.*, **55**, 68 (1983).
360. C. H. Lochmuller, D. B. Marshall, and D. R. Wilder, *Anal. Chim. Acta*, **130**, 31 (1981).
361. C. H. Lochmuller, D. B. Marshall, and J. M. Harris, *Anal. Chim. Acta*, **131**, 263 (1981).
362. C. H. Lochmuller, A. S. Colborn, M. L. Hunnicutt, and J. M. Harris, *Anal. Chem.*, **55**, 1344 (1983).
363. D. K. Ryan, and J. H. Weber, *Anal. Chem.*, **54**, 986 (1982).
364. O. S. Wolfbeis and E. Urbano, *Anal. Chem.*, **55**, 1904 (1983).
365. C. J. Halfman and A. S. Schneider, *Anal. Chem.*, **54**, 2009 (1982).
366. P. M. Roemelt and W. R. Seitz, *Environ. Sci. Tech.*, **16**, 613 (1982).
367. A. J. Lapen and W. R. Seitz, *Anal. Chim. Acta*, **134**, 31 (1982).
368. W. R. Seitz, *Appl. Spectrosc.*, **36**, 161 (1982).
369. R. K. Stevens and J. A. Hodgeson, *Anal. Chem.*, **45**, 443A (1973).
370. W. R. Seitz and M. P. Neary, *Anal. Chem.*, **46**, 188A (1974).
371. L. J. Kricka and G. H. G. Thorpe, *Analyst*, **108**, 1274 (1983).
372. W. R. Seitz, "Chemiluminescence Analysis in Water Pollution Studies," in E. L. Wehry, Ed., *Modern Fluorescence Spectroscopy*, Vol. 1, Plenum, New York, 1976, Chap. 7.
373. J. D. Barry, J. M. Heitman, and C. R. Lane, *J. Appl. Phys.*, **50**, 7181 (1979).
374. H. R. Schroeder and F. M. Yeager, *Anal. Chem.*, **50**, 1114 (1978).
375. J. P. Auses, S. L. Cook, and J. T. Maloy, *Anal. Chem.*, **47**, 244 (1975).

376. K. Matsumoto, K. Fugiwara, and K. Fuwa, *Anal. Chem.*, **55**, 1665 (1983).
377. K. Honda, J. Sekino, and K. Imai, *Anal. Chem.*, **55**, 940 (1983).
378. C. P. Keszthelyi, N. E. Tokel-Takvoryan, and A. J. Bard, *Anal. Chem.*, **47**, 249 (1979).
379. I. Rubinstein, C. R. Martin, and A. J. Bard, *Anal. Chem.*, **55**, 1580 (1983).
380. E. H. Ratzlaff and S. R. Crouch, *Anal. Chem.*, **55**, 348 (1983).
381. K. W. Sigvardson and J. W. Birks, *Anal. Chem.*, **55**, 432 (1983).
382. J. K. Nelson, R. H. Getty, and J. W. Birks, *Anal. Chem.*, **55**, 1767 (1983).
383. A. Fontijn, "Chemiluminescence Techniques in Air Pollutant Monitoring," in E. L. Wehry, Ed., *Modern Fluorescence Spectroscopy*, Vol. 1, Plenum, New York, 1976, Chap. 6.
384. A. M. Winer, J. W. Peters, J. P. Smith, and J. N. Pitts, *Environ. Sci. Tech.*, **8**, 1118 (1974).
385. D. M. Steffenson and D. H. Stedman, *Anal. Chem.*, **46**, 1704 (1974).
386. R. D. Matthews, R. F. Sawyer, and R. W. Schefer, *Environ. Sci. Tech.*, **11**, 1092 (1977).
387. A. Fontijn, H. N. Voltrauer, and W. R. Frenchu, *Environ. Sci. Tech.*, **14**, 324 (1980).
388. A. A. Mehrabzadeh, R. J. O'Brien, and T. M. Hard, *Anal. Chem.*, **55**, 1660 (1983).
389. J. H. Richardson, *Meth. Enzymol.*, **66**, 416 (1980).
390. K. Miyaishi, M. Kunitake, T. Imasaka, T. Ogawa, and N. Ishibashi, *Anal. Chim. Acta*, **125**, 161 (1981).
391. T. Hirschfeld, *Appl. Spectrosc.*, **31**, 245 (1977).
392. J. D. Ingle, *Appl. Spectrosc.*, **36**, 588 (1982).

CHAPTER

3

TRACE ANALYSIS BY INFRARED SPECTROSCOPY

A. LEE SMITH

Dow Corning Corporation
Midland, Michigan

1.   INTRODUCTION

The use of IR (infrared) spectroscopy for trace analysis is a comparatively recent development, given impetus by the availability of the more energy-efficient Fourier transform and laser spectrometers. Dispersive spectrometers, which have been valuable analytical tools for chemists since 1941, are handicapped by low efficiency in the intrinsically energy-poor IR region. Thus, only in special cases has it been possible to analyze homogeneous samples for constituents present in the parts-per-million range. Microsampling, or analysis of a single small particle, was possible, however, by using beam condensing optics, along with wide slits and slow scanning (to improve the signal-to-noise ratio) and ordinate expansion (1). With good technique, samples as small as one microgram could be analyzed.

With the coming of FT (Fourier transform) IR, many formerly difficult problems became much easier to solve. The reason, of course, lies in the significant energy advantage of FTIR over dispersive IR spectrometers—something on the order of 10 to 40, depending on the wavelength region and the specific configuration of the spectrometers being compared. Currently, good spectra are being obtained on nanogram quantities of material, and this limit may be extended downward in the future.

If one does not have access to an FTIR, however, all is not lost. Modern computerized dispersive spectrometers are capable of a high level of performance, particularly over a limited wavelength range. Thus, much of the following discussion applies equally well to dispersive and FT systems.

In this chapter, it is assumed that the reader has a basic knowledge of IR spectroscopy techniques and instrumentation such as are described elsewhere (2–4). Our discussion therefore includes only minimal descriptions of instruments and their functioning, except for points specifically relevant to trace analysis.

It is also assumed that the analyst already has some training or experience in micro- or trace-analysis procedures. The analyst, not the instrument, is the key factor in success here. Trace analysis, as a first exercise, is not a suitable way to become acquainted with analytical chemistry. As the American Chemical Society Committee on Guidelines for Data Acquisition in environmental chemistry has pointed out, accurate analysis cannot be based solely on the ability of an instrument to respond to small concentrations of analyte in a sample—properly trained and skilled chemists are also essential (5).

Predicting needs for trace analyses is difficult, but some of the applications of current interest are: environmental chemistry and pollutants; surfaces and interfaces; semiconductor materials; and polymers. These topics are discussed in Section 4. In a field showing such rapid progress as IR spectroscopy, however, a review such as this can never be entirely up to date. For current awareness, the biennial reviews in the journal *Analytical Chemistry* are highly recommended.

1.1. What Does "Trace" Mean?

The word trace can connote almost any concentration below 1% or so, depending on the problem and the perspective of the user. In this chapter, it can usually be taken to mean parts per million or lower concentrations. It does not mean "detectable but not measurable"; trace analysis must include at least semiquantitative assessment of the substance sought, although in many cases the error of measurement will be relatively large. Fortunately, great accuracy is seldom needed in trace analysis. The difference between 0.08 and 0.06 ppm is not likely to be very significant, even though the relative error is 25%.

Micro analysis is a special case of trace analysis, defined here as analysis of a single small particle, or a minor constituent in a heterogeneous mixture. *Trace* analysis, on the other hand, refers to analysis for a minor component in a homogeneous mixture. The approach to these two situations is usually quite different.

The problem of trace analysis near the limit of detection may be important in some cases (such as environmental analyses). It will now be considered.

1.1.1.   Limit of Detection (LOD) and Limit of Quantitation (LOQ)

The limit of detection for an element is defined as the concentration that the
analyst can determine to be statistically different from an analytical blank (6).
Although this definition is valid when applied to IR spectroscopy, the criteria
for detection are quite different from those used with elemental analysis. With
emission spectroscopy, for example, the detection of a signal at a particular
precise wavelength is certain evidence for the presence of an element known to
emit at that wavelength. With IR spectroscopy, many different compounds can
absorb at any given wavelength, and identification is made on the basis of the
complete spectral pattern rather than one or two absorptions. Thus, the presence
of a single band is rarely definitive for a compound (except in high-resolution
spectroscopy of gases), so we need to impose some additional constraints on the
word "detection." These constraints vary with the problem and the compound
sought, and are not amenable to precise definition. The reasons will become
apparent from a study of Fig. 1.

Several quite different problems can exist with trace analysis:

Case I. The analyte is known (one or more substances).

1.   Is the analyte present? (Usually some limit such as the LOD must be
     specified.)
2.   If present, what is its concentration?

Case II. The analyte is unknown.

1.   Identify the unknown material(s).
2.   Identify and quantify the unknowns.

Case I has been addressed by the ACS Committee on Environmental Improvement as it relates to environmental analysis (7). They define three levels of
uncertainty (Fig. 2): high uncertainty (of detection); less certain quantitation;
and the region of quantitation. If the gross analyte signal is S_t, the field blank

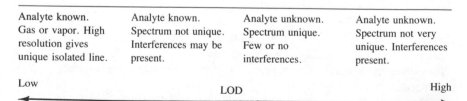

Fig. 1.   Possible trace analysis situations, illustrating the variation in limit of detection for different
kinds of problems.

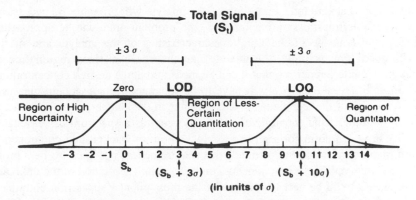

Fig. 2. Relationship of LOD and LOQ to signal strength. The LOD is located 3σ above the measured average difference between the blank (S_b) and total (S_t) signals. The LOQ is 10σ above S_b. Reprinted with permission from ACS Committee on Environmental Improvement, *Anal. Chem.*, **55**, 2210. Copyright 1983 American Chemical Society.

or background is S_b, and σ is the variability (standard deviation) of the field blank, then the limit of detection is:

$$S_t - S_b \geqq k\sigma \tag{1}$$

The recommended value of k is 3 for limit of detection and 10 for the limit of quantitation.

If the detector is an infrared spectrometer and unknown interfering materials may be present, we need to be more conservative. If the analyte has at least 3 absorption bands, and at least one of these is at a reasonably unique position, the LOD may be taken as 10σ and the LOQ as 20σ for at least 2 of the hands. Clearly, under these circumstances a good deal of judgment enters the decision as to what is a unique band and what is not; the above values for k are suggested as a guideline. In the less usual case of detecting a vibration–rotation line in the high-resolution spectrum of a gas, the ACS Committee's criteria may be adequate. It would be well, however, to confirm the identification or quantitation on a second line from the same compound.

1.1.2. The Effect of Background

Dispersed materials can be further classified as having a high, low, or variable transmission background. The approach taken in each of these situations depends on the definition of the problem. *Detection* or *quantitation* of a known compound (is compound XYZ present in this material at the 0.1-ppm level?) in a high transmission background situation is likely to be relatively simple. With a low transmission background, separation of the analyte will probably be necessary. For a variable transmission background, the problem often can be approached by taking advantage of the unique characteristics of the analyte and of the background. For example, the analyte may have a band in a region free of background interference, so that it can be measured at the desired concentration. Or, more likely, the analyte will have to be separated for subsequent measurement, by one of the techniques discussed later in the chapter.

On the other hand, *identification* usually requires a spectrum of reasonable quality in which no major regions are blocked out by background components. With high transmission background, this requirement is not likely to be a problem, but with variable or low transmission background, separation of the unknown component(s) will be necessary. One of the most popular separation and quantitation methods, gas chromatography (GC), has been successfully combined with an excellent identification method (IR spectroscopy) to form a powerful and useful tool. This topic is discussed more fully in Section 3.

1.2.   Choice of a Method

Many approaches to trace analysis are used. For dispersed (homogeneous) samples, separation–determination methods such as GC–MS (gas chromatography—mass spectrometry), spectrophotometric methods, and chemical methods are useful. For micro samples, optical and electron microscopy, electron microprobe, Raman, IR, and many others are employed. For analysis of materials on surfaces, techniques such as XPS (X-ray photoelectron spectroscopy), SIMS (secondary ion mass spectrometry), and X-ray fluorescence are often used in addition to IR. It is important to know the capabilities and limitations of each of these methods in order to choose the most effective approach to a problem. Sometimes, however, sophisticated instrumentation is not available, so one does one's best with the instrument at hand. If there is a choice of instruments, best results are usually obtained by synergistically combining the information obtained from several of them. Most instruments of this type give complementary data, and they should not be thought of as "competing." We see some examples later in this chapter where combining IR with another technique was essential to the solution of a difficult problem.

A comparison of the characteristics of various microspectroscopic methods is given in Table 1. Clearly, the strength of IR spectroscopy lies in its universality and uniqueness, rather than its sensitivity (8).

1.3.   Sampling

Samples are assumed to represent a larger population or distribution of the analyte. It is therefore essential that the sampling program be carefully thought through before starting the analytical process, to assure that the analytical results

Table 1.   Optical Microbeam Methods[a]

	UV–Visible Absorption	Fluorescence	Raman	IR Absorption	Light Scattering
Nominal working wavelength	~3500 Å	~3500 Å	~3500 Å	~10 μm	~3500 Å
Detection Limits					
Size	~2500 Å	~100 Å	~500 Å	5 μm	~100 Å
Concentration	10–100 ppm	1 ppb	>1000 ppm	>1000 ppm	1 ppm
Quantity	10^{-17} g	10^{-21} g	10^{-13} g	10^{-12} g	10^{-19} g
Universality	w/ Reagents	No	Almost	Yes	No
Qualitation:					
Recognition	w/ Reagents	Often	Yes	Yes	Rarely
Identification	No	Often	Yes	Yes	No
Quantitation	1%	1%	2%	1%	2%

[a]Reproduced with permission from K. F. J. Heinrich, Ed., *Microbeam Analysis—1982,* San Francisco Press, Box 6800, San Francisco, CA 94101-6800.

are meaningful. For difficult problems such as environmental analysis for low levels of hazardous materials, the plan should include the following elements (5,7):

1.   A careful definition of the objectives of the analysis, based on the intended and probable use of the results.

2.   A proper statistical design, including selection of sampling sites, the number of sampling sites, the number of samples, the timing of the sample collection, and the acceptable level of fluctuation in the results due to heterogeniety.

3.   Protocols for collecting, labeling, preserving, and transporting samples to the analysis site.

4.   Training of personnel in the specified procedures. Trial runs with known concentrations of analyte may be necessary to identify problem areas. In wastewater analysis, for example, solutions or suspensions of organic materials may show significant loss of analyte from plating out on the glass walls of the sample container. Or sometimes the container itself (or bottle cap liner) contaminates the sample. When working with trace levels of material in solutions, one cannot assume that the sample container is inert. Occasionally, it is necessary to exclude

light or air from the sample, and the effectiveness of this process should be tested by using known standards. Sample handling should be minimized, and it should be subject to rigorous procedures that have been tested and proven effective.

1.4.   Control of Contaminants and Artifacts

In situations where a single fingerprint or particle of dust can overwhelm the sample being analyzed, the need for careful control of contaminants is obvious. Not so obvious are ways to carry out this control.

It is important for the chemist doing trace analyses to realize that nature is working against him or her. We know, for example, that an absolutely pristine surface exists only momentarily after its formation unless one is working in a very high vacuum. Within seconds, it is covered with water, and within minutes, with a layer of hydrocarbon. The analyst himself may unintentionally add to the contamination problem. Smoker's cough, an oily skin, or even a bad case of dandruff may be the cause of much futile and unproductive effort, in which the analyte is partially or completely obscured by the contaminant. More likely, however, a tendency to take shortcuts and a failure to appreciate the difficulty of achieving and maintaining a clean analytical system will be responsible. Such problems, fortunately, can be controlled by taking appropriate precautions, including training of the analyst in the use of microscopical techniques.

Solvents are often necessary and can be used successfully in trace analysis, but one should be wary of procedures where appreciable quantities of solvent are evaporated, because residues often occur unexpectedly. Contamination from apparently clean glassware and containers is also a problem. A report on contamination control in infrared trace analysis has been published by Chen and Dority (9). Solvent residue contamination of chemical isolates from pesticides,

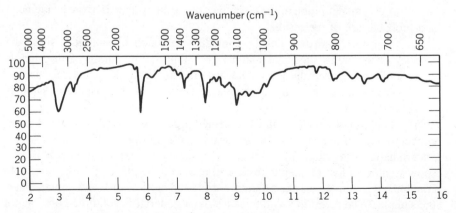

Fig. 3.   Residue from 13-mL glass-distilled acetone evaporated to dryness (9).

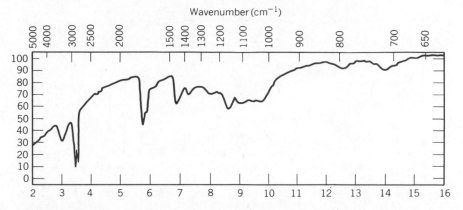

Fig. 4. Residue from a supposedly clean and dry test tube (9).

food additives, and drugs was demonstrated. The analytical method used was to evaporate a small amount of sample-containing solvent under N_2 or a vacuum, and then to thoroughly mix the residue with 5 mg KBr. A microdisc was prepared and the spectrum recorded with the aid of a beam condenser. Various sources of contamination were found. Even "distilled-in-glass," reagent grade, and spectroquality grade solvents were found to have nonvolatile residues (Fig. 3). The recommended procedure for minimizing contamination from this source was to use the minimum quantity of the best grade solvent available (determined by test runs using the same quantity as in an analysis).

Laboratory glassware sometimes carries a thin layer of soap or detergent (Fig. 4) The authors recommend cleaning glassware as follows:

1. Heat in nitric acid near the boiling point for several hours.
2. Rinse thoroughly with distilled water.
3. Boil in distilled water to remove traces of nitrate ion.
4. Rinse thoroughly in 95% ethanol.
5. Dry completely, using a hot plate.

Plastic laboratory ware (except possibly Teflon®) should be avoided. These authors also point out that minute quantities of impurities in some solvents, even though they may be volatile, can react with some samples to give unexpected results. For example, phosgene or HCl in $HCCl_3$ can react with aliphatic amines.

Silicone fluids or greases sometimes find their way adventitiously into solvent or sample (10). Because their absorptions are so intense, silicones may be visible way out of proportion to their concentration. Their low surface tension makes it easy for them to spread or creep on all types of surfaces (11). A spectrum of

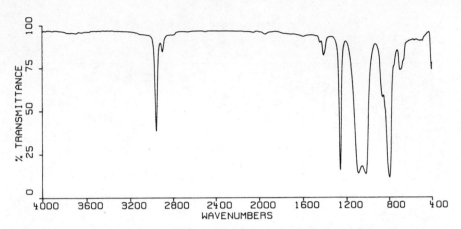

Fig. 5.   Polydimethylsiloxane, 100 cst viscosity.

a typical silicone fluid (polydimethylsiloxane) is shown in Fig. 5. Silicone stop-cock grease should not be used on apparatus involved with trace analysis.

Phthalate esters, often used as plasticizers in flexible plastic tubing, are also strong IR absorbers and are easily extracted. The spectrum of such an extract is shown in Fig. 6.

The sampling, handling, and storage of materials for trace analysis is discussed by Moody (13).

Launer (12) has listed a number of artifacts that sometimes appear in spectra, and his list is given in Table 2.

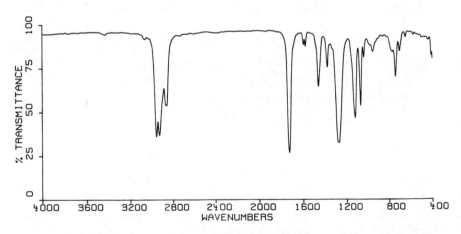

Fig. 6.   Extract from plastic tubing.

Table 2. Common Spurious Absorption Bands[a]

Approximate Frequency (in cm^{-1})	Wavelength (in μm)	Compound or Group	Origin
3700	2.70	H_2O	Water in solvent (thick layers)
3650	2.74	H_2O	Water in some quartz windows
3450	2.9	H_2O	Hydrogen-bonded water; usual in KBr disks
2350	4.26	CO_2	Atmospheric absorption
2330	4.3	CO_2	Dissolved gas from dry ice
2300 and 2150	4.35 and 4.65	CS_2	Leaky cells
1996	5.01	BO_2	Metaborate in the halide window
1400–2000	5–7	H_2O	Atmospheric absorption
1820	5.52	$COCl_2$	Decomposition product in purified $HCCl_3$
1755	5.7	Phthalic anhydride	Decomposition product of phthalate esters or resins
1700–1760	5.7–5.9	C=O	Bottle-cap liners leached by sample
1720	5.8	Phthalates	From plastic tubing
1640	6.1	H_2O	Entrained in sample, or water of crystallization
1520–1620	6.2–6.6	(structure: C—C with two O, carboxylate)	Reaction product of alkali halide windows or KBr pellet with organic acid
1520	6.6	CS_2	Leaky cells
1430	7.0	CO_3^{-2}	Contaminant in halide window
1360	7.38	NO_3^{-}	Contaminant in halide window
1270	7.9	$SiCH_3$	Silicone oil or grease
1110	9.0	?	Impurity in KBr for disks
1000–1110	9–10	SiOSi	Glass; silicones
980	10.2	K_2SO_4	From double decomp. of sulfates in KBr pellets
935	10.7	$(CH_2O)_x$	Deposit from gaseous formaldehyde
907	11.02	CCl_2F_2	Dissolved Freon-12
837	11.95	$NaNO_3$	(See 1360 cm^{-1})
823	12.15	KNO_3	From double decomp. of nitrates in KBr pellets
794	12.6	CCl_4 vapor	Leaky cells
788	12.7	CCl_4 liquid	Incomplete drying of cell or contamination

Table 2.   *(Continued)*

Approxmiate Frequency (in cm^{-1})	Wave length (in μm)	Compound or Group	Origin
720 and 730	13.7 and 13.9	Polyethylene	
728	13.75	Na_2SiF_6	SiF_4 + NaCl windows
667	14.98	CO_2	Atmosphere
	Any	Fringes	If refractive index of windows is too high, or if a cell is partially empty, interference fringes may appear

[a]From Launer (12) and Smith (4).

The possibilities for instrumental artifacts are almost endless; some of the more common ones are considered at the appropriate place in our discussion. One has a good chance of avoiding most of these kinds of problems if he or she (1) has a thorough knowledge of the principles and operation of the spectrometer, (2) has a good understanding of the sampling technique, and (3) uses both as intelligently as possible.

2.   INSTRUMENTAL CONSIDERATIONS

Only brief descriptions of instrument systems are given here. Many detailed and excellent descriptions of all types of spectrometers are available elsewhere. Thus, we highlight only features important to the practice of trace analysis and refer to the literature (2–4,14) for more complete discussions.

The general problem faced by workers in the infrared region is that of low energy. Much effort has been devoted to designing spectrometers that maximize energy transmission. Fourier transform IR spectrometers have achieved popularity because they are more energy-efficient than dispersive spectrometers, yet many spectroscopists find themselves limited by energy restrictions even with an FTIR system. It therefore becomes important to set one's objectives carefully, to plan strategies that make the best use of available equipment, and to use the most effective techniques with a full knowledge of their strengths and their shortcomings.

2.1.   Detectors

The detector plays a key role in the operation of an IR spectrometer. It is important that its characteristics meet the requirements of the spectrometer, or inferior results will be obtained. In trace analysis, particularly, awareness of the detector characteristics can help the analyst optimize his system for maximum sensitivity.

Infrared detectors are classified as *thermal* detectors (thermocouples, bolometers, pneumatic detectors) or *photon* detectors (photoconductive, photovoltaic, or photoemissive devices). Photon detectors usually require cooling to the temperature of liquid N_2 or liquid He. Because energy in the IR region decreases as wavelength increases, the detection of photons becomes more and more difficult as we go farther into the region. Thus, detector selection is determined by the wavelength region in which one must work. The subject is discussed in detail by Kruse (15).

Dispersive spectrometers usually use thermocouple detectors, which have relatively slow response (0.05 to 0.1 s) but respond uniformly to all wavelengths. Fourier transform spectrometers commonly use DTGS (deuterated triglycine sulfate) pyroelectric bolometer detectors. They have the fast response, large dynamic range, and linearity of response that are important for FTIR instruments. Cooled MCT (mercury cadmium telluride) photon detectors are a factor of 2 to 10 more sensitive, but the most sensitive MCT unit has its low-frequency cutoff at about 750 cm^{-1}.

For purposes of comparing detectors, several figures of merit have been proposed. Currently, the most widely used unit is called $D*$ (dee-star), and it is defined as the root mean square signal-to-noise ratio (S/N) in a 1-Hz bandwidth per unit radiant power.

$$D* = \frac{(A_d B)^{1/2}}{P} \left(\frac{v_s}{v_n} \right) \qquad (2)$$

where A_d is the detector area in cm^2; B is the electrical bandwidth in Hz; P is the incident radiant power; and (v_s/v_n) is the rms (root-mean square) S/N voltage ratio. It is assumed that the field of view is hemispherical (2π steradians). The units of $D*$ are cm Hz$^{1/2}$/W. With this definition, the better the detector, the higher the value of $D*$.

A related parameter, $D**$, is appropriate to background-limited detectors and eliminates the need to specify the field of view. If the aperture through which radiation reaches the detector has the half-angle θ, $D** = D* \sin \theta$ and, for a hemispherical field of view, $D** = D*$. The units of $D**$ are cm Hz$^{1/2}$ ster$^{1/2}$/W. Fig. 7 shows $D**$ for a number of thermal and photon detectors. Operating temperature is shown after the type designation (16).

An older figure of merit, the noise-equivalent power or NEP, is sometimes encountered. It can be thought of as the smallest power detectable with S/N = 1. The reference bandwidth, the detector area, and the field of view must be specified. It is related to $D*$ by

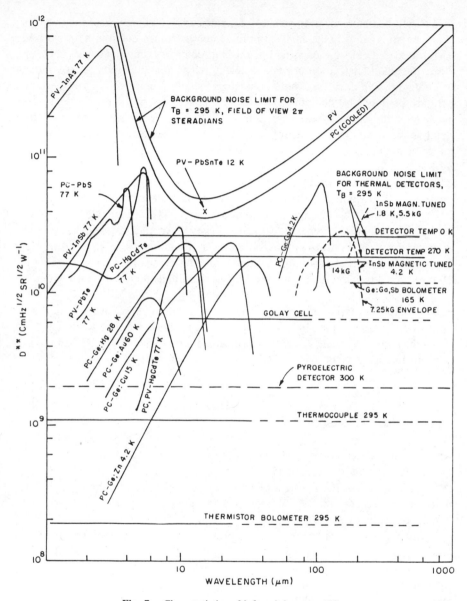

Fig. 7. Characteristics of infrared detectors (16).

$$NEP = \frac{(A_d B)^{1/2}}{D^*} \tag{3}$$

The units of NEP are watts, and the better the detector, the *lower* the NEP. Another term sometimes used is detectivity, which is $(NEP)^{-1}$. Thus it is related to, but not the same as, D^*. For a comprehensive discussion of detectors and their characteristics, the reader is referred to the book by Keyes (17).

In IR trace analysis, two situations are common: (1) the energy passing through the sample is adequate, and small differences between two sets of scans (or two beams) are being measured; or (2) energy is limited, i.e., the S/N is low. In both cases, the detector noise level is an important consideration. For thermocouple and bolometer detectors, Johnson noise or thermal motion of electrons in the receiver element, along with statistical thermal variations in the element and background radiation, limit the S/N. These sources are usually responsible for the noise observed in the spectral record. This noise, unlike shot noise in photomultiplier devices, is independent of the signal level.

It should be noted that the S/N for any detector is directly proportional to the incident power, and inversely proportional to the square root of the target area. For micro samples, it may therefore be advantageous to use a detector with a smaller-than-normal target area. A detector target measuring 50×50 μm should give a 20-fold increase in S/N over a 1×1 mm^2 target. In any spectrometer, to get the best S/N it is advantageous to fill the detector target as nearly as possible, and to use the highest possible radiation flux.

2.2. Dispersive Spectrometers

A typical design for a dispersive spectrometer is shown in Fig. 8. Radiation from the source is divided into two equivalent beams and sent by front-surface mirrors through the sample compartment. A rotating sector mirror alternately directs the sample and reference beams into a common path through the entrance slit and into the monochromator. After the radiation is dispersed by the grating, it passes through the exit slit and is focused on the detector. Because the sample and reference beams alternate at a fixed frequency, usually 10 to 15 Hz, any difference in intensity between the two beams caused by absorption of the sample at that wavelength results in an alternating signal from the detector. (If both beams are exactly balanced, there is, of course, no signal.) In the optical null spectrometer, this signal activates a servo motor which moves an optical wedge into or out of the reference beam in such a manner as to cancel the difference between the two beams. The motion of the wedge is recorded as a function of wavelength, and thus produces the spectral record. Some spectrometers digitize both the wavelength and transmission coordinates so they can be stored in a computer and used to regenerate or manipulate the spectrum.

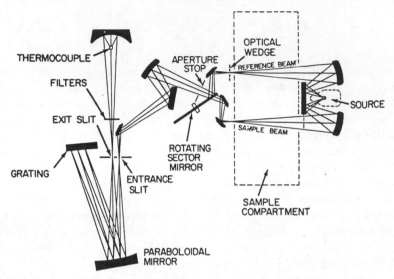

Fig. 8. Optical schematic for a typical dispersive spectrometer. (Courtesy of the Perkin-Elmer Corp., Norwalk, CT.)

Ratio recording spectrometers, instead of using an optical wedge, amplify the electrical signals from the detector and ratio the signals electronically to produce the spectrum.

In both types of spectrometers, the operating parameters can be varied over a wide range of values to suit the needs of the current problem. The variable parameters are:

1. The slit width, which determines the resolution.
2. The scan speed, which must be compatible with the pen response speed.
3. The gain of the pen servo loop, which determines the noise level of the spectrum.

Any *two* of these parameters may be independently chosen, but the third is then fixed by that choice. The relationship between the variables is

$$\text{resolution} = \frac{1}{w} = ct^{1/4} \, (S/N)^{-1/2} \tag{4}$$

where w = slit width; t = response time of the pen servo loop; S/N = the signal-to-noise ratio; and c is a constant. This equation defines the "trading rules" for spectrometer operation. Optimization of the spectrometer for various purposes and use of the trading rules are discussed elsewhere (1,4). Users of dispersive spectrometers should be thoroughly familiar with these optimization procedures before attempting trace analyses.

As an illustration, suppose we are looking for a trace material by analysis of an extremely dilute solution (identical reasoning applies to analysis of a material adsorbed on a surface, or a trace constituent by attenuated total reflectance (ATR). We expect that a 100-fold ordinate expansion will be necessary to detect the analyte. If we simply expand the ordinate scale by a factor of 100 without a reduction in the noise level, we will not have improved the sensitivity over that of a normal scan. We therefore must increase the S/N, either by widening the slits or lengthening the scan time. We note that S/N varies as the square of the slit width, but only as the square root of the scan time. It is generally better to achieve the required gain in S/N by sacrificing resolution than by using inordinately long scan times. (In this example, we would have to decrease the scan rate by a factor of 10^4; if a normal scan takes 20 min, the expanded ordinate scan would take 140 days!) Ideally, we should open the slits by a factor of $100^{1/2}$, or 10, and decrease the wedge servo loop gain by a factor of 100. Dispersive spectrometers do not usually have the capacity to open their slits by this factor, however. (Also, opening the slits beyond the point where the source image completely fills the detector target is futile.) Therefore, we need to use a combination of parameter adjustments. We may open the slits by a factor of 2 (4-fold noise reduction); use a digital smoothing routine (4-fold noise reduction);

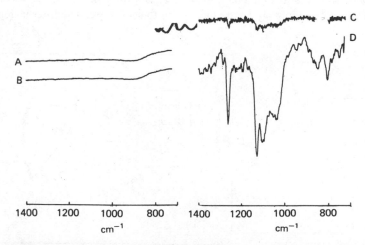

Fig. 9. Illustration of spectrum accumulation by multiple scans: A, reflectance spectrum of metal surface; B, same surface with estimated 20-Å-thick layer of silicone oil; C, result of 15 accumulations of B; D, spectrum C expanded 10X and smoothed. Reprinted with permission from J. P. Coates, *Amer. Lab.*, **8**(11), 67 (1976). Copyright 1976, International Scientific Communications, Inc.

increase the scan time by a factor of 4 (2-fold noise reduction); and accept 3-fold higher noise level in the expanded scan. An example of what can be done with computer-assisted dispersive spectroscopy is shown in Fig. 9. These techniques are discussed in detail later in this chapter, but it should be noted that some dispersive spectrometers (depending on their design, age, and condition) may lack the wavelength and transmittance reproducibility necessary to take full advantage of these techniques. Mechanical parts such as cams and sliding electrical contacts become worn, and under such circumstances wavenumber reproducibility may be degraded. Ordinate reproducibility is affected by inertia, friction, and overshoot in the optical wedge mechanism, and by lack of energy at low transmittance values.

It is important to understand that at low transmission values of the sample, very little energy is available to activate the optical attenuator servo loop. Thus, transmission errors in the 1 to 10% T range can be relatively large. Similarly, any absorbing solvent, when present in both beams, removes energy from the system and the instrument will not respond to sample absorptions in such regions. This statement applies to all spectrometers, including FTIR instruments.

2.3. Dielectric Filter Spectrometers

Dielectric filters can be produced to have reasonably narrow band pass in the IR region with good rejection of unwanted frequencies. Such filters are very efficient; they do not absorb energy, they only transmit or reflect it. A useful invention, which is available commercially, is a filter in which the wavelength passed varies continuously across the face of the substrate. The filter is deposited

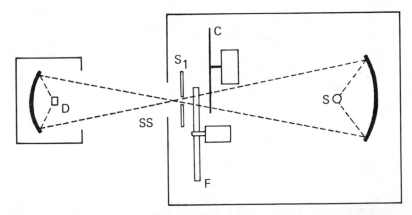

Fig. 10. Schematic drawing of a dielectric filter spectrometer. Radiation from the source S is chopped at C, passes through a slowly rotating variable circular filter F, a slit S_1, a sample space SS, and falls on the detector D.

on a disc, and the spectrum is scanned by slowly rotating the disc. Such a filter forms the basis for a high-energy monochrometer which is incorporated into a simple spectrometer (Fig. 10). The bandpass is variable from 40 to 150 cm^{-1} at 2800 cm^{-1}, or 1.5 to 3.5% of the wavelength setting. Thus, the resolution is low, but the S/N is on the order of 10^3. Such an instrument is capable of very accurate quantitative analysis (19).

The IR filter photometer, when combined with an all-reflecting microscope, forms the basis of an IR microspectrophotometer that can obtain transmission spectra on specimens as small as 20 × 20 μm (20). As shown in Fig. 11, radiation from the source (a Nernst glower) is directed through a sector wheel and variable wavelength filter into the reflecting microscope. There it is focused, with the aid of an auxiliary optical microscope, onto the sample, which is held on an adjustable stage. A variable aperture is used to define the area to be sampled. After passing through the sample, the radiation is focused on a cooled MCT detector. A computer is used to control the scan rate (filter wheel rotation) and to process the detector signals. As an illustration of performance, the spectrum of a single strand of hair is shown in Fig. 12. The instrument functions in

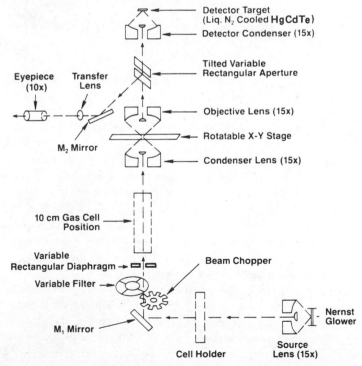

Fig. 11. Schematic of an IR microspectrophotometer. (Courtesy Nanometrics Corp.)

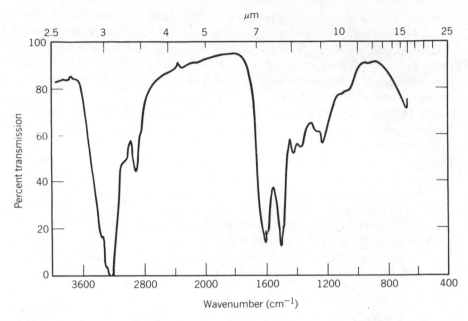

Fig. 12. IR spectrum of a single strand of hair as obtained by a microspectrophotometer. (Courtesy Nanometrics Corp.)

transmission only—if the sample is on a nontransmitting substrate, it must be remounted on IR-transparent material.

2.4. Fourier Transform Infrared Spectrometers

The Michaelson interferometer forms the basis for FTIR spectrometers (Fig. 13). The only moving part is a mirror in one arm of the interferometer, whose position is monitored by a laser and (usually) a white light reference. Intensity measurements of all frequencies are taken simultaneously as a function of mirror position. The resulting interferogram contains information about all the frequencies in the spectrum, but it must be transformed to the frequency domain by a Fourier transformation in order to present the information in the form of a spectrum. This process is discussed in detail elsewhere (2).

Fourier transform IR has a significant energy advantage over dispersive spectrometry, on the order of a factor of 10 to 40, depending on the resolution and the spectral region being observed (21,22). One part of this gain comes from Felgett's advantage, which accrues because all frequencies are measured simultaneously instead of sequentially, as in dispersive spectrometers. Thus, much more energy is available in unit time (the calculated advantage is $\sqrt{N}$ where N

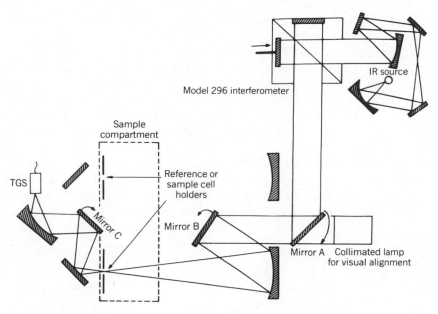

Fig. 13.   Optical schematic of an infrared interferometer, the Digilab FTS-14 (21).

is the number of resolution elements in the spectrum). Jacquinot's advantage results because a circular entrance aperture rather than an entrance slit is used. The third advantage of the interferometer (not related to energy) is called Conne's advantage, and it refers to the increased frequency accuracy (on the order of 0.001 cm^{-1}) which results from laser indexing of the moving interferometer mirror. This kind of accuracy is needed for absorbance subtraction work.

As with dispersive spectrometers, one can trade the energy advantage of the FTIR instrument for shorter scan time or higher resolution, and the trading rules are analogous but not identical (23). For any spectrometer, S/N at a given resolution increases as the square root of measurement time. With FTIR, measurement time is usually varied by changing the number of scan cycles that are signal-averaged to produce the interferogram. Resolution is changed by adjusting the retardation, or length of travel of the moving mirror. The trading rules take different forms, depending on whether *constant throughput* or *variable throughput* is being used as the resolution is changed. In the former case, the entrance aperture is not changed and increasing resolution by a factor of 2 requires only a 4-fold increase in scan time (a dispersive spectrometer requires a 16-fold increase). At some point, however, for higher resolutions, the entrance aperture must be reduced in size (this aperture is analogous to the entrance slit of a dispersive spectrometer). In this case, a two-fold increase in resolution requires a 16-fold increase in measurement time, exactly the same as for a dispersive spectrometer.

Put another way, if we are operating an interferometer at its maximum throughput for low-resolution measurements, we can reduce the time spent scanning by degrading resolution, but the gain will be a factor of 4 less in measurement time than we would get with a dispersive spectrometer. However, if we wish to double the resolution and need to concurrently decrease the solid angle of the IR beam by halving the area of the entrance aperture, we must increase the scan time by a factor of 16.

FTIR spectometers have other unique characteristics which must be understood for their most effective use. The dynamic range of the interferogram signal is very large at the centerburst where all frequencies are in phase; the S/N ratio at zero retardation may be in excess of 10^4. The implication is that the digitizer

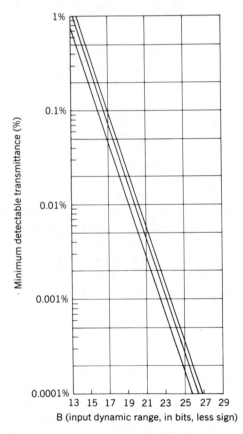

Fig. 14. Smallest detectable transmittance change (in %T) as a function of available input dynamic range for high throughput FTIR (24).

must have a precision of at least 2^{15} bits. In addition, other noise sources and the requirements of signal-averaging have led Foskett (24) to conclude that a *minimum* register length of 27 bits is necessary for measurement of trace materials in high-throughput samples (i.e., S/N $\geq 10^4$). Fig. 14 gives the relationship between minimum detectable transmittance change and input dynamic range. Various ways of meeting this requirement are used by spectrometer manufacturers, such as double-precision arithmetic in the centerburst region, signal averaging, 32-bit computer words, and the like. Griffiths (2) has proposed a dual-beam configuration in which only the *difference* between two interferograms is digitized and transformed. With this arrangement, the dynamic range requirement is much less. This topic is explored more fully elsewhere (2,24,25).

In high background (low transmission) situations, such as microanalysis through a small aperture, the problem is to get enough signal to attain a spectrum having reasonable S/N. The use of cooled detectors, beam condensing optics, and long spectrum accumulation times are helpful. However, all detectors are limited in their detection of radiant power by noise in the detector itself and in the electronic amplifiers, by fluctuations in the background radiation reaching the detector, and by fluctuations in the signal. Many detectors function very close to this limit (see Fig. 7). Detector noise is discussed at some length by Kruse (15).

Small errors and instabilities in one vendor's interferometer spectrometers have been identified and discussed by de Haseth (26). These instabilities are extremely small and it is unlikely that they have any appreciable effect on the average spectrum. However, those working near the limits of the interferometer spectrometer should be aware of them.

2.5. Lasers

In many respects, lasers might seem to be ideal sources for IR trace analyses. They are extremely intense, they are essentially monochromatic, and they can be tailored to cover the entire IR region. They also have some serious short-comings: tuneability is either nonexistent or very limited (depending on the type); stability and reliability have until recently been less than might be desired; and experimental arrangements can get quite complex. Tuneable lasers include semiconductor diode (SD), spin-flip Raman (SFR) and parametric oscillator (PO) types. Their characteristics are discussed elsewhere (27,28). Very briefly, the most popular of the "tuneable" lasers, the SD type, are made from lead salts of the type $Pb_{1-x}Sn_xTe$, $PbS_{1-x}Se_x$, or $Pb_xCd_{1-x}S$. They are broadly tuned by adjusting their chemical composition; each composition has an emission range about 40 cm^{-1} wide. Not all frequencies in the band are available, however; the nominal emission wavelength segment, which is about 1 cm^{-1} wide, is fixed by the diode temperature. The frequency within that segment is then chosen by adjusting the diode current. Thus, one must have a battery of SD lasers available to cover even a relatively short segment of the IR spectrum. This device is excellent for scans over short wavenumber ranges, such as might be required for high-resolution detection of a vibration–rotation line in a gas sample. Resolution is on the order of 10^{-4}cm^{-1}. The entire IR region can be covered (3–30 μm), and a device that emits 20 μW or more at 28 μm has been reported (29). A simple HF monitor that utilizes an SD laser operating at 400 cm^{-1} is shown schematically in Fig. 15. A balloon-borne diode laser spectrometer that uses a retroreflector suspended 0.5 km under the gondola (for a total path of 1 km) has been described (30). This device is designed to measure atmospheric contaminants in the sub-parts-per-billion range. Applications of diode lasers to chemical analysis have been reviewed by Butler et al. (31). Minimum detectable concentrations for several gases, using a multi-pass low-pressure gas sample cell and second derivative detection, are shown in Table 3.

Nontuneable CO or CO_2 lasers have also been used for trace analysis. In this case, one must have either an accidental coincidence of one of the laser lines with an absorption line of the analyte gas, or use Stark-effect modulation to bring a nearby analyte line into coincidence with the laser line. Examples of the former include use of CO_2 lasers to monitor flue gas for NH_3 (33), and ethylene mass flow near a petrochemical factory (34,35). Lines from CO and CO_2 lasers have been used with Stark-effect modulation to detect vinyl chloride, vinylidene chloride, acylonitrile, and methanol (36) with detection limits in the sub-ppm range. A list of gases suitable for quantitative Stark spectroscopy is given in Table 4.

Although many gas-analysis systems used in the field for atmospheric analysis

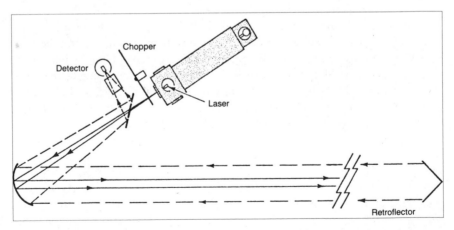

Fig. 15. Open-air monitor for hydrogen fluoride. A sensitivity at 0.5 ppm/m at 400 cm^{-1} is achieved (29).

Table 3. Calculated Minimum Detectable Concentrations for a Tunable Diode Laser System[a]

Pollutant	Approx. cm^{-1}	Sensitivity (ppb)
SO_2	1140	3
O_3	1050	0.5
N_2O	1150	2
CO_2	1075	300
H_2O	1135	50
NH_3	1050	0.05
PAN	1150	0.3
CH_4	1300	0.03
SO_2	1370	0.3
NO_2	1600	0.02
NO	1880	0.03
CO	2120	0.01
CO_2	2350	0.001

[a]From Reid et al. (32).

Table 4. Representative Gases Suitable for Quantitative Stark Spectroscopy[a]

Molecule	Laser	Dipole Moment
Acetonitrile	CO, CO_2	3.92
Acrylonitrile	CO_2	3.87
Allyl chloride	CO_2	
Formaldehyde	CO, CO_2	2.33
Methacrylonitrile	CO, CO_2	3.69
Methyl chloride	CO_2	1.87
Methanol	CO_2	1.70
Tetrahydrofuran	CO_2	1.63
Toluene	CO	0.36
Trichloroethylene	CO_2	
Vinyl chloride	CO, CO_2	1.45
Vinylidine chloride	CO_2	1.34
Nitrobenzene	CO	4.22
Nitric oxide	CO_2	0.15
Ozone	CO_2	0.53
Carbonyl sulfide	CO_2	0.71

[a]From Sweger and Travis (36).

199

utilize either enclosed absorption cells or retro-reflectors, range-resolved measurements are also possible using a high-power pulsed laser with Mie scattering as the reflecting device. Such LIDAR (light detection and ranging) systems are often operated in a differential mode (DIAL, or differential absorption lidar) in which two closely spaced laser lines—one at the analyte absorption frequency and the other not—are measured alternately. The detector signals are amplified, integrated, and used to develop a topological map of the analyte concentration (35). A discussion of various error sources in DIAL—inadequate knowledge of the absorption coefficient, differential spectral reflectance, background interference, and signal fluctuation errors—is given by Menyuk and Killinger (37).

3. TECHNIQUES OF TRACE ANALYSIS

3.1. Selection of a Technique

Many techniques have been adapted to detection of trace components and materials, and to detail them all would be a lengthy and difficult task. Almost all macro sampling techniques have been tried on a small scale. Thus, selection of the optimum approach is not always easy, and several tries may be necessary before the best method is discovered. The selection process may perhaps be made a bit easier if one is aware of the alternatives and the strengths and weaknesses of each.

3.2. Dispersed Samples

We have distinguished between *dispersed* (homogeneous) samples and *micro* (heterogeneous) materials. Micro samples are highly concentrated but of microscopic size. Only one particle may be available. With dispersed samples, plenty of material is usually available—it is just very dilute. Dispersed materials can further be considered as having a high, low, or variable transmission background. Typical examples are: high T, hydrocarbons in air; low T, organic contaminants in drinking water; variable T, additives in gasoline.

The statement of the problem is critically important. Detection of a *known* compound, or its quantitation, is quite different from finding and identifying one or more unknown trace contaminants. Examples of each type of problem will be cited in the section on Applications.

3.3. Separation Methods

Methods of sample preconcentration and enrichment, and separation of interferents, are discussed in Chapter 1 (Sections 2.2 and 2.3). Compensation for interferents is described in Section 2.4 of that chapter. Concentration or isolation of a known or suspected impurity is an important and in many cases mandatory part of the analytical procedure. Distillation, solvent evaporation, crystallization, solvent extraction, gas and liquid chromatography, paper and thin-layer chromatography, cryogenic and charcoal trapping—all have been used to separate minor constituents for IR analysis. In carrying out such procedures, one assumes the risk of loss or contamination of the sample. It is important to minimize the risk, and one way of doing this is by undertaking several "dress rehearsals," using a synthetic sample that approximates the unknown. In all cases, at least one blank run should be carried through concurrently with the sample. Possible artifacts originating with solvents or glassware have been discussed in Section 1.4. Several examples of trace analysis by IR spectroscopy using preconcentration and separation techniques, along with computerized spectroscopy, are described by Hannah et al. (38).

3.4. Infrared Sampling Techniques

In the discussions that follow, we assume that the reader is familiar with the principles of the common IR sampling methods. For those who are not, detailed sources are available and should be consulted (3,4). We highlight the special aspects relating to trace and microanalysis, and attempt to provide information that will permit the analyst to make a reasonable choice for approaching his or her own problem. Ultimate sensitivity is always a point of interest; numbers quoted are approximately correct at this writing. They are usually not easily achieved, however; the reader should be warned not to expect to duplicate these sensitivities without considerable experience and effort. Undoubtedly, better sensitivities will be reported in the future, but with greater and greater difficulty as limits become lower.

3.4.1. Attenuated Total Reflectance (ATR)

This popular sampling method is commonly used for difficult materials such as rubber, fibers, paper, or powders. The theory and practice are well covered in the monograph by Harrick (39). Less well known is the fact that ATR has the potential of being one of the most sensitive methods for trace analysis (40). If the sample can be spread in a thin layer over the surface of an ATR element (deposited from solution or suspension, for example) to form a layer less than one micron thick, it can be used very efficiently. The effective depth of penetration depends on the wavelength, and for thin samples, all of the material is used by the radiation (Fig. 16). The effective sample area (Fig. 17), which is the area contributing to the detected signal, should be determined, and only this area used to deposit the sample. To do this, the ATR accessory is fixed in place

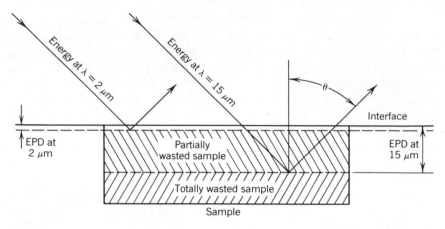

Fig. 16. Effective penetration depth and its variation with wavelength in ATR (41).

and adjusted for maximum transmission. The effective sample area is found by moving an opaque object downward across the face of the element until the spectrometer shows attenuation of the radiation. This procedure is repeated on the lower part of the entrance face, and on the upper and lower parts of the exit face.

The ATR element should be thin, to achieve a large number of reflections; its refractive index should be not too high and close to that of the sample; and the angle of incidence chosen should be just above the critical angle. Figure 18 shows the CH stretching bands of a monolayer (1–2 μg) of calcium stearate adsorbed on a sapphire plate. Much less material could have been detected by using techniques such as repetitive scanning and ordinate expansion.

Jakobsen (42), using FTIR, shows spectra of various thicknesses of stearic acid deposited on a KRS-5 ATR crystal (Fig. 19). He also used ATR to determine spectra of thin films on metal surfaces, by sputtering a thin (50–100 Å) film of

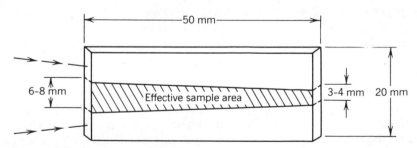

Fig. 17. Effective sample area in ATR. Only the radiation reaching the detector contributes to the ATR spectrum of the sample (41).

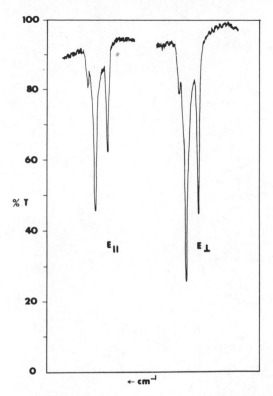

Fig. 18. The CH stretching bands of a calcium stearate monolayer deposited on a 0.5-mm sapphire plate at 45° (about 100 reflections). The relative intensities of the polarized spectra show that the hydrocarbon chains stand roughly at right angles to the absorbing surface (41).

iron onto a germanium ATR crystal and then adsorbing the sample onto that surface. The metal film is thin enough to be penetrated by the radiation.

In striving for utmost sensitivity, however, one ultimately reaches some sort of equilibrium with the environment. A clean ATR crystal exposed to laboratory air for only a few minutes showed the spectrum of Fig. 20. Based on stearic acid intensities, it is estimated to represent less than 0.1 monolayer of hydrocarbon.

3.4.2. Chromatography/IR

Gas Chromatography. Early attempts at identifying gas chromatography (GC) fractions by the use of IR spectroscopy relied on trapping the GC effluent and subsequently scanning it. Strategies such as directing the GC effluent through a capillary tube cooled with dry ice, or holding a salt plate in front of the effluent

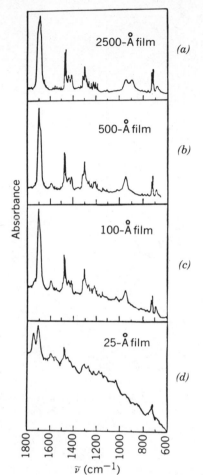

Absorbance

2500-Å film *(a)*

500-Å film *(b)*

100-Å film *(c)*

25-Å film *(d)*

$\bar{\nu}$ (cm^{-1})

Fig. 19. Internal reflection spectra of stearic acid films: (*a*) 2500-Å film; (*b*) 500-Å film; (*c*) 100-Å film; (*d*) 25-Å film (42).

stream to condense the emerging fraction, were sometimes successful with fractions of sufficient size. Another approach, used with samples of low volatility, was to trap the fraction on powdered KBr, which was then pressed into a pellet (43). Samples as small as 50 mµ could be identified in this way (44). Small tubes containing activated charcoal have also been used (45). The trapped eluate is desorbed with CS_2 and scanned using a microcell. The trapped-fraction method has the advantages of being inexpensive and simple, and it permits leisurely or repetitive scanning of the spectrum. Further, the physical state of the sample is such that spectra in the reference library can be used to identify the sample, with no allowance needed for a change in state. Disadvantages include the need for developing a good handling technique, limited sensitivity for small fractions,

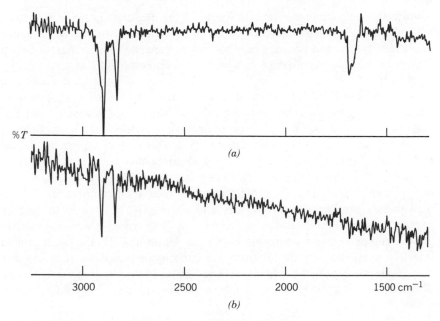

Fig. 20. ATR crystal exposed to laboratory air, showing the result of several thousand co-added scans: (*a*) stearic acid; (*b*) components absorbed from air. The adsorbed material is estimated to be 0.01–0.1 monolayer (42).

danger of contamination, and handling difficulties if the sample has a large number of fractions. For occasional identification of a limited number of major fractions from a packed-column chromatograph, however, this technique is still viable.

Another type of trapped fraction technique that is somewhat easier to carry out and has potentially greater sensitivity is the stopped-flow or vapor-trap method (46,47,48). Here, the GC effluent is directed through a short transfer line to a heated IR gas cell. The flow of carrier gas through the system is stopped when a fraction from the sample is judged to have reached the IR cell, and the spectrum is scanned as usual. A special chromatograph, optimized for IR sampling, can also be used (49,50). Large sample injections (up to 100 μL) and higher carrier gas pressure (120 N/cm^2 or 180 psig) give good separation and sensitivity for minor components. Chromatographic resolution is not degraded by stopping the flow.

Collecting spectra "on the fly" offers substantial advantages to the analyst: the collection proceeds rapidly, no manipulation of small samples is required, chances for contamination are reduced, and spectra on a large number of fractions may be collected in a relatively short time. It should be noted that spectra are not directly comparable to reference spectra in the file, especially for hydrogen-bonded species; a special reference library of vapor spectra scanned at the appropriate temperature and pressure may have to be generated or purchased. Group frequency data for vapor-phase spectra have been compiled and discussed by Nyquist (51).

It is possible to obtain on-the-fly spectra using a dispersive spectrometer especially designed for rapid scanning (52), but the energy advantage of FTIR makes it a clear choice for this type of analysis. Sensitivities for fractions of only a few nanograms have been reported (53). This type of sensitivity did not come easily, however; many technological advances have been brought together to achieve the state of the art. Advances in chromatography, detectors, cell design, and computerized spectrometer operation have all contributed.

Although excellent sensitivity was achieved in GC/FTIR with the use of packed columns (Fig. 21 and reference 53), capillary columns were felt to offer greatly improved chromatographic resolution. Unfortunately, the small sample injection necessitated by the relatively low capacity of such columns meant that only major components could be detected by GC/IR. Much effort has been devoted to solving this problem, both by improving the sensitivity of GC/IR and

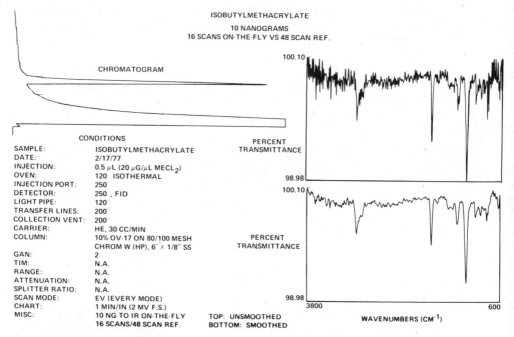

Fig. 21. FTIR spectrum of 10 ng *iso*-butylmethacrylate scanned "on the fly." Top spectrum is unsmoothed; bottom spectrum is smoothed (53).

by devising columns that retain their efficiency with larger sample injections. Thus, a 40-m × 0.5-mm i.d. WCOT (wall-coated open tubular) column coated with an 0.8-μm film of SE-52 stationary phase in combination with a 1-mm × 30-cm light pipe and an MCT detector gave excellent results for 8 μg/component (54). Another publication (55) demonstrates detection limits of under 40 ng/component with good chromatographic resolution, using WCOT columns. Fused silica columns have also been used with GC/IR (56,57). On-column injection of 0.5 μL of a 10% solution of jet fuel in HCCl$_3$ was separated on a 60-m × 0.33 mm i.d. fused silica column having a 1-μm chemically bonded film of DB-5. A nonane standard was easily detected at the 40-ng level; a 400-ng component of the jet fuel, identified as m-xylene, gave a very satisfactory spectrum. A good review of capillary GC/FTIR has been given by Griffiths et al. (58).

Various components of the interface and the spectrometer have also been the subject of intense developmental efforts.

The cooled narrow-range MCT detector has a specific detectivity (D^*) an order of magnitude greater than that of the TGS pyroelectric bolometer (although the wavelength range covered is substantially less, a real disadvantage with some samples). Second, the construction of gas cells or "light pipes" has been improved by producing a uniform coating of highly reflective gold on the interior surface (59). These cells may be optimized to give the longest path and minimum volume, while retaining adequate transmission (60,61). Typically, a 2-mm i.d. × 50-cm long light pipe transmits as much as 25%. Any design, of course, is a compromise, since the optimum dimensions and volume of the cell depend on the volume of the chromatographic peak being sampled. Resolution is degraded if the volume of the peak is much greater than the volume of the cell, and sensitivity is lost if the peak volume is significantly less than the cell volume. Different light pipes are often used for packed column and for capillary chromatography. It is important that the chromatographic resolution not be degraded by the light pipe plumbing, and a GC detector placed at the light pipe exit should give the same chromatogram as one placed at the GC column exit. Sometimes, makeup helium is added between the GC column and the light pipe to prevent degradation of resolution.

One needs also to be concerned about having the transmittance of the light pipe too high; a narrow band MCT detector is extremely sensitive and is easily overloaded (driven beyond its linear range), and the S/N of the interferogram can exceed the dynamic range of the analog–digital converter. If these situations occur, artifacts will appear in the spectrum (58).

Finally, the data system has to be capable of storing the interferogram of each fraction as it appears in the effluent. With a capillary column, chromatographic peak widths of only a few seconds or less are common, and some samples contain several hundred peaks. Spectral resolution no worse than 8 cm^{-1} is needed by the analyst, so for a 3700 to 750 cm^{-1} spectrum at least 2048 interferogram points per spectrum must be stored. Even with the fast Fourier transform, significant time is required to convert the interferograms to spectra. Fortunately, very fast array processors can do the transform in less than a second.

It is not, of course, necessary to store or transform every interferogram, as chromatograms usually contain more spaces without peaks than with peaks. The problem then becomes one of the data system recognizing the presence of a GC peak and triggering the storage mechanism. Then, when the peak has passed, storage is discontinued until the next peak arrives. In practice, however, it is useful to have an occasional background scan interleaved with the peaks in order to make a more accurate correction for any CO_2 and water vapor in the spectra.

Originally, a signal from a GC detector was used to activate the storage system, but this arrangement has obvious disadvantages, particularly with a destructive detector such as a flame ionization detector (FID). A better way is to examine each interferogram and compare it with an interferogram of an authentic background. Two ways of doing this are presently available. In the first, a low-resolution spectrum is computed from the first 512 data points in the interferogram, and ratioed against a reference spectrum of the same resolution taken at the start of the run. This calculation is very fast, so the spectra are available in real time. A difference in absorbance between the calculated spectrum and the reference spectrum indicates that an absorbing fraction is passing through the light pipe. These differences can be sought at up to 5 windows in the spectrum, usually short regions that contain characteristic absorptions for functional groups such as CH, C=O, COC, or aromatic rings. Sensitivity can be set so that any selected difference will trigger interferogram storage. Transformation of the interferogram to 8 cm^{-1} resolution spectra is undertaken after the run is completed. This system has the advantage of giving some chemical information about each chromatogram peak (62).

The second system for real-time chromatogram reconstruction from IR data utilizes the interferograms. One popular system is the Gram–Schmidt vector orthogonalization method, which depends on the fact that each point in the interferogram contains information about all the sampled frequencies in the spectrum. A small region of the interferogram, usually near the centerburst, is selected and n consecutive points in the interferogram are chosen and treated as an n-dimensional vector. The "basis vector" is chosen to represent a "normal" interferogram; i.e., an interferogram obtained with no sample in the light pipe. The basis vector is subtracted from subsequent vectors, and any residual difference represents absorption from material in the light pipe. This difference is recorded as a function of time. The resulting "chromatogram" is a good representation of the passage of sample components, but intensities may differ from those of a conventional chromatogram of the same sample. This result is to be expected, for the total absorption of the material in the light pipe is not the same

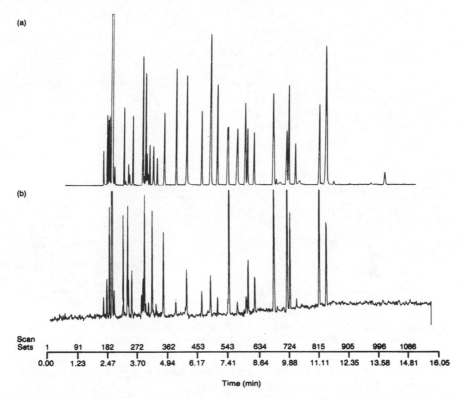

Fig. 22. Comparison of (*a*) a FID response with (*b*) a Gram–Schmidt reconstruction from the IR response. Reprinted with permission from Griffiths et al., *Anal. Chem.*, **55**, 1361A. Copyright 1983 American Chemical Society.

as the response of, say, a FID (cf. Fig. 22). The Gram–Schmidt method is said to be faster and to give higher S/N chromatograms than the real-time spectrum calculation method (63,64).

A novel matrix isolation approach to GC/IR has been proposed by Reedy et al. (65,66). The chromatograph effluent is mixed with a suitable gas (usually argon) and passed through a molecular jet separator, where most of the helium carrier gas is removed. The stream then passes into a vacuum chamber where it impinges on the mirror-like surface of a slowly revolving disc held at 14 K (Fig. 23). The argon condenses to form the matrix, and fractions from the chromatograph are also condensed as isolated molecules within the matrix. Matrix isolation techniques are not new, but they have never been applied to GC. Up to 5 hours' worth of eluting samples can be collected on the disc without overlap. After the matrix has been formed, the disc can be scanned, spot by spot, with the IR spectrometer (special transfer optics including a beam condenser are used).

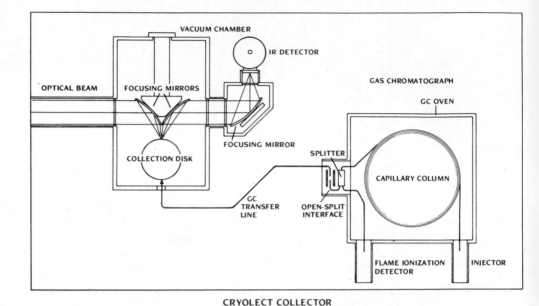

CRYOLECT COLLECTOR

Fig. 23. Schematic diagram of matrix isolation apparatus. Reprinted with permission from Bourne et al., *Amer. Lab.*, **16**(6), 90 (1984). Copyright 1984 International Scientific Communications, Inc.

Figure 24 shows matrix isolation spectra of acetone and *p*-dimethylaminoben-zaldehyde. Scan time was 1 min in both cases.

Several rather stringent conditions must be met for optimum results. The objective is to minimize the area over which each fraction is spread, so that spectral absorption will be as strong as possible, and yet not overlap different fractions even when they elute only seconds apart. Also, the correct ratio of matrix gas to sample must be maintained; if the sample becomes too concentrated, its molecules will no longer be isolated and will show intermolecular interactions. Finally, the heat of condensation of the matrix gas must not unduly warm the condensing surface, nor can the vacuum be degraded appreciably. In practice, the flow of matrix gas is on the order of a few cm^3 min^{-1}.

Although spectroscopic data are not collected in real time, the matrix isolation method has some real advantages. First, it is extremely sensitive—good spectra can be obtained on 5 ng of material, and recognizable spectra on subnanogram fractions. Because the matrix can be held indefinitely, extended signal-averaging can be used to enhance the S/N of minor component spectra. Optical throughput for the system is high (~50%), so more energy is available than with light pipes. Samples, after exiting the chromatograph, are held at low temperatures where no decomposition or rearrangement can occur. Finally, restrictions on spectral

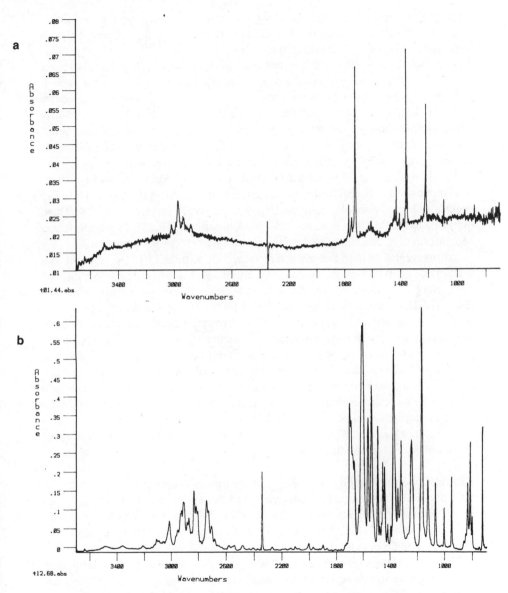

Fig. 24. Matrix isolation spectra of (*a*) 2 ng acetone, and (*b*) 20 ng *p*-dimethylaminobenzaldehyde (66).

211

range caused by, for example, detector response, can easily be circumvented. One can even return to a previously scanned fraction and rescan it using a different resolution, S/N, or wavelength range.

Disadvantages of the method include requirements for expensive and complex cryogenic equipment (now commercially available as a single unit, however), and the need for a special spectrum reference library. Matrix isolated molecules have very sharp absorptions, and the frequencies and intensities do not necessarily coincide with those of spectra taken on condensed phases at room temperature. Thus, a reference library of spectra scanned under matrix isolation conditions is virtually required for rapid identification of condensed fractions.

It has been realized for some time that GC/IR and GC/MS give complementary information (54, 67). Polar compounds give strong IR absorptions and distinctive spectra, whereas nonpolar compounds are more easily identified by MS. Thus, direct-linked GC/IR/MS systems have been described (68,69), in which IR and MS information is accumulated simultaneously as the peaks elute.

An integrated system that combines GC/FTIR with GC/FTMS (electron ionization *and* chemical ionization collected alternately) was found to give improved compound identification over the less complex systems previously described (70). The large amounts of data generated by this system make it important to minimize human intervention. Identification depends heavily on automated library searches, thus emphasizing the need for extensive and good quality reference libraries. Even when explicit identification is not achieved, however, molecular weight information is available from the chemical ionization spectra, and can be linked with the best matches from the IR search for further consideration. In a test of a 17-component mixture, 16 constituents were positively identified. The entire experiment took less than 3 hours.

The use of tandem GC/IR/MS promises to be a valuable approach for analysis of complex organic mixtures.

Liquid Chromatography (LC/IR).   Many mixtures cannot be fractionated by GC. Some or all of the components may not have sufficient vapor pressure, they may degrade at the temperatures necessary for volatilization, or they may react with the column stationary phase or other materials in the chromatograph.

If the sample is amenable to analysis by liquid chromatography or LC (the term liquid chromatography as used here includes gel permeation chromatography, high-performance liquid chromatography, column chromatography, and more), IR can give a more or less complete analysis of the eluting fractions. Sensitivity is somewhat poorer (by a factor of 10–100) for LC/IR than for GC/IR as of this writing, but this situation will no doubt improve as the technique is further optimized. Obtaining IR spectra on LC fractions has one major difficulty that is not met in GC/IR, however; LC fractions are always dilute solutions in a solvent that invariably obscures a portion of the spectrum. Two means of overcoming this difficulty have been tried, neither of which is completely satisfactory. Either the sample is examined in the solvent, which places severe limitations on both the chromatography and the spectroscopy, or the fraction must be separated from the solvent so that it can be scanned without solvent interference.

Which method is chosen depends on the purpose of the analysis. For monitoring a component or fraction having a defined functional group, the direct approach of scanning the sample in the solvent works well, if a chromatographic solvent having suitable transmitting windows can be found. No single solvent is completely transparent over the entire 4000–400 cm^{-1} range, and even the best IR-transmitting solvents are opaque in some spectroscopically important regions (71,72). The more polar the solvent, the fewer are the windows, so unless one is exceptionally lucky, one can seldom identify a complete unknown from its LC/IR transmission spectrum. Fortunately, in many cases (such as polymer systems) the general nature of the components is well known and the choices for the "unknown" fraction are quite limited.

Cell design is subject to infinite variation. Often a conventional microcell can be adapted to LC/IR use. Cell design is discussed by Vidrine (72). He recommends that:

1. Dead volume in the cell should be minimized.
2. Cell volume should be smaller than the volume of the expected chromatographic peaks, to retain chromatographic resolution.
3. The exit tube should be at the top of the cell, so any bubbles are quickly purged.
4. The exit tube should be larger than the inlet tube, and should not have significant flow resistance to cause pressure build-up in the cell.

For highly absorbing solvents, such as water, a low-volume ATR cell having a novel cylindrical element can be used (73).

An IR filter photometer (19) has been adapted as an LC/IR detector. Only a single wavelength (at low resolution) is used, but it can easily be changed to detect different functional groups. If detection at several wavelengths is desired, however, the chromatograph run must be repeated several times. The instrument has a high energy throughput and is very stable, so it can reliably detect small changes in absorbance.

Fourier transform IR spectrometers, because of their rapid scanning capabilities, can be used to accumulate data over the entire spectral range in real time (except, of course, in regions of solvent absorption). The chromatograph effluent flows through a microcell of appropriate thickness (usually 0.1 to 1 mm). Temperature-sensitive or volatile components are preserved, and fractions can

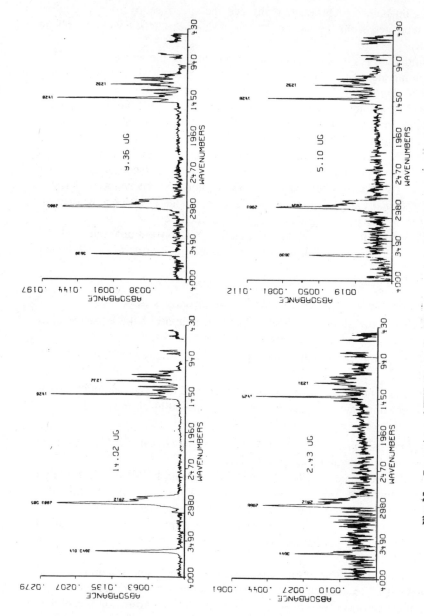

Fig. 25. Comparison of FTIR spectra for different sample weights of 2,6-di-*tert*-butylphenol: 1-mm × 100-cm Polar Amino Cyano (PAC) column, 20 mL/min^{-1}, 5 mL injected, 0.2-mm flow cell. Reprinted with permission from Brown and Taylor, *Anal. Chem.,* **55**, 1492. Copyright 1983 American Chemical Society.

be collected for further processing. Chromatograms for up to 5 spectral regions can be displayed in real time, as for GC/IR. Absorbance subtraction can be used to remove solvent bands (74), although for wavelengths where bands have absorbance greater than 1, the S/N will be seriously degraded; and for bands having absorbance greater than 2, no meaningful information can be obtained (see also Section 3.6.3). Further subtraction of eluted fractions can show structural or functional group variations. "Flowthrough" LC/IR is discussed at some length by Vidrine (72,74).

The use of microbore (1-mm i.d.) LC columns with flow cell detection has some advantages over normal LC/IR (75). First, the approximately 20-fold increase in eluate concentration over analytical scale work improves the detectability of the eluting fractions. Second, the low solvent consumption (1 ml/run or less) makes the use of expensive solvents such as $DCCl_3$ feasible, if their IR transparency offers an advantage. Third, the high efficiencies of microbore columns make possible the separation of complex mixtures.

On-the-fly spectra of 2,6-di-*tert*-butylphenol obtained with the system described by Brown and Taylor (75) are shown in Fig. 25. A comparison of injected minimum detectable quantities (IMDQ) for the microbore column and for an analytical column are shown in Table 5. The detection limit for different functional groups depends on the absorption coefficient for that group, but the relative sensitivity of the two columns should be approximately constant, and it is. The microbore system is approximately 8 to 10 times more sensitive than the conventional analytical system. Further development of this technique would seem to be warranted.

For identifying completely unknown components, the solvent must be removed without destroying the integrity of the sample. Of the various approaches tried, the most successful to date involves evaporating the solvent from each fraction

Table 5. Injected Minimum Detectable Quantities, Analytical and Microbore Columns, for 2,6-Di-*tert*-butylphenol[a]

IR Band	Origin	Noise Level	IMDQ[b] Analytical (μg)	Microbore (μg)	Ratio
3640	OH stretch	0.0007	12.87	1.67	7.71
2960	CH_3 stretch	0.0008	10.52	1.32	7.90
1426	CH_3 bend	0.0004	6.04	0.63	9.59
1232	CO stretch	0.0002	6.38	0.76	8.39

[a]Reprinted with permission from Brown and Taylor, *Anal. Chem.*, **55**, 1492. Copyright 1983 American Chemical Society.
[b]IMDQ takes as 3 × noise level.

on a bed of powdered KCl or KBr, which is then scanned using diffuse reflectance (76). The first such system described in the literature consisted of a carousel having a series of cups on its perimeter, each holding about 40 mg powdered KCl, which was programmed to receive fractions from the HPLC. After the sample was deposited, the cup was moved to a new position where the solvent was evaporated, and then into the spectrometer sampling position where the diffuse reflectance spectrum was scanned. Later variations of this apparatus have also been constructed. Sensitivity is on the order of 10 to 100 ng for a well-designed system with efficient collection optics.

Thin Layer Chromatography (TLC).   The concept of identifying compounds by IR after separating them using TLC is an appealing one, but its achievement has not yet been completely successful. The principal obstacles are the small sample sizes and the strong IR response of typical TLC absorbants. Griffiths and co-workers (77,78,79) have experimented with special TLC plates made of a 0.1-mm layer of Al_2O_3 or SiO_2 on AgCl substrate. These plates could be used directly in transmission measurements after the sample spot was developed by exposure to iodine. A $4 \times$ beam condenser was used. Samples were ratioed to a spectrum taken on an adjoining segment of the plate. Plates were treated with mineral oil or Fluorolube to reduce scattering by the absorbant. Programmed multiple development TLC (79) gave better sensitivities, by a factor of approximately 5, and as little as 10 ng of a (known) strong IR absorber could be detected. Identification of an unknown material would have required considerably more sample.

The *in situ* approach suffers from several obvious shortcomings. Only a limited region of the spectrum ($1300–2000$ cm^{-1}) is available because of the strong absorption of SiO_2 and Fluorolube below 1300 cm^{-1}. TLC plates must be specially prepared on an AgCl substrate; commercial plates cannot be used. If the analyte does not have strong, distinctive bands in the $1300–2000$ window, sensitivity may be inadequate. Treatment of silica gel with a solvent tends to alter its surface characteristics, leading to frequency shifts and intensity changes.

These drawbacks led Fuller and Griffiths (80) to a second approach: that of eluting the sample from the silica gel before IR scanning. They report far superior results when the spot containing absorbent was scraped from the plate and extracted with solvent. The extract was centrifuged and filtered through a sintered glass filter to remove the SiO_2 particles. The solution was then evaporated on powdered KCl in a DRIFTS cup (see Section 3.4.3). A complete spectrum, free of interferences, was obtained for 100 ng of a strongly absorbing sample (Fig. 26).

One disadvantage of this method is apparent: success is strongly technique-oriented, and opportunities for contamination are great. A second problem is mentioned by the authors: occasional solvent interference is noted, apparently resulting from a hydrogen bonding donor–acceptor complex between sample

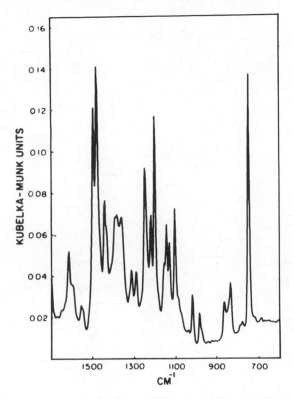

Fig. 26. DRIFTS spectrum of 100 ng Indophenol Blue, obtained from a TLC plate (see text) (80).

material and solvent. This problem could place severe limitations on the choice of eluting solvent.

Supercritical Fluid Chromatography (SFC). Certain classes of compounds that have low volatility or are sensitive to the high temperatures of GC can often be successfully chromatographed using SFC with much greater resolution than can be obtained with LC (81). Obviously, in some cases at least, IR detection and characterization of the eluting fractions would be desirable. Such detection has been reported by Shafer and Griffiths (82), who used a wide-bore fused silica column (60 m, 0.33-mm i.d. WCOT column with a 1-μm layer of DB-5) and supercritical CO_2 as the mobile phase. Supercritical CO_2 is quite transparent in the IR down to about 800 cm^{-1}; its strong absorption at 2300 comes in a region that is generally of little interest in IR spectroscopy. The absorption does vary with pressure, however, as shown in Fig. 27. The result of injecting 3 μg of each of 3 components is shown in Fig. 28.

The use of the preferred small capillaries, 50-μm i.d., for which a 100-nL

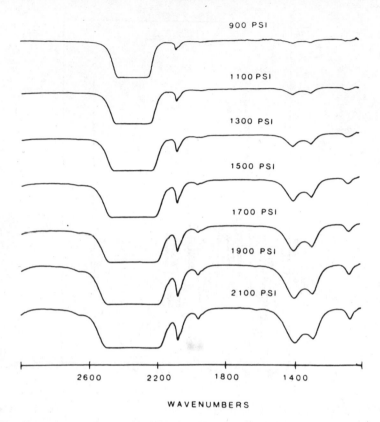

Fig. 27. Transmittance of supercritical CO_2 in a 10-mm cell at various pressures. Reprinted with permission from Shafer and Griffiths, *Anal. Chem.*, **55,** 1939 (1983). Copyright 1983 American Chemical Society.

injection is typical, gives high chromatographic efficiency but unfortunately does not allow FTIR detection because of the small fraction weights (83).

3.4.3. Diffuse Reflectance

Diffuse reflectance is not a new technique—it has been used for years in the ultraviolet and visible regions for which highly reflecting integrating spheres are available. A special FTIR spectrometer that incorporated an integrating sphere as a part of its optical train has been described (84) but it required long measurement times because of poor energy efficiency and was not commercially successful. In 1978, Fuller and Griffiths (85) described a novel optical accessory that utilized an ellipsoidal mirror to collect radiation that was diffusely scattered from a 4-mm cup containing the powdered sample (Fig. 29). These workers also

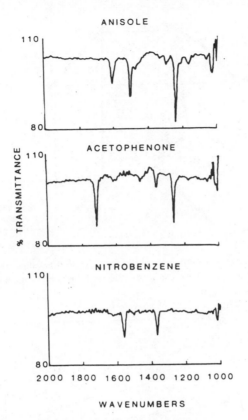

Fig. 28. Spectra of separated components obtained by SFC/IR, 3 μg each. Reprinted with permission from Shafer and Griffiths, *Anal. Chem.*, **55**, 1939 (1983). Copyright 1983 American Chemical Society.

found that powdered KCl scattered light quite efficiently, and that relatively small amounts of sample mixed with the KCl gave excellent diffuse reflectance spectra. Best results are obtained if the KCl is ground to have a particle size of less than 10 μm (Fig. 30). Solid samples can be run directly, but should be ground, preferably to less than 75-μm particle size. Better spectra are sometimes obtained if the powdered sample is mixed with 80–90% powdered KCl. Solutions can be evaporated onto the KCl powder, but nonpolar solvents should be used. Polar solvents tend to change the surface characteristics of the KCl and are sometimes difficult to eliminate completely. Sensitivity of the method can be improved, sometimes dramatically (depending on the sample) by compressing the KCl powder to a point somewhat short of pressure that could cause fusion and give a transparent pellet (86). Better reproducibility of band intensities is also claimed. The diffuse reflectance method is quite sensitive; less than 200 ng

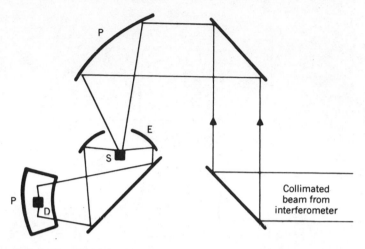

Fig. 29. Optical schematic of the diffuse reflectance attachment. The ellipsoid E collects scattered radiation from the sample S, and focuses it on the detector D, which is not coplanar with the other optics (80).

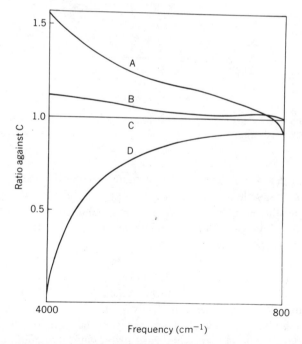

Fig. 30. Reflectance spectra for powdered KCl having particle sizes (A) < 10 μm; (B) 10 < d < 75 μm; (C) 75 < d < 90 μm; and (D) d > 90 μm, ratioed against (C). Reprinted with permission from Fuller and Griffiths, *Anal. Chem.*, **50**, 1906. Copyright 1978 American Chemical Society.

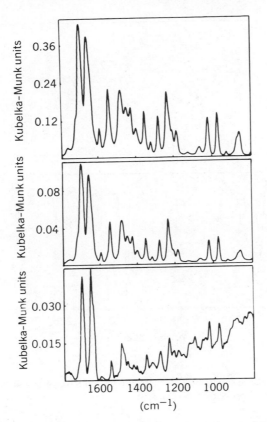

Fig. 31. Diffuse reflectance spectra of (*top*) 11 μg, (*middle*) 256 ng, and (*bottom*) 11 ng of caffeine in KCl versus a KCl reference (80).

of moderate absorbers give reasonable spectra (Fig. 31), and only 2 ng of a strong absorber gave some detectable bands (80). The name "DRIFTS" was coined for the technique; it is an acronym for diffuse reflectance infrared Fourier transform spectroscopy.

The use of diffuse reflectance, however, is not limited to FTIR. Hannah and Ancreon (87) have shown that good spectra can be obtained with a commercial diffuse reflectance unit and dispersive spectrometer. Only 500 ng of phenacetin gave an excellent spectrum (Fig. 32), and a useable spectrum was obtained from 50 ng. This technique would seem to have considerable potential for trace analyses, but the sensitivity should be optimized by careful optical alignment, use of sensitive detectors, and care in achieving optimum particle size for the sample and KCl (or KBr).

It is even possible to do quantitative work and apply absorbance subtraction

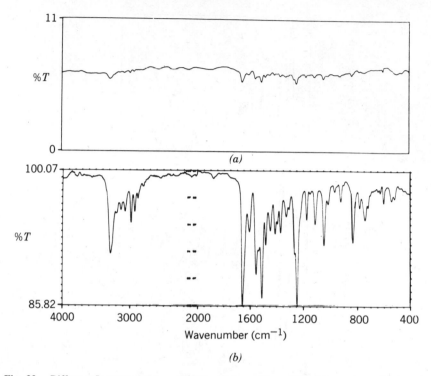

Fig. 32. Diffuse reflectance spectrum of 500 ng phenacetin on KBr. (*a*) raw data; (*b*) expanded, smoothed, and background corrected data (87).

routines. Kubekla–Munk theory (which relates sample concentration and scattered radiation intensity) predicts a linear relationship between the reflectance intensity observed for a band and its absorption coefficient, *provided* the scattering coefficient remains constant. The scattering coefficient depends on both particle size and particle size distribution, so it is important to control these factors if one wishes to undertake quantitative or scaled subtraction procedures. It was found that accurate subtractions over the entire spectrum were not possible for samples that were not diluted with KCl (85).

3.4.4. Emission

Emission techniques can sometimes be applied to samples that cannot be studied in absorption, but such situations are relatively uncommon. Alternative methods are usually easier to carry out and are generally more sensitive as well. However, spectra can be obtained by emission from materials when temperature is only a few degrees above or below ambient, provided conditions are right. Best results are obtained from a larger temperature differential, however, provided the sample does not decompose or sublime. Condensed-phase samples must not only be very thin but must also be in the right range of thickness (88). The sample temperature should be uniform; nonuniformity has large and unexpected effects on the spectra (89). At least two measurements (sample, and a blackbody reference) need to be made; if a cooled detector is used, four measurements are required (sample and blackbody at two different temperatures) (90,91). With interferometer spectrometers, reflection from the beam splitter back to the sample, which is often deposited on a highly reflecting metal surface, can result in multiple passing of the radiation. The effect is to introduce artifacts such that some fraction of the intensity at frequency v appears at $2v$, $3v$, etc. in the spectrum. This effect is difficult to eliminate, but Chase (91) has found that a powdered KBr substrate for the emitting sample gives better results than a reflecting metal substrate (Figs. 33 and 34). The spectrum shown in Fig. 34 was

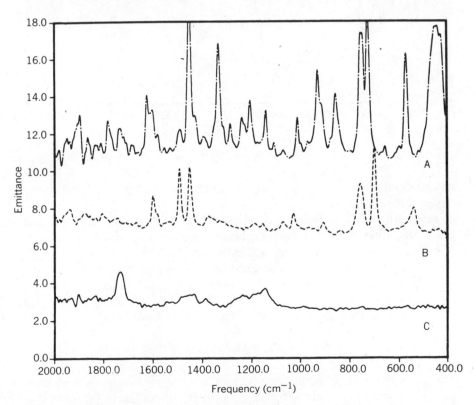

Fig. 33. Emittance spectra of (A) 15 μg carbazole, (B) 5 μg of polystyrene; and (C) 2 μg of poly(methylmethacrylate) on a reflecting gold surface (91).

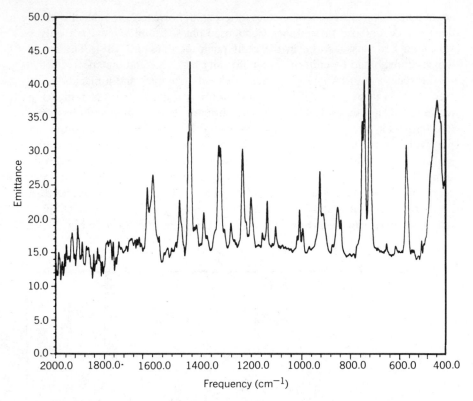

Fig. 34. Emittance spectrum of 2 μg of carbazole deposited on KBr powder (91).

obtained with 2 μg of carbozole; one can infer that 200 ng would still give a recognizable spectrum.

Fewer difficulties are encountered with obtaining emission spectra from heated gases (88), although some intensity anomalies are likely to be seen, apparently because of temperature variations within the sample.

Lauer et al. (92,93) have used IR emission spectroscopy to study the behavior of a thin lubricant film under load. In another study, residual lubricant on steel tire cords could be detected efficiently by emission (94). In that work, a viewing angle of 70°, and temperatures of 80 to 190°C, were used.

3.4.5. KBr Pellets

This popular sampling method has been adapted to microanalysis, and commercial equipment is available for pressing pellets as small as 0.5 mm in diameter. Nevertheless, the sensitivity of this method is not as good as that of several other methods and it has some other rather limiting characteristics. It does, however, fill an important niche in the repertoire of micro sampling methods for samples in the range of 0.05 to 100 μg. Many micro samples studied at these levels, especially in the lower part of the concentration range, are dissolved in a suitable solvent which is evaporated on the KBr powder. Sometimes lyophilization, or freeze-drying of a water solution of the sample, is practiced. After the solvent is evaporated, the powder is pressed into a disc for placement in the spectrometer. A beam condenser is advantageous and is mandatory with smaller pellets. Hirshfeld (95) has derived the optimum diameter of a FTIR microcell (micropellet), and recommends using the strongest possible focus micro sampling optics.

The limitations of the KBr pellet technique should be understood and accepted before it is used for microanalysis. They are discussed elsewhere (4), so are reviewed only briefly here.

1. Many materials yield spectra that are strongly dependent on the sample history. Polymorphs may not give consistent results, and their spectra often depend on the solvent from which they are precipitated, the pressure used in forming the pellet, and other factors which are poorly understood. Phenols and organic acids show spectral distortions, apparently because of adsorption onto the alkali halide particles. Hydrochloride salts of amines often undergo partial or complete anion exchange, which may give large spectral changes.

2. Whenever a solvent is evaporated, impurities and contaminants are likely to appear, which sometimes mask the sought-for constituent (reference 9 and Section 1.4). Some solvents, such as acetone, are difficult to eliminate completely from the KBr powder. The use of blank runs is essential to detect any such problems.

3. Samples showing even slight volatility can be lost during preparation of the pellet. Griffiths and Block (96) studied the recovery of methyl phosphonate monoesters (B.P. 200°C at 1 mm) after preparing KBr pellets of successively lower concentration by serial dilution using KBr powder. Simply exposing the powder mixture to the atmosphere led to decreased absorbance of the POC stretching band such that 90% of the original 1-μg sample was lost in 2 min and 97% was lost in 20 min. No way of preventing this loss could be found. Another convincing demonstration of volatility loss is given by King (97), who tried evaporating CS_2 solutions containing different concentrations of 2,6-dimethoxyphenol (B.P. 263°C) onto KBr powder, which was then pressed into 0.5-mm discs. He found that more dilute solutions showed large losses of analyte. Also, baseline excursions and irregularities limited ordinate expansion. He concludes that 0.05 μg is the smallest amount of material that can be transferred and analyzed by this method.

For appropriate samples, however, the KBr method is relatively fast and easy. A system for concentrating the analyte from solvent solution (as from TLC adsorbant), the Wick-Stick,® has been described (98). A triangular piece of porous KBr, prepared from KBr powder, is held upright in the center of a small vial with the apex of the triangle near the top of the vial. The solvent solution is placed in the vial and a cap containing a vent hole placed on the vial. Capillarity draws the solution up the Wick-Stick, where the solvent evaporates, leaving the sample on the upper part of the KBr triangle. Two or three additional passes with solvent effectively concentrate the analyte at the tip, which can then be broken off and pressed into a micropellet. With TLC samples, the TLC adsorbant containing the sample can be scraped off the plate and placed in the vial. The solvent then carries the sample to the tip of the KBr triangle, leaving the adsorbent behind. No additional filtering is necessary. The solvent used must, of course, be very pure. Samples containing 10–50 μg of material give satisfactory spectra when pressed into a 1.5 mm pellet.

Thus, the sensitivity of the KBr pellet method is limited, not so much by intrinsic shortcomings of the sampling method but rather by the difficulty of manipulating samples smaller than 0.05 μg and keeping them free of contamination. For nonvolatile solid samples in a solution of volatile solvent, this seems to be the lower limit of sample size for reliable work (99).

3.4.6. Long Path Sampling

When one is dealing with a high transmission background, or when a variable transmission background has spectral "windows" in which the components of interest can be detected, no separation of analyte from the matrix is necessary. Sample path lengths of 10 to 1000 or more times normal can be used to achieve the needed sensitivities.

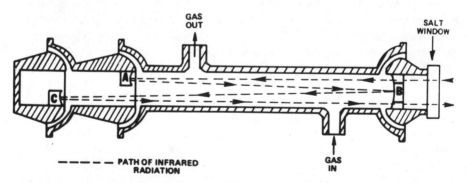

Fig. 35. Multipass cell with ball joint mounting of mirrors. Reprinted with permission from P. L. Hanst, *Appl. Opt.,* **17,** 1360 (1978).

Table 6. Estimates of Pollutant Detectability Limits[a]

Pollutant	Best Measurement Frequency (cm^{-1})	Detectability Limit c_{min} in Billionths of an Atmosphere (ppb)	Remarks
Acetylene	735	0.02	Strong absorber.
Ammonia	965	0.2	
Carbon monoxide	2180	1.0	The atmospheric carbon monoxide concentration never falls below about 80 ppb.
Ethylene	950	0.2	
Formaldehyde	2765	1.0	This estimate is for 1-cm^{-1} resolution. Higher resolution may further increase the detection sensitivity.
Formic acid	1100	0.8	
Hydrogen chloride	2820	1.0	This estimate is for 1-cm^{-1} resolution. Higher resolution may further increase the detection sensitivity.
Hydrogen sulfide	1300	50	Weak spectrum.
Methane	3020	0.6	The atmospheric methane concentration never falls below 1400 ppb.
Nitric acid	880	0.6	
Nitric oxide	1900	2.0	
Nitrogen dioxide	1615	0.2	
Ozone	1060	0.6	
Peroxy acetyl nitrate	1160	0.2	
Peroxy benzoyl nitrate	990	0.2	
Phosgene	850	0.1	
Propylene	915	1.0	
Sulfur dioxide	1360	0.2	

[a]Using 50 passes through a 10-m cell, with 98% mirror reflectivity; sample pressure, 1 atm. From Hanst et al. (104).

Transparent gases such as air are most often studied in a multireflection cell of the type designed by White (100). Such cells can be constructed so as to vary the number of reflections and thus the path length. Because of reflection losses at the mirror surfaces, however (about 2% for gold-coated mirrors), the number of reflections cannot be increased indefinitely; a point is reached where the decrease in S/N more than cancels the gain from additional path length. Also, for large numbers of reflections, small changes in the lateral adjustment, caused by temperature drifts, can cause large changes in path length. Another trade-off exists between source magnification and maximum path length. Cell mirror configurations, and the important requirement of matching the cell to the spectrometer, are discussed by Hanst (101). Hanst also has devised a microvolume multipass cell that uses glass ball joints for adjusting the number of reflections

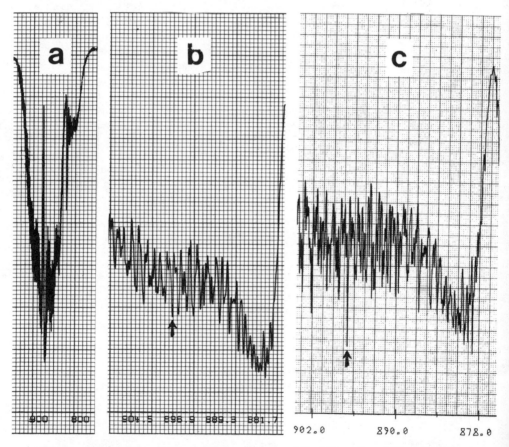

Fig. 36. High-resolution spectra of ethylene oxide. (*a*) and (*b*), 0.25 cm^{-1} resolution; (*c*) 0.04 cm^{-1} resolution. Note the appearance of the 897.069 cm^{-1} band at higher resolution (106).

(Fig. 35). This cell, when used at 32 passes, had a path length of 115 cm but a volume of only 3 cm^3. About 15 ng of ammonia gave 3% absorption at 935 and 970 cm^{-1}.

At the other extreme, a 22.5-m-long cell has been described that gives an overall path length of 1 km (103). For a 10-m cell with 50 passes (0.5 km) Hanst et al. (104) calculated detectable limits for a series of pollutants shown in Table 6. One advantage of confining the sample to a cell is that the contents can be subjected to sunlight, UV radiation, etc., and the response of the pollutants therein monitored (105).

A method of increasing sensitivity and selectivity in gas analysis is to use higher resolution scans. Line width in a vibration–rotation band of a typical gaseous molecule at normal temperature and pressure may be on the order of

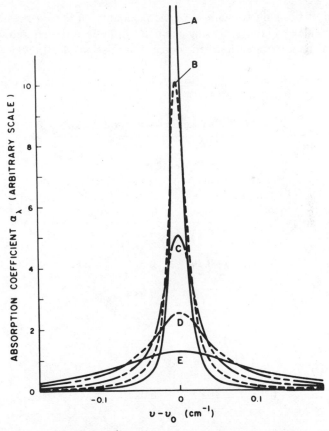

Fig. 37. Line profile as a function of pressure. Curves A–E correspond to pressures of 45, 95, 190, 380, and 760 Torr, respectively (107).

0.1 cm^{-1}. Spacing of the individual lines in a band depends on the moments of inertia of the molecule. Whereas considerable improvement in sensitivity for light molecules may result from scanning at higher resolution (see Fig. 36), for heavy molecules the gain will be minimal. Reducing the pressure, however, sharpens the lines (Fig. 37) because of reduced collisional (pressure) broadening, and one may actually achieve better sensitivity at a lower pressure. Pressure reduction may be combined with longer path length to achieve best sensitivity. Further pressure reduction is, of course, ineffective once the Doppler width of the line has been reached (Doppler widths are usually on the order of 0.001 cm^{-1}). Also, the spectrometer must be able to take advantage of the line narrowing (only a few FTIR spectrometers can reach resolutions of 0.05 cm^{-1}, but this figure is easily surpassed with diode lasers). The subject of line widths, spectral resolution, and reducing interferences from overlapping absorptions is discussed by Hanst (101).

Instead of bringing the sample to the laboratory, it is also possible to take the measurements in the field using a remote retroreflector, a remote source such as the sun, or a portable infrared source, or emission measurements from the sample.

Atmospheric pollutants can also be concentrated by condensing the air and

Table 7.   Trace Gases in Ambient Air[a]

Date and Weather	Fluoro-carbon-11 (ppb)	Fluoro-carbon-12 (ppb)	Carbon Tetra-chloride (ppb)	Carbonyl Sulfide (ppb)	Acet-lyene (ppb)	Paraffinic Carbon from 3.4-μm band (ppb)
Jan. 15, 1975 Clear, breezy	0.19	0.23	0.10	NM	2.5	60
Jan. 23, 1975 Overcast	0.36	0.26	0.11	0.20	4.2	100
Feb. 25. 1975 Clear, breezy	0.12	0.20	0.09	0.24	1.6	30
May 9, 1975 Overcast	0.17	0.22	0.10	NM	1.7	60
May 15, 1975 Overcast, raining	0.18	0.37	0.08	0.20	3.2	90
May 22, 1975 Hot, sunny and humid	0.26	0.44	0.08	NM	NM	150

[a]ppb = parts per billion (10^{-9} atm partial pressure); NM = not measured. From Hanst et al. (108).

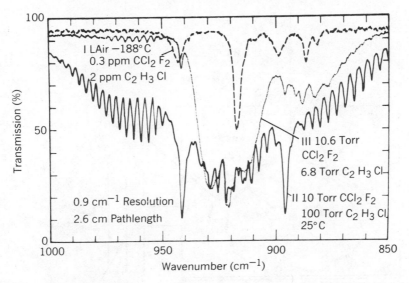

Fig. 38. IR spectra of 0.3 ppm CCl₂F₂ and 2 ppm C₂H₃Cl. (I) dissolved in liquid air; (II) and (III) Same two components in the gas phase. Reprinted with permission from Freund et al., *Anal. Chem.*, **50**, 1260. Copyright 1978 American Chemical Society.

subjecting the condensate to functional distillation. Concentration factors of 5000 or more have been achieved (101,108). Liquified O_2 and N_2 were fractionated off, and the CO_2 was absorbed with NaOH. Table 7 shows the compounds measured and their typical concentration at Research Triangle Park, NC, at the time of the test.

A technique for actually using liquid air or other liquified gases as solvents for IR analysis has been suggested by Freund et al. (109). The gas is cooled until it liquifies, whereupon the trace impurities are determined by IR. Band shapes for light molecules are considerably simplified (Fig. 38) and sensitivity is good, since optical path lengths of 1–10 cm can be used.

Only a few liquids are sufficiently transparent that they can be sampled at path lengths of 1–10 cm. $SiCl_4$ (used in semiconductor and optical fiber manufacture) is one such material. A stainless-steel cell with AgCl windows has been described (110) for analysis of $SiCl_4$. Concentrations as low as 0.07 ppm (Si)OH, 1.4 ppm HCl, and 1.6 ppm SiH could be found using a 10-cm cell.

3.4.7. Photoacoustic Spectroscopy

Although PAS (photoacoustic spectroscopy) is not generally thought of as a technique for trace analysis, it has been found useful in surface studies, such as for supported metal catalysts and molecules adsorbed on powders. The principle

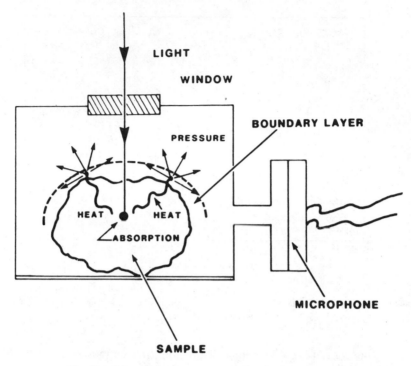

Fig. 39. Schematic representation of a PA cell for solids (112).

is rather simple; modulated radiation impinges on the sample which is enclosed in a cell containing a microphone (Fig. 39). Those frequencies that are absorbed by the sample give an active signal at the modulation frequency, as a result of heat absorption and release in the form of a pressure wave or acoustical signal. The depth of penetration of the radiation depends on the "optical thickness" of the sample, its heat capacity, and its thermal conductivity (111). It has the advantage of requiring no sample preparation (although finely divided solids give better results than coarse granular or undivided ones). PAS has been used to study coal surfaces, inorganic salts, pesticides on clay, catalysts, and powders (112). An efficient PAS cell that incorporates temperature and atmosphere control can be used for studies of reactive chemical systems (113). PA spectra of silica powder before and after treatment with $TiCl_4$ vapor are shown in Fig. 40.

3.4.8. *Reflection–Absorption*

This rather specialized technique is particularly useful for studying films or adsorbates on polished metal surfaces. Two approaches are possible; which one

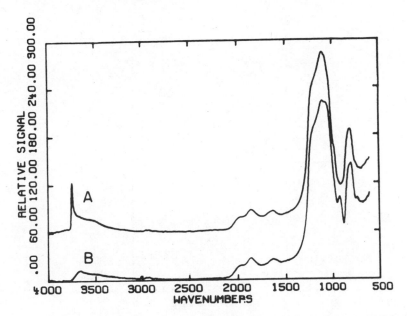

Fig. 40. Mid-IR PAS spectra of silica powder (A) before and (B) after reaction with TiCl₄ vapor. Reprinted with permission from Kinney and Staley, *Anal. Chem.*, **55**, 343. Copyright 1983 American Chemical Society.

is used depends on the sample thickness. For coatings with a thickness in the same range as a normal transmission sample (0.01 mm), a simple specular reflection arrangement with the beam perpendicular to the metal surface gives a satisfactory spectrum. With extremely thin samples (less than the wavelength of the radiation) this technique is ineffective, because an electric standing wave field has a node approximately at the reflecting surface. The field strength at the node is zero, and therefore the radiation cannot interact with molecules at the surface. If, however, the radiation encounters the surface at a glancing angle (80–89° angle of incidence, measured from the normal to the surface) the node is not at the surface and IR spectra can be obtained on a very thin layer, even a partial monolayer, of molecules. One possible experimental arrangement is shown in Fig. 41 (114).

The number of reflections must be carefully chosen to optimize band intensities, since energy is lost at each reflection. For some substrates, one or two reflections is optimum. Absorption intensity is about 10–50 times stronger for the same thickness of sample than in normal transmission spectroscopy. Only vibrations having their transition moment perpendicular to the metal surface are detected, so information about the orientation of the adsorbed molecules can be inferred from their spectra. The theory of RAIR has been given by Greenler

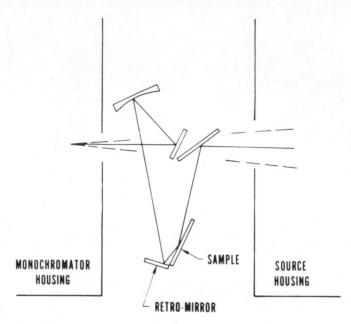

Fig. 41. Optical schematic of a single-reflection sampling device (114).

(115), and techniques have been discussed by Greenler (115), Tompkins (114), and Allera (116).

Applications to chemisorption of gases on metals, studies of catalysis, and oxidation-corrosion of metals have been described (117). Blanke et al. (118) have studied the orientation of stearate films on gold and copper substrates. They describe a modified dielectric filter spectrometer especially constructed to study emission spectra of surface species, but the sensitivity was not very good. Reflection–absorption measurements could be made, however, using a polarizing modulator. Polarization modulation was also used by Dowrey et al. (119) with an FTIR spectrometer. The improvement in S/N for a 10-Å-thick film of cellulose acetate on polished copper as compared with a normal RAIR spectrum is demonstrated in Fig. 42.

3.4.9. Solutions

The use of dilute solutions for recording IR spectra has long been standard practice in some laboratories (3,4). Solutions have the advantage of giving reproducible spectra, having little perturbation by external factors, on many kinds of samples, including solids and volatile liquids. The disadvantages are that no single solvent gives an unobscured view of the entire spectrum; some solutes

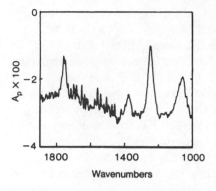

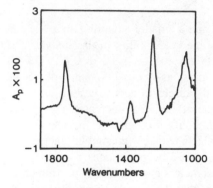

Fig. 42. (*top*) Normal RAIR spectrum versus clean copper; (*bottom*) Polarization-modulation spectrum of same samples (see text) (119).

react with the solvent; and not all samples are soluble. From those materials that are soluble, however, high-quality spectra that can be used for quantitative analysis can usually be obtained.

For analysis of small samples (0.5 μg or more), microcells of 2–3 μL capacity and 0.5–1 mm path length combined with a beam condenser are useful, and give better results than micropellets (97). Some problems may be encountered with dispersive spectrometers because of the heating effect of the IR sample beam, which can cause the solvent to boil out of the cell. FT spectrometers have less effect because the most energetic portion of the radiation is not present in the sample beam.

Griffiths and Block (96) state that the optimum method for studying samples of finite volatility is in solution in a microcell. They show a spectrum of 70 ng ethyl phosphonate, run as a 0.05% solution in CCl_4. The total volume of solution was 0.2 μL.

In designing an experiment, the analyst should consider the optimum solution concentration and cell size very carefully. On the one hand, a very dilute solution in a cell of full beam cross section could be selected. At the other extreme, a more concentrated solution contained in a thicker cell of extremely small cross section might be used. The latter approach is preferred, for reasons that will become apparent. (See also Section 3.6.)

Let us consider an example. To detect an absorption band of 0.01% T, we need a S/N ratio of 10^5:1, in order that the band be 10 times the noise amplitude. A well-tuned dispersive spectrometer may have S/N for normal scans of 10^3/1 (0.1% noise). To detect the trace constituent, then, the S/N must be improved by a factor of 100. This can be done by opening the slits by $\times 10$; by lengthening the scan by $\times 10^4$; or by some combination thereof. FTIR instruments can more easily achieve high S/N ratios, but they, too, are limited in ordinate expansion. With a high transmission background, it is easy to exceed the linear dynamic range of the detector and of the A/D converter. Unless some interferogram noise is digitized, S/N will *not* increase as scans are co-added (see Section 3.6.1 and Chapter 7 of reference 2).

By using a more concentrated solution and a smaller sampling area, we can improve both the S/N of the spectrum and the detectability of the analyte. The optimum S/N is achieved when the analyte absorbance is $1/e$ or 0.43 (37% T) (3); a range of 0.2–0.7 (20–60% T) is almost as good. Often it will not be possible to reach this optimum range, but the direction is clearly indicated. The next question is how far to go in the trade-off between increased absorbance resulting from the same amount of sample being in a smaller cross section of the beam, and vignetting of the beam, which reduces transmitted energy and limits ordinate expansion. We assume first that a beam condenser is used.

The answers differ for dispersive spectrometers and for FTIR spectrometers. With dispersive instruments, it is important that the sample image fill the maximum *width* of the slits. Energy lost at the top and bottom of the slits is essentially wavelength independent and can be compensated for, using the trading rules. If the full width of the slits is not filled, however, the energy loss will be wavelength dependent and will cause a sluggish or dead pen in some areas, as well as possible background irregularities. Thus, the resulting spectrum may be badly distorted and therefore useless for any purpose. This factor places a limitation on the minimum microcell size, and best results will be obtained if the sample aperture in the microcell matches the beam image at the slit (i.e., rectangular).

For FTIR instruments, the beam is round and therefore the increased sampling area should be round or square, but not rectangular. The optimum size for the cell aperture has been calculated by Hirschfeld (95) as

$$r_{min} = r_o \left[(1/4)(1/SNR)(A/\Delta A) \right]^{1/4} \tag{5}$$

where r_{min} is the optimum r corresponding to the minimum weight of sample; SNR the signal to noise ratio in the 100% line; and $A/\Delta A$ the ratio of absorbance to absorbance noise for the sample absorptions. As an example, if $r_0 = 2.0$ mm, SNR $= 10^4$/1, and $A/\Delta A = 20$, detection limits in the nanogram range can be achieved for strong absorbers, and $r_{min} = 0.3$ mm. Hirschfeld summarizes by saying, "To optimize microsampling detection limits in an FTIR instrument, one must thus use the strongest possible focus microsampling optics, and use a cell whose clear aperture is that of Eq. (5)."

3.5. Microsampling

The major limitations of most traditional IR microsampling procedures have been listed by Coates (18). They are low energy throughput, which degrades S/N; distortion of the sample beam, which can cause artifacts in the spectrum; and localized heating, which may cause losses or changes in the sample. To these we add a fourth: difficulty in handling small samples without losing or contaminating them.

Obtaining an IR spectrum on a single fiber, a microscopically small paint chip, or a tiny spot of discoloration on a semiconductor device presents a challenging problem. Manipulation of such samples should be minimized to avoid contamination or loss of sample. However, some handling will undoubtedly be necessary, and it is important that the analyst possess the skills necessary to work effectively with microsamples.

Such skills might well be learned in the microscopy laboratory and developed by making several trial runs on known samples which have a character similar to that of the unknown. Humecki et al. (120) suggest using the stereomicroscope and sharpened tungsten needles to place particles less than 200 μm in size on a salt window. The manipulation of nanogram-sized samples for IR spectroscopy is also discussed by Cournoyer et al. (121). They recommend that the sample be prepared in such a way that it is only a few μm thick. Many sample preparation techniques are possible; the one chosen is determined by the nature of the sample and by the preferences and skills of the analyst. The sample is more often too thick than too thin, so one generally tries to make it as thin as possible. It can be placed on a microscope slide and a fine-pointed probe tip rolled over it, or it can be pressed between two microscope slides. It is then peeled from the slide with a solvent-washed razor blade and mounted. Or, the sample may be pressed between two freshly cleaved thin salt plates, and gentle pressure applied with a probe. When the plates are separated, the sample usually adheres to one plate.

To mount the sample, Cournoyer et al. suggest placing the sample side of the salt plate against the aperture disc so that the sample sits in the aperture rather than some distance above it. The salt plate may be secured to the aperture disc with a bit of soft wax, applied with a fine-pointed probe.

A heavily filled polymer fragment can be subjected to a micropyrolysis procedure to separate the polymer from inorganic pigments or fillers. The procedure is described by Humecki (120) (Reproduced from K. F. J. Heinrich, Ed., *Micro-*

beam Analysis—1982, San Francisco: San Francisco Press, Box 6800, San Francisco, CA 94101-6800):

Borosilicate capillaries of 0.2 to 0.5 mm i.d. and 3 to 6 cm long are used. A short length of borosilicate glass wool is twisted and inserted into the tube while it is observed under the microscope. The glass wool is fused to the inside by gently playing of the flame of a hobbyist minitorch near, but not at, the end of the tube. The glass wool is trimmed to form a brush. A drop of ethylene chloride is applied to the opposite end of the tube and the brush end is touched to an absorbant surface to draw out the solvent. This cleaning operation reduces background contamination.

A few sample particles are pushed into the open end of the capillary tube with a tungsten needle or wire to a depth of 2 to 4 mm. The tube end is sealed with a torch and the flame is played gently over the area containing the sample until sample pyrolysis is complete. When the heated end has cooled, it is broken off and a drop of solvent is applied while the capillary is held in a horizontal position. The capillary is brought into position over the salt plate and tipped so that the liquid washes over the pyrolysis condensate and flows into the brush. The brush is touched lightly to the salt plate. As the solvent evaporates, the brush confines the liquid to an area less than 200 μm in diameter by capillary action.

The capillary brush technique can be used for depositing any liquid in a restricted area. To restrict spreading of the liquid or to minimize the production of interference fringes in the spectra, a pinhole well is sometimes made in a salt plate. The well is made by gently twisting a tungsten needle into a salt window. The needle must have a stout point, not the fine delicate point used for small particle handling. Typically, a hole with a depth of up to 50 μm is produced. The powdered salt from the piercing operation is pushed back into the hole and tamped down, and the liquid sample or solution is then deposited from the brush in the powder filled well. Scattering of the energy by the powdered salt is not a problem because the powder is wetted by the liquid to produce a transparent or almost transparent sample site.

Good spectra have been obtained from only a few micrograms of a carbon-black-filled polymer with this procedure.

A beam condenser or IR microscope is essential for maximizing throughput. The sample must be masked using a pinhole or other aperture to prevent unwanted source radiation from overwhelming the absorption from the sample. Infrared microscopes are available for both transmission and reflectance studies.

3.5.1. *Transmission Measurements*

The optimum size or thickness for a microsample may often not be available— one takes what one can get—but it is important to understand the limitations imposed by the size and shape of the sample. Normally, for the best IR spectrum, one works with samples that are 0.001–0.01 mm in thickness, that is, 1–10 μm. The smallest diameter sample that can be dealt with comfortably is about 25 μm, although samples as small as 10 μm can be accommodated if one is willing to accept some fall-off of intensity with wavelength because of diffraction effects (at 10 μm, the sample diameter is less than the wavelength of radiation over part of the spectrum). Thus, a 1 × 10 × 10 μm sample is about the minimum size that can be scanned, and then only with a great deal of difficulty. If the particle has a density of 1.5, its total mass is 0.15 ng, which represents the approximate lower limit for this type of sample. The spectrum of a 0.9-ng sample of cellulose acetate, as obtained by Cournoyer et al. (121), is shown in Fig. 43. The sample contains 10% triphenyl phosphate, which is not visible in this spectrum but can be seen in a spectrum obtained from 96,000 scans co-added (approximately 40 h scanning time).

Usually, the sample will not have optimum dimensions; it will more likely be an irregular or spherical particle, or a fiber. If it is a 30-μm-diameter sphere, for example, it is wide enough, but may be too thick to provide a recognizable spectrum. Certainly, a particle larger than 100 μm will not give a useable result. Similarly, fibers may have a cylindrical shape that is at once too narrow and too thick for the desired purpose.

It is sometimes possible in these cases to flatten the sample by pressing it between the polished and hardened ends of steel pins, or by placing it directly in a diamond cell, where it can be compressed between the windows to a suitable thickness.

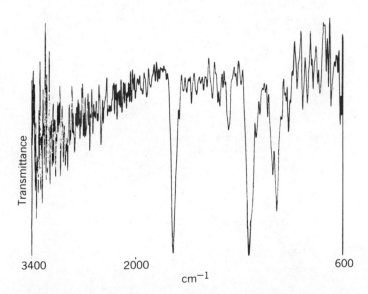

Fig. 43. Spectrum of 0.9 ng cellulose acetate containing 10% triphenyl phosphate in a 50-μm aperture, scanned 2000 times at 8 cm⁻¹ resolution. Reprinted with permission from Cournoyer et al., *Anal. Chem.*, **49**, 2275. Copyright 1977 American Chemical Society.

When a beam condenser is used, the sample can be mounted on a precision aperture (these are available commercially at reasonable cost in sizes ranging from 1 to 100 μm), perhaps backed with a thin sheet of KBr that is glued to the aperture disc. The disc may be placed in a micropellet holder for ease of handling.

Infrared microscopes are available for use with samples too small for beam condensers. We have already described an integrated microscope-filter photometer for transmission measurements (Section 2.3). In addition, other microscopes are available that fit into the sample compartment of a conventional dispersive or FT spectrometer (122,123). One microscope that can be used in either the transmission or reflectance mode has a 32 × all-reflecting Cassegranian objective (122). A separate optical viewing system permits masking of the sample area at a secondary focus, precluding the need for placing apertures or masks on the sample itself. The sample also may be photographed through the optical viewer. The sample image is focused on a cooled MCT detector having a 0.25-mm-diameter target. The small detector element gives reduced noise (see Section 2.1).

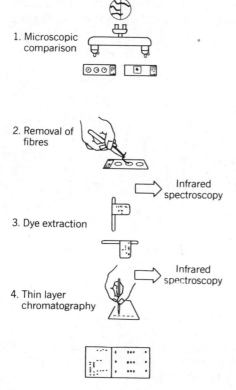

1. Microscopic comparison

2. Removal of fibres

Infrared spectroscopy

3. Dye extraction

Infrared spectroscopy

4. Thin layer chromatography

Fig. 44.   Fiber examination procedure (124).

Spectra of samples as small as 10×10 μm can be recorded, but better results are possible with slightly larger samples (20×20 μm).

Techniques for preparing microscopic samples of textile fibers for forensic identification are discussed by Cook and Paterson (124). The general procedure for fiber examination is outlined in Fig. 44. If the fiber is lightly dyed or undyed, it may be examined after the initial microscopic comparison. Otherwise, the dye is first extracted with a suitable solvent and identified by TLC. Two methods of handling fibers are discussed. In the first, about 3 μg of fiber is dissolved in 2–3 mL of an appropriate solvent (see Table 8). The solution is placed on a surface giving easy release (such as silane-treated glass) and the solvent evaporated. The fiber is sandwiched between powdered KBr charges in a 1-mm KBR pellet die and pressed to give a transparent pellet. A $6 \times$ beam condenser is used with a dispersive spectrometer. In the second method, the fiber is flattened between steel dies using 8 tons pressure before being cut into 0.25-mm lengths and, with the aid of a dissecting needle, transferred to a 0.25-mm aperture cut in a piece of 0.1-mm lead foil. A small amount of powdered KBr is placed over the sample and pressed at 4 tons pressure for 15 s. The lead-foil method was found to be more sensitive and gave fewer problems than the cast-film method, but demanded a high degree of manipulative skill. The authors comment that most persons could produce acceptable results on 1-μg samples within a day, but more practice was required to produce spectra of a consistently high quality.

For particulate or fibrous samples weighing 1 μg or more, the diamond cell provides a convenient method of mounting. A high-pressure diamond microcell is available commercially. The sample is mounted on one of the diamond faces

Table 8. Preparation of Solvent Cast Films from Fibers[a]

Fiber Type	Solvent	Evaporation Time (70°C Oven)	Solvent Removal
Polyacrylonitrile	Dimethylformamide	From 20 min[b]	Boil film in water, then dry in 70°C oven
Modified polyacrylontirile	Dimethylformamide		
Polyamide	m-Cresol	From 45 min[b]	Place film in cold ethanol, then dry cold
Polyester	m-Cresol	From 45 min[b]	
Cellulose acetate	Acetone	5 min	No further treatment needed
Cellulose triacetate	Chloroform	40 min	

[a]From Cook and Paterson (124).
[b]Depends on film thickness.

and the cell is then closed and compressed. The objective here is not to pressurize the sample, but rather to squeeze it to a useable thickness for transmission spectroscopy. So-called Type II-a diamonds are quite transparent in the finger-print region, although because of the small aperture, a beam condenser must be used with the cell. Under these circumstances, transmission of the cell may be only 5–10%, which means that the cell must be carefully aligned in the sample beam, and trading rules must be used to set scanning parameters properly. The diamond cell has been especially useful for forensic spectroscopy (99). Samples of paint chips, plastic fragments, foam rubber, and single fibers have been found to yield excellent spectra with minimal sample preparation (Fig. 45). The use of diamond cells has been reviewed by Ferraro and Basile (125).

Samples can also be prepared by conventional microtoming. Thus, Shearer (126) has reported the analysis of layers found in a small paint chip taken from a fifteenth-century painting. The chip was mounted in epoxy and microtomed so as to diagonally cut across the paint layers, thus exposing the maximum surface of each layer for analysis. The sample was positioned using the optical microscope, and the focused and diaphragmed IR beam transmitted through each layer in turn. Good quality spectra were obtained which were used to identify not only the binder but also the pigment in each layer.

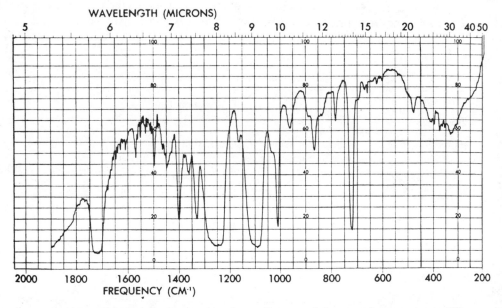

Fig. 45. Diamond cell spectrum of a polyester fiber, recorded on a grating spectrometer (99).

3.5.2.   Reflectance

If the microsample is present on a reflecting substrate, specular reflectance in a reflecting microscope can be used. Because the radiation passes through the sample twice, the sample should be only half as thick as for a normal transmission measurement. Reflection from the front surface of the specimen may distort the spectrum.

3.6.   Computerized Spectroscopy for Trace Analysis

Dedicated microcomputers are a necessary and integral part of most FTIR spectrometers and provide an opportunity for the analyst to perform arithmetic operations such as addition, subtraction, smoothing, and the like on the spectra. The fact that dispersive spectrometers, when properly interfaced to a computer, can carry out the same operations, is sometimes overlooked. Our discussion, therefore, generally applies to both types of instrumentation, with the caveat that FTIR often can do more spectacular manipulations because of its higher S/N ratio and better wavelength reproducibility.

3.6.1.   Signal Averaging (Coaddition)

We have noted that S/N for a spectrum is proportional to the square root of the measurement time. For the FTIR, where the scan rate of the moving mirror is fixed by limitations of the detector, the only practical way of doing longer scans is to scan repetitively and add the results (usually, to save doing multiple transforms, interferograms are added and the final result is transformed).

It should be noted that if S/N at the centerburst is high enough to exceed the dynamic range of the analog-to-digital converter (ADC) that is used to digitize the interferogram, signal averaging will *not* increase S/N. The least significant bit of the ADC must be smaller than the noise level, that is, noise must be present in the digitized interferogram. If it is not, the signal is said to be *digitization noise limited*, and a *decrease* in signal can lead to increased sensitivity (2,58,96). Thus, for high transmission background situations, one must take care not overload either the detector or the ADC, and in some cases it may be necessary to stop down the entrance aperture or use other methods to reduce the energy flux.

3.6.2.   Smoothing

Random fluctuation in the spectrometer signal, or noise, arises from many sources, including thermal motion of electrons in the recorder elements and statistical thermal fluctuations in the circuitry (15). Smoothing is the process of reducing noise by averaging the spectrometer response across a limited wavelength interval.

Dispersive spectrometers that record spectra in real time use an electrical RC network and/or mechanical filtering to reduce higher frequency noise in the pen servo loop. Several filter configurations are available (127); they all distort the spectrum to some extent, such that resolution becomes poorer (peak intensities of sharp bands becomes less) and there may be a shift in peak position in the direction of scan (Fig. 46). Such effects are minimized by reducing the scan rate, and in fact the maximum scan rate is determined by the extent of noise filtering. It is usually recommended that the scan rate be no greater than 0.2 half-band widths per response period. (The pen response speed, of course, is a measure of the degree of filtering. A fast response implies little filtering, while a slow response implies more filtering.)

Computerized spectrometers utilize digital filtering, and the smoothing can be accomplished in a slightly different way. Usually a quadratic least-squares smoothing technique is used (129). Because raw data are available on both sides of the center point of the smoothing interval (not just on one side, as in a scanning dispersive spectrometer), twice as many points are available for smoothing. Distortions on a dispersive spectrometer are much less than are found with an RC network filter, and may in fact, be negligible as long as the smoothing interval is restricted. Using a large number of smoothing points gives a distorted spectrum, however. Best results are obtained if the data are digitized at high

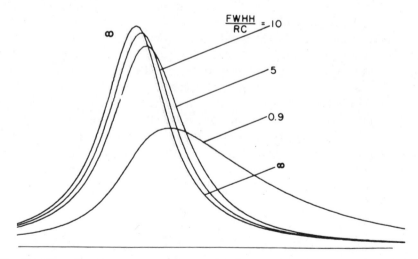

Fig. 46. Distortion in peak intensity and position caused by filters (128). Effect is shown for several values of FWHH to RC.

densities (points close together) and no more than one inflection is included in the smoothing interval. In FTIR spectrometry, only two data points are taken for each resolution element, so smoothing has a more pronounced effect on the spectrum. Griffiths (130) recommends that, rather than use smoothing, one measure the spectrum at lower resolution with a larger number of scans.

In other words, nothing is gained by including more points than are used to define the band of interest. The use of wider intervals tends to degrade resolution and distort the spectrum. With a FTIR spectrometer at 4 cm^{-1} resolution, for example, a band 10 cm^{-1} wide should have no more than a 3-point smooth. The effect of the number of smooth points on a moderately sharp band is shown in Fig. 47. A more complete discussion of filtering and smoothing of spectra is given by Willson (128).

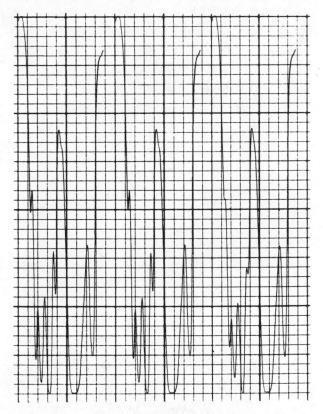

Fig. 47. Smoothing effects. (*Left*), No smoothing; (*center*), 13-point smooth; (*right*), 25-point smooth.

3.6.3. *Absorbance Subtraction*

The technique of spectral stripping or absorbance subtraction is a computerized method of removing the spectrum of a material from that of a mixture. It was first reported for FTIR systems in 1975 (131); prior to that time, the same operation could be accomplished optically in double-beam dispersive spectrometers, by using the proper thickness of the pure component in the reference beam. Because the thickness had to be exactly correct to achieve complete cancellation, the technique was not easy to carry out. With computerized instruments, the proper scaling factor can easily be selected. In theory, then, absorbance subtraction with a computerized spectrometer ought to provide a quick, sensitive method for detecting small differences between samples, and for quantitatively recording the difference.

In practice, all sorts of gremlins lie in wait to cause problems for the naive or inexperienced (132). One finds, on consulting the literature, that pursuit of these gremlins has provided many a happy hour for spectroscopists, and indeed

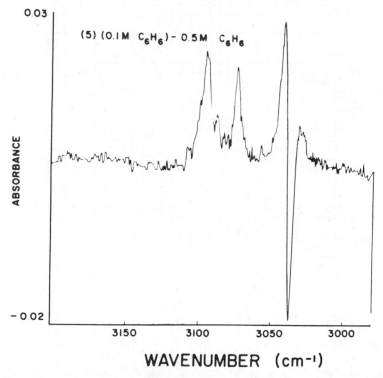

Fig. 48. Residual perturbation spectrum of C_6H_6–CCl_4 solution resulting from a concentration change from 0.1 to 0.5 M (133).

some fairly rigid requirements must be met for success with this technique.

Among the worst of the problems is the failure of Beer's law, especially at higher concentrations. Figure 48, taken from Ref. 133, shows the residual spectrum from two solutions containing the same number of molecules but having different concentrations and path lengths. Two of the bands have changed intensity with the concentration change; the third has shifted in wavenumber and gives a derivative-like curve. Gas spectra (Fig. 49) show analogous results because of the difference in line broadening at different pressures.

Instrumental requirements for absorbance subtraction are discussed by Hirschfeld (134), from which Figs. 50 and 51 are taken. It is apparent that excellent wavenumber reproducibility, especially at higher background absorbances, is essential, because small band shifts cause spurious peaks. Hirschfeld has also shown that samples that vignette the sample aperture can cause wavenumber shifts (135), and he makes the point that both of the samples being compared must have the same or nearly the same (within 1 or 2%) vignetting properties. A high S/N ratio is also essential, and Figs. 50 and 51 can be used to set an upper limit on the tolerable background absorbance for identification or detection of the analyte. An automated procedure for successive subtraction of known spectral components is described by Gillette and Koenig (136).

Another source of artifacts is effects from the instrument line function (137). Results are also sensitive to the apodization function (138).

As suggested above, intermolecular interactions are a principal source of

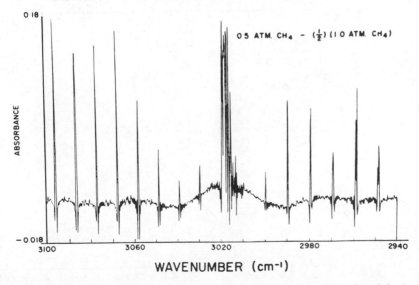

Fig. 49.   Residual perturbation spectrum of CH_4 resulting from a pressure change from 0.5 to 1 atm (133).

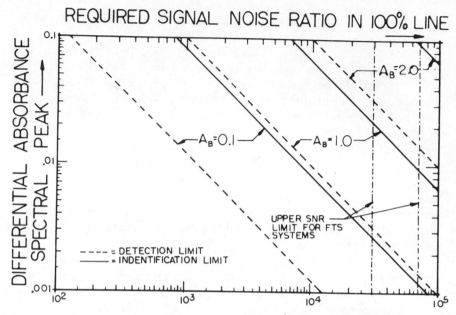

Fig. 50. Required S/N ratios for absorbance subtraction detection and identification at 0.1, 0.2, and 2.0 background absorbances (134).

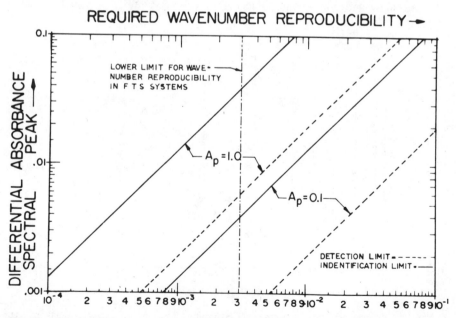

Fig. 51. Required wavenumber reproducibility for absorbance subtraction detection and identification for 0.1 and 1.0 background absorbance peaks (134).

subtraction artifacts, and the artifacts can in fact be used to study such interactions. However, one must be careful to distinguish between true interactions
and optical phenomena arising from dispersion effects (139). The apparent anom-
aly results if the two pure components have different refractive indices, because
of the pronounced effect of the real part of the refractive index on a strong
absorption band (Fig. 52). Small shifted residual peaks can in this case be
attributed solely to optical effects (Fig. 53).

The requirement of high S/N means that one must be very cautious about
working with spectra having absorption bands of intensity greater than $A = 0.7$.
Regardless of the type of spectrometer one is using, when the sample beam
energy drops below about 10% T ($A = 1$), the instrument response becomes
significantly poorer, and good subtractions become difficult or impossible to
achieve. Figure 50 may be used to estimate the S/N ratio requirements for
different values of background and sample absorbance. Clearly, bands hidden
under strong solvent absorptions (as in HPLC/IR) cannot be visualized by absorbance subtraction procedures.

Thus, while absorbance subtraction can be an extremely useful technique, it
must be approached with caution. The spectra that are to be subtracted must be
scanned under as nearly the same arrangements as is possible; that is, using the
same cell, identical positions of the samples in the beam, low concentrations if
solutions are used, identical sample temperatures and all instrumental conditions,
including purge rates, the same. Better results are obtained if the scaling factor

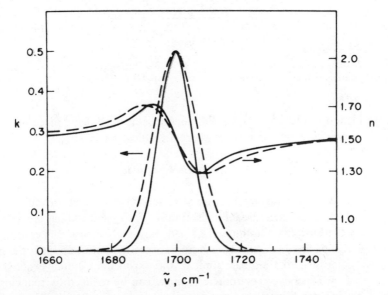

Fig. 52. Absorbance (k) and dispersion (n) curves for a hypothetical oscillator of 1700 cm^{-1} with
a Gaussian line shape, for 15-cm^{-1} line width (solid) and 20-cm^{-1} line width (dashed) (139).

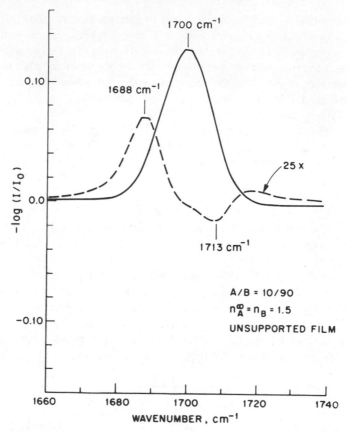

Fig. 53. Absorbance spectra for an unsupported 10/90 A/B film with $N_a = N_b = 1.5$, for mixture spectrum (solid) and difference spectrum (dashed) ($\times$ 25) (139).

is close to unity (132,133). For solid samples, the thickness, placement, and orientation are factors critical to the success of the operation.

3.6.4. *Ordinate Expansion*

The practice of expanding the transmission scale is extremely useful in trace analysis, as it can produce useable absorptions from bands which are virtually invisible in the base line. Nothing is gained, however, unless S/N is such that the analyte bands are distinguishable from background noise, and it is the noise level that determines the ultimate limit of the expansion. We have noted previously that in dispersive spectrometers, noise can be reduced by reducing the gain in the pen servo loop, and restoring energy by opening the slits. In any

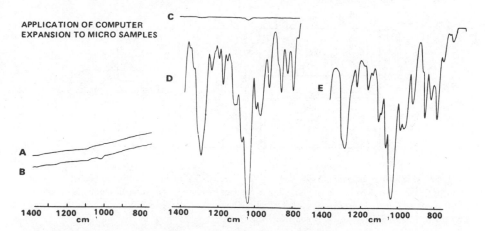

APPLICATION OF COMPUTER
EXPANSION TO MICRO SAMPLES

Fig. 54. (A) Background trace from ATR element; (B) 1 μg Birlane deposited on ATR element; (C) Computer-calculated difference (B − A); (D) Spectrum (C) expanded 200-fold and smoothed; (E) Reference ATR spectrum of Birlane. Reprinted with permission from J. P. Coates, *Amer. Lab.*, **8**(11), 67 (1976). Copyright 1976 International Scientific Communications, Inc.

spectrometer, S/N can be increased by longer data collection time, T, with S/N proportional to $\sqrt{T}$ (except as discussed in Section 3.6.1). Some additional gain in S/N can be had by smoothing, if the loss in resolution is not critical. Thus, expansion factors of several hundred are sometimes possible (see Fig. 54). As one might anticipate, however, an upper limit exists for expansion, even with FTIR. The noise in the detector, digitizer, and computer will, if sufficiently amplified, eventually overwhelm a weak signal. With an upper limit for S/N of 10^4 to 10^5 for FTS systems (Fig. 50), the theoretical expansion limit to attain a 10% T signal is around 10^3 to 10^4.

3.7. Correlation Spectroscopy

This technique has been applied to trace gas analysis to achieve enhanced sensitivity in qualitative analysis and increased precision in quantitation. It is applicable either to absorption or emission spectra. In effect, what is done is to relate the entire spectrum, rather than a single peak, to a reference spectrum of the analyte. This operation can be done optically or mathematically, depending on the instrument configuration. It is claimed that reliable detection and quantitation can be obtained at S/N levels less than unity. A nondispersive gas filter analyzer such as is shown in Fig. 55 provides an example of an optical correlation spectrometer. Here the radiation passes alternately through a reference cell that contains a high concentration of target gas, and a sample cell and a neutral density filter that have about the same overall transmission as the reference cell.

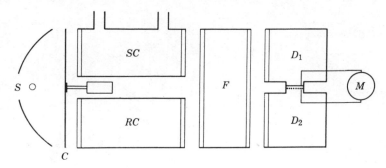

Fig. 55. Gas filter analyzer. (*S*) IR source; (*SC*) and (*RC*) sample and reference cells; (*F*) filter cell; (*D₁*) and (*D₂*) detector cells containing a condenser microphone between them; (*M*) readout. From *Applied Infrared Spectroscopy*, A. L. Smith. Copyright © 1979 John Wiley & Sons. Reprinted by permission of John Wiley & Sons, Inc.

The detector is a microphone in a chamber that contains an optimum amount of the target gas. When no target gas is present in the sample cell, the two beams are balanced and there is no signal at the detector. A small amount of target gas in the sample cell will cause a reduction in transmission in the sample path, but will not affect the beam passing through a reference cell. The detector thus sees a signal whose magnitude depends on the concentration of gas in the sample cell. Absorbers other than the target gas do not cause any signal at the detector, unless there is an accidental overlap of absorptions between the target gas and the extraneous material. In this case, the interfering gas can be placed in the filter cell where the troublesome wavelengths will be subtracted from both beams equally.

Correlation interferometry can be done with the FTIR spectrometer without optical modification. The interferogram, or a portion of it, is compared with a correlation function that resembles the interferogram of the pure analyte gas. Correlation spectroscopy is discussed in detail by Wiens and Zwick (140,141).

Another mathematical approach to correlation spectroscopy is the least-squares method of Haaland and Easterling (142). The digitized information from a computerized dispersion or FTIR spectrometer is correlated with the spectra of the analyte through one of several possible least squares approaches, the particular method chosen depending on the behavior of the baseline of the sample and reference spectra. The authors claim sensitivity enhancements of 5–12 in detection limits over conventional IR gas analysis methods.

4. APPLICATIONS

4.1. Atmospheric and Environmental Analysis

Because IR spectroscopy is so highly specific and is adaptable to the analysis of many types of materials, much effort has been devoted to overcoming its relatively poor sensitivity and lack of portability. As a result of the success of these efforts, the techniques outlined in this chapter have given the analyst an important tool for his analytical toolbox, concurrent with a rapidly expanding interest in, and need for, sensitive, accurate, and specific means of monitoring environmental contaminants. Although laboratory analysis gives better control over many analytical variables, field analysis is often necessary, and portable IR spectrometers make such analyses possible.

4.1.1. Breathing Zone Analysis

Several options are available for atmospheric monitoring of breathing zones in an industrial setting. To monitor the concentration of organic solvents, the Occupational Health and Safety Administration (OSHA) and the National Institute of Occupational Safety and Health (NIOSH) recommend use of trapping tubes containing activated charcoal, through which a known volume of air is drawn at a fixed rate (143). The organic vapors are trapped by the charcoal, and are subsequently desorbed and analyzed by GC/MS or IR spectroscopy. Use of CS_2 as both an extracting medium and an IR solvent has been reported (144). Best recoveries were obtained when both the charcoal and the CS_2 were cooled with liquid N_2 before extraction. The 150-mg charcoal charge was extracted with 0.5 mL CS_2, which was then placed in a microcell having a 6-mm path length and a minimum dead volume. Total cell volume was 0.43 mL. Normal samples containing 10–60 μg of adsorbed material could be analyzed qualitatively and quantitatively. Mixtures could also be analyzed, and if desired, the CS_2 solution could later be analyzed by GC also. A few cases of reaction on the charcoal were reported, suggesting that some preliminary trials might be appropriate.

Localized air analysis for industrial hygiene monitoring of factory environments can be accomplished with simple portable IR spectrometers (145,146,147). Sensitivities in the sub-ppm range are attained for some contaminants. Qualitative analysis on a limited scale is also possible.

The 8-h maximum allowable average exposure for nickel carbonyl (an intermediate in the synthesis of acrylate monomers and in nickel refining processes) has been set at 1 ppb by OSHA. An FTIR method has been developed which relies on long-path sampling and absorbance subtraction to detect this toxic

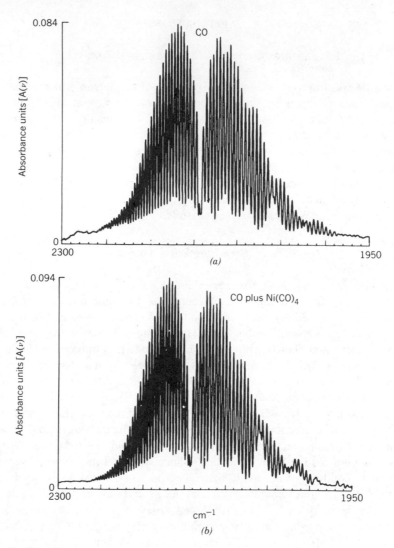

Fig. 56. (*a*) CO reference spectrum; (*B*) CO containing 0.6 ppb Ni(CO)$_4$; (*c*) result of subtracting (*a*) from (*b*) showing Ni(CO)$_4$ band at 2057 cm^{-1} (148).

material in the presence of interfering CO at the 1000-ppm level (Fig. 56). A Teflon-lined 10-m-path multipass sample cell was used. The spectrometer was operated at 2 cm^{-1} resolution with a cooled MCT type detector. Because of the large dynamic range requirements, double precision software (32-bit word length) was necessary (148).

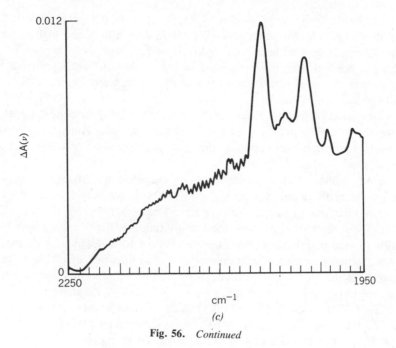

Fig. 56. *Continued*

4.1.2. Flue Gas Analysis

A CO_2 laser analyzer for NH_3 in flue gas exploited an accidental overlap between the R(18) emission line of a $^{13}C^{16}O_2$ laser and the $\nu_2 aQ(6.6)$ absorption line of $NH_3(33)$. The laser beam passed through the flue duct and any absorption was detected by comparison with a coincident beam from a second $^{13}C^{16}O_2$ laser R(14). Signals from the two lasers were separated electronically by chopping them at different frequencies. Interferences from H_2O and CO_2 in the flue gas were negligible at the concentrations measured (75–500 ppm V).

4.1.3. Laboratory Atmospheric Analysis

A condensation method for concentrating air pollutants is described by Hanst (108), wherein air was condensed in a 3.5-L cryocondenser. After the condenser was filled, O_2 and N_2, still at 78 K, were pumped off. This condensation–concentration process was typically repeated 16 times—a four-hour procedure. The concentrated residue was finally volatized into a long-path IR cell. Cells of

10 to 120 m in length were used with dispersive or FTIR spectrometers. Concentration factors on the order of 3000 to 5000 were found, although factors of up to 10^6 were reported later (101). Most materials, with the exception of CO, CH_4, and possibly NO, were retained in the residue without significant loss. Sensitivities down to partial pressures of 10^{-11} atm were found, with typical results as noted in Table 7.

A limitation of this procedure is the possible interaction between analytes: ozone and NO, HCl and H_2O, HCl and NH_3, and the like. However, for hydrocarbons, chlorinated hydrocarbons, and nonacid carbonyl compounds, it appears to give valid results.

Use of a tuneable diode laser source in conjunction with a multi-pass gas cell gave sensitivity for SO_2 of 0.2 ppb (149). Laboratory studies of photochemical reactions in the atmosphere have been described by Hanst and Gay (104). The reactor (Fig. 57) was used as a multipass IR cell with a path length of 504 m, and permitted studies of ppm levels of formaldehyde, chlorine, and NO_2 in 1 atm of air under the influence of a controlled flux of UV light. An IR spectrum of the reaction products is shown in Fig. 58, and their concentrations as a function of time are shown in Fig. 59. When NO_2 was in excess over Cl_2, O_3, CO, HNO_3, N_2O_5, HCl, and nitryl chloride ($ClNO_2$) were found. When O_2 was the major species, the principal reaction product was peroxy nitric acid ($HOONO_2$). The half-life of this material was about 10 mins. Formyl radicals (HCO) reacted with O_2 to give CO and HO_2, the latter then adding to NO_2 to give $HOONO_2$.

Many of the atmospheric pollutants of most interest are nonseparable and/or transitory. These include O_3, H_2O_2, peroxyacetyl nitrate, HCHO, NO, NO_2, N_2O, and NH_3. Here, measurements must be made directly on the atmosphere,

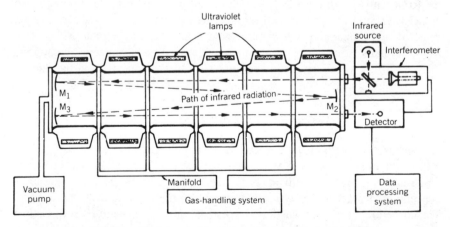

Fig. 57. Photochemical reactor and IR detection system (105).

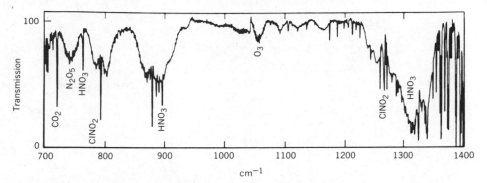

Fig. 58. Products of the reaction of 3.6 ppm NO₂, 2.4 ppm H₂CO, and 1.0 ppm Cl₂ in 1 atm of dry air photolyzed 12 min in the glass reaction chamber. The IR absorption path was 504 m (105).

either with a long-path cell through which ambient air is circulated, or with a remote source, retroreflector, or LIDAR system.

Since the supply of atmosphere is essentially unlimited, very large cells can be used. Multireflection cells having path lengths up to 1 km have given sensitivities in the low ppb range for many pollutants (150). Sub-ppb concentrations of most pollutants are said to be detectable with 50 passes through a 10-m cell (104); these levels are well below the usual ambient levels and thus should provide adequate sensitivity for most measurements. Hanst (101) has reviewed trace pollutant gas analysis in the atmosphere.

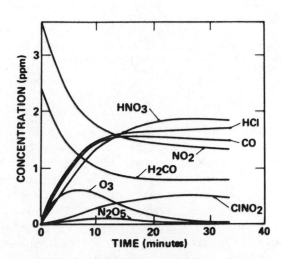

Fig. 59. Changes in concentration of reactants and products during irradiation of 3.6 ppm NO₂, 2.4 ppm H₂CO, and 1 ppm Cl₂ in 1 atm of dry air (105).

Table 9.   Sensitivities[a] of Analytical Lines[b]

Molecule	Frequency (cm^{-1})	Sensitivity (ppm)
O_3	1025.08	33.0
	1026.52	29.0
	1031.00	21.0
	1034.84	23.0
	1038.53	25.0
	1047.61	21.0
N_2O	1278.11	11.0
	1292.29	10.0
	1293.89	8.9
	1295.48	9.1
	1297.05	9.4
CO	2111.55	5.1
	2115.63	4.4
	2119.68	4.0
	2123.70	4.0
	2127.69	4.3
	2154.60	4.6
CH_4	1283.45	27.0
	1303.71	18.0
	1306.11	14.0
	1327.06	14.0
	1332.71	15.0

[a]Concentration that produces 0.01 absorbance for a 1-m path, neglecting the effects of interferences.
[b]Reprinted with permission from Golden and Yeung, *Anal. Chem.*, **47**, 2132. Copyright 1975 American Chemical Society.

Choice of the absorption line for analysis of a given material is not always obvious. It should, of course, be as intense as possible, but it should also be relatively free of interference from other substances that may be present. The best lines for IR analysis of common pollutants have been listed by Golden (151), who systematically calculated the extent of interference for six molecules and critically evaluated the results. A temperature of 25°C, pressure of 1 atm, and resolution of 0.01 cm^{-1} were assumed. Sensitivities of the selected lines for O_3, N_2O, CO, and CH_4 are given in Table 9, and the contribution to the total absorbance by each species at the best analytical wavelength is given in Table 10.

Table 10.   Percent Contribution to Total Absorbance at AL Frequencies[a]

	AL frequency (cm^{-1})	H_2O	CO_2	O_3	N_2O	CO	CH_4
O_3	1025.08	0.0	0.0	99.9	0.0	0.0	0.0
	1026.52	0.0	0.0	99.9	0.0	0.0	0.0
	1031.00	0.0	0.0	100.0	0.0	0.0	0.0
	1034.84	0.0	0.0	100.0	0.0	0.0	0.0
	1038.53	0.0	0.0	100.0	0.0	0.0	0.0
	1047.61	0.0	0.0	100.0	0.0	0.0	0.0
N_2O	1278.11	33.6	0.0	0.0	63.9	0.0	2.5
	1292.29	33.2	0.0	0.0	65.6	0.0	1.3
	1293.89	32.9	0.0	0.0	64.1	0.0	3.0
	1295.48	34.4	0.0	0.0	62.7	0.0	3.0
	1297.05	35.3	0.0	0.0	62.4	0.0	2.2
CO	2111.55	1.3	68.9	0.6	0.0	29.2	0.0
	2115.63	1.2	65.2	2.4	0.0	31.2	0.0
	2119.68	1.1	63.4	3.6	0.0	31.9	0.0
	2123.70	1.4	63.7	3.0	0.0	31.8	0.0
	2127.69	1.0	67.2	1.5	0.1	30.2	0.0
	2154.60	0.7	74.1	0.0	0.2	25.1	0.0
CH_4	1283.45	22.2	0.0	0.0	2.7	0.0	75.1
	1303.71	16.9	0.0	0.0	2.5	0.0	80.6
	1306.11	18.3	0.0	0.0	3.0	0.0	78.7
	1327.06	18.6	0.0	0.0	0.0	0.0	81.4
	1332.71	22.1	0.0	0.0	0.0	0.0	77.9

[a]Reprinted with permission from Golden and Yeung, *Anal. Chem.*, **47**, 2132. Copyright 1975 American Chemical Society.

4.1.4.   Field Studies

Measurement of gases emitted from a factory or refinery is important from the viewpoint of reducing pollution and also of reducing economic losses from process vents and leakages. Both FTIR and laser spectrometers have been used for this purpose. Gas filter correlation spectroscopy has been used to measure CO from auto emissions in urban air, in the 0.02–20 ppmv range (152). Provision to correct for a small water interference was necessary.

A CO_2 laser spectrometer with retroreflectors was used to perform mass flow measurements on ethylene emitted from a petrochemical factory (Figs. 60, 61)

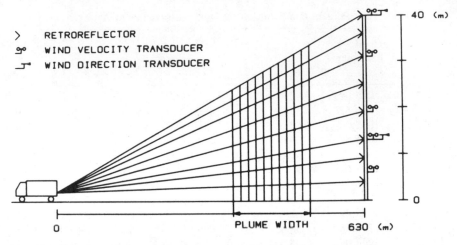

RETROREFLECTOR
WIND VELOCITY TRANSDUCER
WIND DIRECTION TRANSDUCER

40 (m)

0

0   PLUME WIDTH   630 (m)

Fig. 60. Side views of the experimental set-up used in the field experiment. The measurement volume is divided into segments for mass flow calculations.Reprinted with permission from Persson et al. *Appl. Opt.*, **21**, 4417 (1982).

(34). Total path length was about 1.2 km. Measurements were made at two wavelengths, one where ethylene has a large absorption cross section (the P-14 line), and the other at a wavelength of low absorption (the P-20 line). Intensities were time-averaged and compared, concentrations calculated, and wind velocity factored in, to obtain the total ethylene mass flow (found to be 120 kg/h in this case; cf. Fig. 62). This figure agreed with a mass balance deficiency calculated at the plant.

A mobile LIDAR system has also been used to measure ethylene distribution

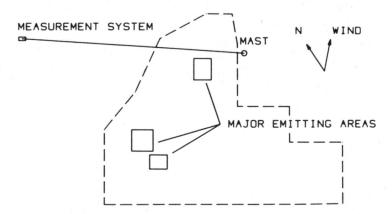

MEASUREMENT SYSTEM

MAST

N   WIND

MAJOR EMITTING AREAS

Fig. 61. "Top" view of factory area, showing sensor and plant locations. Reprinted with permission from Persson et al. *Appl. Opt.*, **21**, 4417 (1982).

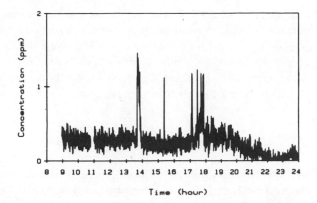

Fig. 62. Measured ethylene concentrations on Sept. 4, 1979. Reprinted with permission from Persson et al. *Appl. Opt.*, **21**, 4417 (1982).

over a refinery (35). This type of system is not, of course, tied to a retroreflector, so it is more flexible in both horizontal and vertical measurements. Concentrations of 10–220 ppb ethylene were found and plotted on a contour map of the area.

A specially modified interferometer spectrometer has been described for detecting and characterizing trace gases in a localized cloud, using emission (153). The cloud temperature must be different from that of the background for this method to succeed. The instrument measures the spacial contrast produced by the cloud by simultaneously observing the cloud and adjacent background. Zachor (154) has tabulated minimum detectable quantities for 14 trace gases analyzed with this system.

A ground-based spectrometer with a portable remote source has been used for remote sensing of source emissions, such as atmospheres over ponds, jet engine enhaust, oil refineries, and stack gases (155). The spectrometer, equipped with a 30-cm-diameter $f/5$ collecting mirror, is housed in a van, and is also capable of remote emission spectroscopy. Measurements are made at high resolution (0.5, 0.25, and 0.125 cm^{-1}) to minimize interferences and to improve detectability. Concentrations as low as 0.01 ppm of some pollutants could be detected.

Methane in air at the 1-ppm level was measured with an airborne nondispersive gas filter correlation analyzer (156) equipped with a multipass cell of 8.15 m total path.

Trace species in the stratosphere have been measured by a balloon-borne tuneable laser spectrometer in which four diode lasers with a path length of 1 km were designed to provide a detection limit of approximately 0.1 ppbv for NO, NO_2, HNO_3, O_3, and HO_2NO_2 (30). Lines used were: NO_2, 1604.172; NO, 1906.14; and HNO_3, 1722.00.

Nordstrom et al. (157) used FTIR with the sun as a source to observe CCl_2F_2 in an atmospheric transmission window. It was necessary to compute synthetic solar spectra to correct for pressure, temperature, and concentration variations of both the analyte and interfering substances such as H_2O, CO_2, O_3, CO, and NO_2 in the various layers of atmosphere. The final result indicated a concentration of 110 ppt of CCl_2F_2 in the Earth's atmosphere, that is, a density of 3×10^9 molecules cm^{-3}.

4.2. Soil and Water Pollutant Analysis

One of the principal engines driving the development of trace analysis methods has been the interest in determining and monitoring environmental contaminants. In fact, this field is now highly specialized, having its own practioners, literature, and, of course, myths, misconceptions, and poor science. Because of the medical, legal, and economic implications of the analytical data, it is most important that environmental data be soundly based and supportable. Toward achieving that end, the ACS Committee on Environmental Improvement has compiled a set of guidelines for conducting analytical measurements of environmental samples (7). These guidelines cover planning, quality control, verification and validation, precision and accuracy, sampling (including statistical design of the sampling process), measurements, and documentation and reporting. Much futile effort could be saved if environmental chemists followed these guidelines; they should be carefully studied by the analyst before any work is undertaken. The guidelines are particularly important for those analyses near the limit of detection, or analyses whose results are likely to have far-reaching economic, legal, or psychological implications.

Rarely will IR be the sole choice for an analytical method, but it will often be used in combination with GC, LC, and GC/MS (67). Whereas GC/MS is more sensitive than GC/IR, it is well known that MS has its own limitations. It usually cannot, for example, distinguish between positional isomers. Also, when two components of a mixture have the same retention time (a situation that occurs more frequently than is realized), GC/MS tends to give erroneous results. In one study, it was estimated that this situation would occur for 40 compounds of a 200-component mixture (158). Thus, GC/MS and GC/IR should be considered as complementary methods which can provide the best range of absolute identification of components. Since IR gives more definitive results with polar compounds and MS is better for nonpolar materials, the combination should give the best of both worlds. These techniques have in fact been combined in one set-up (68,69) and tandem GC/IR/MS will probably be a routine technique in the near future (70).

One of the first environmental problems to which IR was applied was detection of organic contaminants in waste water. A large concentration factor could be achieved by extracting a substantial volume of water with a small amount of an incompatible organic solvent, sometimes helped by salting-out the organic materials. The extract was scanned in the IR spectrometer, either directly or after evaporating the solvent. Problems with incomplete extraction and poor recovery, along with the need for better sensitivity, have led to more sophisticated approaches to water analysis. Better liquid–liquid extraction procedures, or use of a trapping resin, gives a larger concentration factor and better recoveries. Combinations of GC, MS, UV, and IR are likely to be used for the final analysis. Although much progress has been made in identifying trace species in water, current results are by no means infallible, and efforts to improve sensitivity and accuracy are continuing.

Analysis of hydrocarbons in water, using IR spectroscopy, UV spectrofluorometry, and GC has been studied by Ducreux et al. (159). They found that best sensitivity is attained by trapping the hydrocarbon on resin. Ten grams of Amberlite XAD-2 macroreticular neutral polystyrene resin was placed in a column and 10 L of water passed through. For IR analysis, the adsorbed hydrocarbons were eluted with 50 mL CCl_4, which was placed directly in a 1 to 5-cm-long quartz cell. The aliphatic hydrocarbon content was related to the CH stretching band at 2925 cm^{-1}.

A similar approach to identifying trace organics in water using FTIR is outlined by Gomez-Taylor et al. (160). Twenty-five liters of water, spiked with 2 mg L^{-1} of several organic chemicals, were passed through a column of Amberlite XAD-2 resin. The organic materials were adsorbed on the resin and subsequently eluted with 100 mL Et_2O, which was then evaporated to 1 mL. A 10-μL sample injected into a packed column GC/IR system gave recognizable spectra. A similar

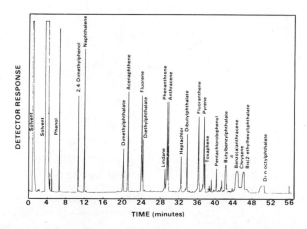

Fig. 63. Chromatogram of separated priority pollutants obtained with a FI detector placed behind the GC/IR light pipe (54).

test of 8 pesticides at the 50-ppb level (50 μg L^{-1}) showed recoveries for most species of 90 to 100%.

A method for analyzing water for anionic surfactants in the presence of interfering materials is given by Hellman (161). Concentrations as low as 40 μg L^{-1} could be analyzed.

Analysis of industrial waste water in the real world is more likely to be approached using a combination of methods. Shafer et al. (67) have detailed the analysis of effluent from a plant that produced nitrobenzene, dichlorobenzene, o-nitrophenol, aniline, and oil additives. The water was extracted with CH$_2$Cl$_2$, which was then evaporated to a concentration of 8 mg/mL of sample. They used GC/FTIR, HPLC/FTIR, and GC/MS. As expected, the methods proved to be

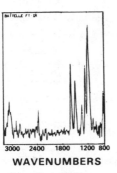

WAVENUMBERS

Scan Set: 876
Identification:
Phenol

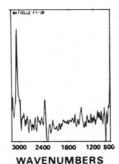

WAVENUMBERS

Scan Set: 1403
Identification:
Naphthalene

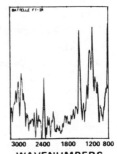

WAVENUMBERS

Scan Set: 1271
Identification:
2,4-Dimethylphenol

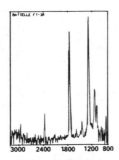

WAVENUMBERS

Scan Set: 2051
Identification:
Dimethylphthalate

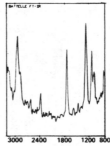

WAVENUMBERS

Scan Set: 4375
Identification: *
Di-n-octylphthalate
(Baseline corrected and smoothed once)

Fig. 64. GC/IR spectra of some priority pollutants separated by WCOT capillary chromatography (54).

complementary, with GC/MS showing best sensitivity, but unable to distinguish isomers of the polysubstituted benzenes. Several GC/IR identifications were tentative, taken alone, but, when combined with GC/MS date, were positive. HPLC/IR was a factor of 20 less sensitive than GC/IR, and was of limited value for this problem (if nonvolatile oligomers or polymers had been present, however, HPLC/IR could have provided valuable information).

A comparison of GC/IR with GC/MS for analysis of a mixture of 21 priority pollutants again demonstrated the complementary nature of these two techniques (54). A wall-coated open tubular (WCOT) capillary column gave adequate separation (Fig. 63). Injection of 8 μg/component for the GC/IR run resulted in reasonably satisfactory spectra (Fig. 64). Identifications based on computer search results are shown in Table 11.

An extract of a soil sample containing hazardous wastes with 44 components was analyzed by both GC/IR and GC/MS (57) on WCOT based SiO_2 capillary

Table 11. Identifications Based on Computer Search Results[a]

Compound	By GC/FT-IR[b]	By GC/MS[b]
Phenol	1	1
2,4-Dimethylphenol	1	1
Naphthalene	2	3
Dimethylphthalate	1	1
Acenaphthalene	1	1
Fluorene	No[c,d]	1
Diethylphthalate	1	1
Lindane	No[c,d]	1
Phenanthrene	No[c]	1
Anthracene	1	4
Heptachlor	No[c,d]	1
Dibutylphthalate	3	2
Fluoranthene	No[c,d]	2
Pyrene	No[c,d]	1
Toxaphene	No[c,d]	1
Pentachlorobiphenyl	No[c,d]	1
Butylbenzylphthalate	1	No
Benzo(a)anthracene	No[c,d]	1
Crysene	No[c,d]	1
Bis(2-ethylhexyl)phthalate	No[d]	1
Di-n-octylphthalate	1	3

[a]From Shafer et al. (54).
[b]Number indicates priority in list of possible compounds named.
[c]No GC/FT-IR spectrum obtained.
[d]Not in GC/FT-IR reference spectral library.

columns. In this study, which used a more sensitive GC/IR interface than was used in (54), 28 compounds were completely identified and 15 partially so by GC/IR. On the other hand, GC/MS gave 13 complete identifications and 23 partial ones. Eight compounds were found by IR that were not detected by MS, and one found by MS and not by IR.

A comparison of packed column GC/FTIR, fused silica capillary GC/FTIR, and fused silica capillary GC/MS on a soil extract consisting of organic materials, gave sensitivities roughly in the ratios 1:(2–8):(80–3200) (162). Success in detecting and identifying pollutants, however, was found to depend strongly on the class of compound, confirming the findings of other workers.

4.3.   Polymers

Trace analysis in polymer chemistry is usually undertaken to solve one of three possible types of problems: microsample identification—either a contaminant in a polymer, or a bit of polymer in or on some other material; the characterization of surface changes on a polymer as a result of oxidation or other chemical reaction; or the study of changes in crystallinity or structure induced by drawing, temperature changes, or some other physical process.

The contaminant problem is more easily solved if the contaminant is distributed throughout the matrix, rather than being an occasional isolated speck. In the case of catalyst residues in high-density polyethylene, simple absorbance

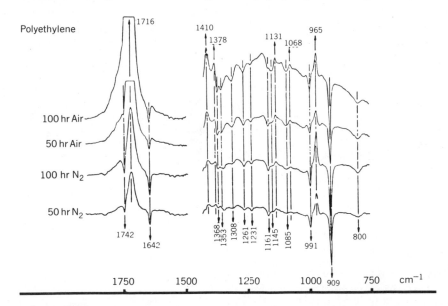

Fig. 65.   Difference spectra for polyethylene films irradiated in N_2 or air for 50 and 100 h (131).

subtraction of 1-mm-thick specimens gave a good measure of the SiO_2 catalyst support in production polymers, referenced to a SiO_2-free polymer. However, an even simpler technique involved measurement of the 470 cm^{-1} silica band above a base line drawn between two spectral minima (163).

A black particle 0.25 mm in diameter found in a molded plastic plate was sampled by microtoming the plastic and then mounting the speck on a micro holder with a suitable mask in a 6× beam condenser. The resulting IR spectrum, while rather nondescript, was a perfect match to that of a dead insect found in the storage bin that held the formulated polymer feed stock (97).

Surface photo-oxidation studies on polystyrene have been carried out using both ATR and transmission spectroscopy. Use of different incident angles with KRS-5 and Ge ATR elements gave a measure of oxidized species as a function of distance from the surface, attributed to slow oxygen diffusion into the interior of the polymer. Some chemical insights as to the nature of the oxidation reaction were also gained (164).

ATR was also used by Hill (165) to study the effect of weathering on plasticized PVC. Reactions of the plasticizer as well as those of the polymer could be monitored as a function of exposure to weather.

The chemical effects of irradiation damage in polyethylene film was illustrated by Koenig (131), using absorbance subtraction (Fig. 65). Growth of the carbonyl band at 1742 cm^{-1} in both air and N_2 is apparent, as is the loss of vinyl end groups at 909 and 991 cm^{-1}. Other changes are also documented. This paper

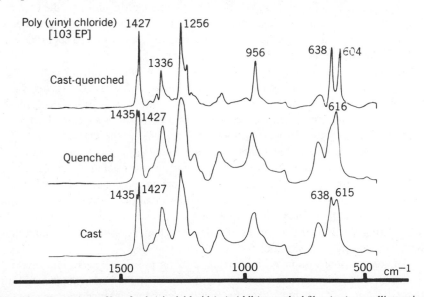

Fig. 66. (*Bottom*) cast film of poly(vinylchloride); (*middle*) quenched film; (*top*) crystalline regions (cast minus quenched film spectra) (131).

also demonstrates the ability of the absorbance subtraction method to extract the spectrum of crystalline polymer from spectra of two partially crystalline materials having different degrees of crystallinity (Fig. 66).

Migration of constituents or surface segregation of polymer components is not uncommon. Both photoacoustic and ATR sampling were used by Gardella et al. (166) to estimate the degree of segregation in a complex biocompatible polymer.

4.4. Semiconductors

Using transmission IR spectroscopy to detect parts per billion of elemental phosphorus, boron, and other impurities in silicon is obviously impossible; yet, in practice, it works quite well. Infrared spectroscopy is in fact the preferred method; chemical procedures and other instrumental methods lack the necessary sensitivity, and electrical measurements are nonspecific and tedious. The unusual characteristics of hyperpure silicon that allow it to function as a semiconductor also work in favor of the IR analytical method. First, silicon is essentially transparent throughout the mid-IR region. Second, the electrical properties of semiconductor silicon depend on the amount and type of impurities (from Groups III and V of the periodic table) substituted in the silicon lattice. The ionization potentials of these impurities are less than 0.1 eV, so they undergo electronic transitions that absorb IR radiation. The most common elements found or introduced as dopants in Si are B, P, Al, As, Sb, In, and Ga. In addition, the elements O and C, which form chemical bonds to Si in the lattice, are of interest because of their effect on the properties of the finished devices. Infrared methods have been developed to monitor all of these elements.

The analyses are not without problems. Silicon has a high refractive index (3.4), so reflection losses are large (31% per surface). Worse, the surfaces must be polished to reduce scattering, and the result, at least for thin slices, is that very intense interference fringes are superimposed on the spectrum. Further, Si has lattice absorptions in exactly the same regions as the carbon and oxygen absorptions. The dopant bands are too wide to detect at room temperature, but at cryogenic temperatures they become very narrow, necessitating high-resolution scans for quantitation. Finally, only the net dopant content can be measured by direct IR spectroscopy; compensated impurities are not measured, so total impurity content is not obtained. Each of these problems has been confronted and at least partially solved, however, so the desired elements can now be quantitated with extraordinary sensitivity.

Silicon is usually sampled as a wafer having polished faces. The thickness is chosen to give reasonable intensity for the bands to be measured; it may range from 0.1 mm to 2 cm, but is usually 0.2–0.5 cm. Interference fringes are often troublesome with the thinner samples. If an FTIR is used, interference fringes may be removed by replacing the secondary interferogram, usually found some distance from the centerburst of the main interferogram, with a straight line segment (167). Or, one can use a resolution setting larger than the fringe spacing (168).

4.4.1. Oxygen and Carbon

Silicon has a 3-phonon lattice band at 1100 cm^{-1} and a 2-phonon band at 610 cm^{-1} which interfere with the frequencies for oxygen (1104 cm^{-1}) and carbon (607 cm^{-1}). The approach used with double-beam dispersive spectrometers was to place a pure wafer of the same thickness in the reference beam to optically compensate for the lattice absorptions. This approach worked well as long as a pure reference specimen was available in the same thickness as the sample. Computerized spectrometers eliminated the identical-thickness requirement, as the spectrum of the reference could be stored in memory. A scaled absorbance subtraction thus became possible and the scaling factor could be adjusted to compensate for the thickness discrepancy.

Adequate sensitivity for O and C can be had with the sample at room temperature; detection of 0.1 ppma has been reported on 0.4-mm-thick wafers (168). Vidrine (169) gives a procedure for achieving 100 ppb sensitivity on a 2-mm-thick wafer with greater speed and accuracy. No reference sample is needed, but because the C absorption is only 4 cm^{-1} wide and lower resolution settings gave nonlinear responses, a resolution of 0.6 cm^{-1} was recommended. It was also found that O and C distributions were not necessarily uniform over the area of the wafer. Sensitivity of the C and O measurements can be increased by a factor of 5 by working at 20 K (168). Regardless of the method used, adequate S/N ratios must be achieved at the point of maximum absorption of the lattice bands, or reproducible quantitative results will be impossible to achieve.

4.4.2. Group III–V Elements

For quantitation of the Group III acceptor (p-type) and Group V donor (n-type) elements, it is necessary to work at temperatures below 77 K, and best results are obtained at sample temperatures below 20 K. Absorptions are electronic transitions from the ground state of neutral impurities to levels below their respective band edges. Only neutral centers absorb IR radiation; the compensated impurities are not IR active.

There is, however, a way to determine compensated impurities. Flooding the Si specimen with photons having energy greater than the Si bandgap energy (1.1 eV) generates excess electrons and holes that neutralize the ionized centers. Thus, simultaneous illumination of the specimen with white light and scanning with the IR spectrometer allows detection of all impurities. Scanning without

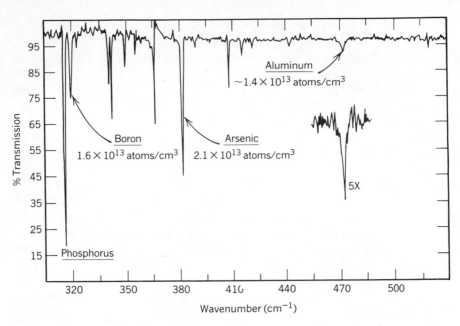

Fig. 67. Total impurity spectrum of 265 Ω-cm n-type Si obtained by the simultaneous illumination method (170).

white light illumination gives only the uncompensated impurities (170). The impurity bands are very narrow (0.6 cm^{-1} for P at 316 cm^{-1}, and 1.6 cm^{-1} for B at 320 cm^{-1}), so a resolution of 0.5 cm^{-1} or better is necessary. A total impurity spectrum is shown in Fig. 67, and the strongest impurity lines are listed in Table 12. Detection limits as a function of baseline noise are indicated in Fig. 68.

Table 12.   Strongest Absorption Lines for Shallow Impurities in Silicon[a]

Impurity	Type	Wavenumber (cm^{-1})	Wavelength (μm)
B	Acceptor	319.7	31.3
Al	Acceptor	471.7	21.2
Ga	Acceptor	548.0	18.2
In	Acceptor	1175.9	8.5
P	Donor	315.9	31.6
As	Donor	382.2	26.2
Sb	Donor	293.6	34.1

[a]From Baber (170).

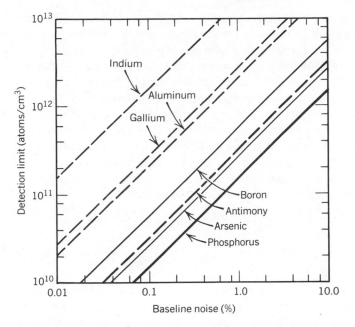

Fig. 68. Detection limits for shallow impurities in 1.0-cm-thick Si. Dashed lines show limits based on estimated calibration factors (170).

For P, a detection limit of 5×10^{11} atoms/cm^3 is equivalent to 0.01 ppba or 0.04 ng for a wafer 5 mm thick and 2 cm in diameter; a truly impressive figure.

Artifacts in the form of weak emission-like bands at twice the frequency of strong absorption bands can appear in the FTIR spectrum. They are attributed to remodulation of the reflected beam from the sample faces (171).

Several examples of the use of infrared microanalysis to solve problems in semiconductor processing are given by Scott and Ramsey (172), who used a transmission IR microscope combined with a microspectrophotometer to examine microscopic contaminants appearing on chips and masks.

4.5. Surfaces and Interfaces

Surfaces are important in chemistry and physics, because in many systems, the surface is "where the action is." Although the word surface is not precisely defined (it is applied to monolayers as well as to layers 100-μm thick), surface analysis is the subject of much activity. Many techniques are used, with some giving elemental composition and others molecular characterization. Electrons, photons, ions, and neutral particles are all used as probes. Depths of penetration depend on the nature of the probe particle and its energy. Each of these methods has its place, but here we discuss only the application of IR spectroscopy to trace analysis on surfaces (most surface analysis is trace analysis almost by definition, since, except for high-surface powders, the surface comprises only a tiny fraction of the bulk of the system). Topics such as catalysis, surfactant effectiveness, adhesion and release, lubrication, corrosion, weathering, electrochemistry, and transport of materials across biological membranes are all surface or interface phenomena that are currently being intensively studied by workers using a variety of techniques, including IR. While our discussion is far from complete, it should be adequate to give the flavor of current work, and perhaps suggest possible applications of IR to new problems.

For current awareness, the *Analytical Chemistry* biennial reviews on surface analysis (173) are recommended. Applications of IR are covered in the review on IR spectroscopy (174). Application of FTIR to surface studies has been reviewed by Jakobsen (42).

4.5.1. Catalysis

As pointed out by Angell (176), IR is popular for the study of catalysts because it permits direct examination of the adsorbed molecules on solid surfaces. Further, it can even give a dynamic picture of that surface while reaction is taking place.

Infrared spectroscopic studies of catalysis generally focus either on model systems such as CO or hydrocarbon on a pristine metallic surface using RAIR,

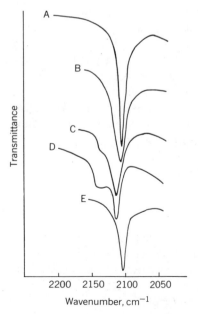

Fig. 69. CO on copper at various stages of oxidation: (A) prior to exposure; (B) after 2 Torr min O_2 exposure; (C) after 7 Torr min; (D) after 18 Torr min; (E) after H_2 reduction (177).

or on powdered supports such as zeolite, clays, or silicas. The latter are usually studied by transmission or DRIFTS techniques.

Example of RAIR studies are given by Thompkins and Greenler (177). Figure 69 shows spectra of CO on copper at various stages of copper oxidation. A peak shift of 18 cm^{-1} and formation of an additional band are noted. Other studies of CO on metal surfaces are reviewed by Bell and Hair (117).

Photoacoustic FTIR spectroscopy (PAS) has also been employed in the study of molecules on silica powder and supported metal catalysts (113). Spectra of palmitic acid neat and deposited on SiO_2 powder (16% by weight) are shown in Fig. 70.

Studies on particle-supported catalysts are often done using thin self-supported pressed discs, which are contained in a special cell that can be evacuated, heated, or cooled. Transmission is generally quite low in some regions of the spectrum because of scattering by the particles of the disc and absorption by the substrate. In other regions, transmission may be quite good. Thus, when an FTIR spectrometer is used, care must be taken to obtain adequate S/N ratios in the regions of heavy absorption and yet not overload the detector or the A/D converter. Methods of dealing with this problem have been discussed by Angell (176). They include use of a sharp cutoff optical filter to blank out the unused spectral

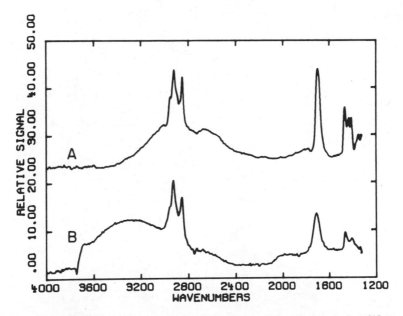

Fig. 70. Mid-IR FTPAS spectra of palmitic acid: (A) pure powder; (B) deposited on SiO_2 powder with an average surface density of 10^{10} mol/cm^2. Reprinted with permission from Kinney and Staley, *Anal. Chem.,* **55,** 343. Copyright 1983 American Chemical Society.

regions, and digital smoothing of the final trace. With dispersive spectrometers, widening the slits to permit more energy transmission accomplishes the same objective (1).

FTIR spectra of CO on silica-supported Pt are shown in Fig. 71 (21). Absorbance subtraction was used to null out interference from the SiO_2 support, giving a clean spectrum of adsorbed CO.

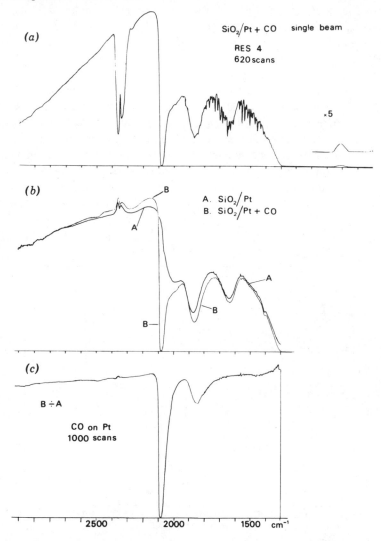

Fig. 71. IR spectra of CO on SiO_2-supported Pr: (a) single beam spectrum; (b) ratioed spectra of (A) SiO_2 support and (B) adsorbed CO + support; (c) residual spectra of CO after subtraction of SiO_2 support (21).

4.5.2. Composites

The interface between glass fibers and plastics in polymer composites has been intensively studied by Koenig et al. using a variety of techniques (131,178–180). The performance of composite materials is greatly enhanced by small quantities of organofunctional silane derivatives such as vinyl triethoxysilane, bonded to the glass surface at the silicon end and to the organic resin at the other. The nature of the bonds to the glass surface is still the subject of some controversy, however. Are these primary chemical bonds, or simply strong hydrogen or van der Waal's bonds that permit the glass surface to slide against the resin and yet maintain adhesion? Another theory proposes that the coupling agent modifies the resin layer adjacent to the glass surface to give a more flexible interface.

The effect of hydrolysis, prior to treatment of glass fibers, on the silane (in this case, γ-methacryloxypropyltrimethoxy silane) is shown in Fig. 72. Spectrum 1 is pure water, spectrum 2 is a 5% solution of the silane, spectrum 3 is the expanded difference, and spectrum 4 is the unhydrolyzed silane. Loss of the strong $SiOCH_3$ bonds at 1190, 1167, and 1089 cm^{-1} is evident, as is the formation of SiOH at 920 cm^{-1}. The silane solution is used to coat the fibers, which are then dried and heated to cure the silane. After drying, the fibers may be ground and sampled in a mull or KBr pellet, studied by diffuse reflectance (181), or simply run as transmission specimens without any special preparation (180). Sometimes, high-surface SiO_2 or powdered glass is used as a model system (178). Cure of the silane can be followed as a function of time and temperature. A study of the glass fiber-coupling agent–resin system (178) showed that the coupling agent–matrix interface is covalently bonded when styrene is the matrix.

Liquid–solid interfaces can sometimes be studied in transmission, although better sensitivity is usually obtained with ATR. In both cases, one's choice of solid is restricted to infrared-transmitting materials. Yang et al. (182) used ATR to study absorption of stearic acid on Ge and Al_2O_3 internal reflection elements. Some information about orientation was obtained with the use of polarized radiation.

ATR, under carefully controlled conditions, has been used to determine the composition of thin surface layers on polymers with the aid of absorbance subtraction (183). The study involved polyurethane polymers made nonthrombogenic by a thin coating of polydimethylsiloxane at the surface that is in contact with blood. An adjustable platform was constructed to mount the ATR accessory firmly and reproducibly at the optimum position in the sample beam of the FTIR spectrometer. The ATR element was chosen so as to minimize angle setting and beam convergence errors, as well as distortions from refractive index changes in the vicinity of strong absorption bands. The authors suggest that a Ge ATR element at an incident angle of 45° is generally appropriate. Germanium,

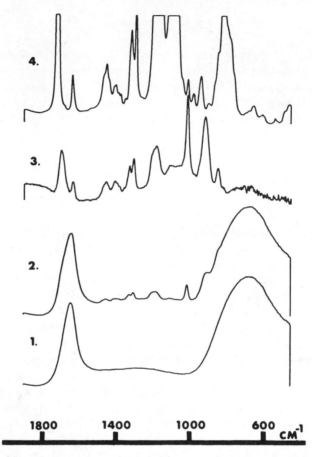

1800   1400   1000   600 CM⁻¹

Fig. 72. IR spectra showing hydrolysis of γ-methacryloxypropyltrimethoxysilane (γ-MPS) (see text) (122).

however, has absorptions below 750 cm^{-1}, and if this region is of interest, ZnSe or KRS-5 elements may be used. Spectra were measured at 2 cm^{-1} resolution using 400 scans and an cooled MCT detector. Spectra obtained with a ZnSe element at 1000 to 650 cm^{-1} are shown in Fig. 73. Absorbance subtraction with the appropriate scale factor revealed the polydimethylsiloxane absorption at 801 cm^{-1}. Concentration was calculated at 1.0%; the calculation was carried out by ratioing the 801 cm^{-1} band to the 1532 cm^{-1} polyurethane band (not shown), and referring to a previously established working curve.

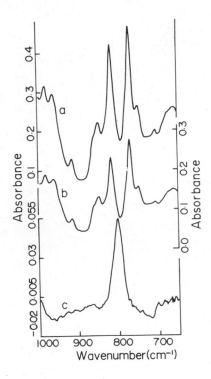

Fig. 73. FTIR-ATR spectra, obtained with a ZnSe element at 51°, for the air-facing surface of: (a) cardiothane 51 film; (b) pure polyurethane film; (c) the difference spectrum (a − b)(183).

4.5.3. Biological Materials

Protein absorption onto surfaces—specifically, blood clotting—has been studied by ATR (184,185). The ATR crystal was Ge, and it was either used bare or coated with a thin (100–1000 Å) film of polymer. Flowing blood from a beagle dog was shunted onto the crystal, and within a few seconds adsorption started. Subtraction of the water spectrum (90% of the signal) gave spectra of the adsorbed species as a function of time (Fig. 74), and subsequent subtraction of one spectrum set from the next gave a picture of the changes that occurred as adsorption progressed. The initial precipitate was found to be rich in albumin and glycoproteins, followed by replacement of the albumin by other proteins. The amount of adsorbed material could be followed by plotting the appropriate absorption band intensity against time (Fig. 75).

Many other biological applications could be listed; however, we refer to the monograph by Parker (186) for a more complete discussion. The use of IR in the biomedical field is now poised for rapid growth.

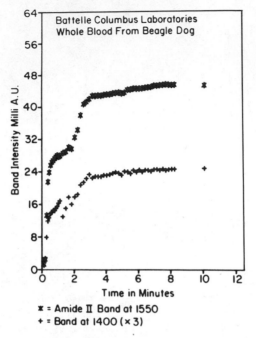

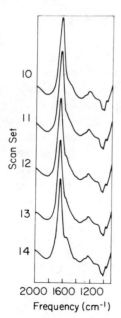

Fig. 74. Blood clotting studied by ATR. First five scan sets after blood entered liquid cell. Water spectrum has been subtracted. Amide I and II bands appear at 1640 and 1550 cm^{-1} (185).

Fig. 75. Plot of amide II (1550 cm^{-1}) and 1400 cm^{-1} bands versus time (185).

278

REFERENCES

1. W. J. Potts, Jr. and A. L. Smith, *Appl. Opt.*, **6**, 257 (1967).
2. P. R. Griffiths, *Chemical Infrared Fourier Transform Spectroscopy*, Wiley, New York, 1975.
3. W. J. Potts, Jr., *Chemical Infrared Spectroscopy, Vol. 1, Techniques*, Wiley, New York, 1963.
4. A. L. Smith, *Applied Infrared Spectroscopy*, Wiley, New York, 1979.
5. ACS Committee on Environmental Improvement, *Anal. Chem.*, **52**, 2242 (1980).
6. G. L. Long and J. D. Winefordner, *Anal. Chem.*, **55**, 712A (1983).
7. ACS Committee on Environmental Improvement, *Anal. Chem.*, **55**, 2210 (1983).
8. T. Hirschfeld, in K. F. J. Heinrich, Ed., *Microbeam Analysis—1982*, Proc. 17th Annual Conf. of Microbeam Analysis Soc., San Francisco Press, 1982, p. 247.
9. J.-Y. T. Chen and R. W. Dority, *J. Assoc. Offic. Anal. Chem.*, **53**, 978 (1970).
10. M. P. Fuller and P. R. Griffiths, *Appl. Spectrosc.*, **34**, 533 (1980).
11. H. W. Fox, P. W. Taylor, and W. A. Zisman, *Ind. Eng. Chem.*, **39**, 1401 (1947).
12. P. J. Launer, *Perkin-Elmer Instrument News*, **13**, No. 3, 10 (1962).
13. J. R. Moody, *Phil. Trans. Roy. Soc. (London) A*, **305**, 669 (1982).
14. J. E. Stewart, *Infrared Spectroscopy: Experimental Methods and Techniques*, Dekker, New York, 1970.
15. P. W. Kruse in R. J. Keyes, Ed., *Optical and Infrared Detectors*, 2nd ed., Springer-Verlag, New York, 1980, p. 5.
16. National Materials Advisory Board, NRC, *Materials for Radiation Detection*, Publication NMAB 287 (National Materials Advisory Board, National Research Council), National Academy of Sciences—National Academy of Engineering, Washington, DC, 1974.
17. R. J. Keyes, Ed., *Optical and Infrared Detectors*, 2nd ed., Springer-Verlag, New York, 1980.
18. J. P. Coates, *Amer. Lab.*, **8**(11), 67 (1976).
19. W. B. Telfair, A. C. Gilby, R. J. Syrjala, and P. A. Wilks, Jr., *Am. Lab.*, **8**(11), 91 (1976).
20. J. N. Ramsey and H. H. Hausdorff, *Microbeam Analysis*, San Francisco Press, 1981, p. 91.
21. D. H. Chenery and N. Sheppard, *Appl. Spectrosc.*, **32**, 79 (1978).
22. P. R. Griffiths, H. J. Sloane, and R. W. Hannah, *Appl. Spectrosc.*, **31**, 485 (1977).
23. P. R. Griffiths, *Anal. Chem.*, **44**, 1909 (1972).
24. C. T. Foskett, *Appl. Spectrosc.*, **30**, 531 (1976).
25. T. Hirschfeld, "Quantitative FT-IR," in J. R. Ferraro and L. J. Basile, Eds., *Fourier Transform Infrared Spectroscopy*, Vol. 2, Academic Press, New York, 1979, p. 193.
26. J. A. DeHaseth, *Appl. Spectrosc.*, **36**, 544 (1982).
27. J. R. Allkins, *Anal. Chem.*, **47**, 752A, (1975).
28. D. H. Whiffen, "Lasers in Infrared Spectroscopy," in A. R. West, Ed., *Molecular Spectroscopy*, Heyden, London, 1977.

29. G. T. Forrest, *Lasers and Applications,* **2**(6), 63 (1983).
30. R. T. Menzies, C. R. Webster, and E. D. Hinkley, *Appl. Opt.,* **22**(17), 2655 (1983).
31. J. F. Butler, K. W. Nill, A. W. Mantz, and R. S. Eng, *Applications of Tunable Diode-Laser-IR Spectroscopy to Chemical Analysis* (ACS Symposium Series, Vol. 85), American Chemical Society, Washington, D.C., 1978, p. 12.
32. J. Reid, B. K. Garside, J. Shewchun, M. El-Sherbiny, and E. A. Ballik, *Appl. Opt.,* **17,** 1806 (1978).
33. A. Stein, T. R. Todd, and B. N. Perry, *Appl. Opt.,* **22,** 3378 (1983).
34. U. Persson, J. Johansson, B. Marthinsson, and S. T. Eng. *Appl. Opt.,* **21,** 4417 (1982).
35. K. W. Rothe, *European Spectrosc. News,* **40,** 22 (1982).
36. D. M. Sweger and J. C. Travis, *Appl. Spectrosc.,* **33,** 46 (1979).
37. N. Menyuk and D. K. Killinger, *Appl. Opt.,* **22**(17), 2690 (1983).
38. R. W. Hannah, S. C. Pattacini, J. G. Grasselli, and S. E. Mocadlo, *Appl. Spectrosc.,* **32,** 69 (1978).
39. N. J. Harrick, *Internal Reflection Spectroscopy,* Interscience, New York, 1967.
40. T. Hirschfeld, *Appl. Spectrosc.,* **20,** 336 (1966).
41. A. C. Gilby, J. Cassels, and P. A. Wilks, Jr., *Appl. Spectrosc.* **24,** 539 (1970).
42. R. J. Jakobsen, in J. R. Ferraro and L. J. Basile, Eds., *Fourier Transform Infrared Spectroscopy,* Vol. 2, Academic, New York, 1979, p. 165.
43. T. H. Parliment, *Microchem. J.,* **20,** 492 (1975).
44. S. K. Freeman, "Gas Chromatography and Infrared and Raman Spectrometry," in L. S. Ettre and W. H. McFadden, Eds., *Ancillary Techniques of Gas Chromatography,* Wiley, New York, 1969, pp. 227–267.
45. D. Goldfarb and C. W. Brown, *Appl. Spectrosc.,* **33**(2), 126 (1979).
46. R. F. Brady, Jr., *Anal. Chem.,* **47,** 1425 (1975).
47. J. E. Crooks, D. L. Gerrard, and W. F. Maddams, *Anal. Chem.,* **45,** 1823 (1973).
48. C. W. Louw and J. F. Richards, *Appl. Spectrosc.,* **29**(1), 15 (1975).
49. M. J. Flanagan, *Appl. Spectrosc.,* **34,** 87 (1980).
50. R. H. Shaps and A. Varano, *Industrial Res.,* **19**(2), 86 (1977).
51. R. A. Nyquist, *The Interpretation of Vapor-Phase Infrared Spectra. Vol. 1. Group Frequency Data,* Sadtler, Philadelphia, 1984.
52. B. Katlafsky and M. W. Dietrich, *Appl. Spectrosc.,* **29,** 24 (1975).
53. D. L. Wall and A. W. Mantz, *Appl. Spectrosc.,* **31,** 552 (1977).
54. K. H. Shafer, M. Cooke, F. DeRoos, R. J. Jakobsen, O. Rosario, and J. D. Mulik, *Appl. Spectrosc.,* **35,** 469 (1981).
55. S. L. Smith, S. E. Garlock, and G. E. Adams, *Appl. Spectrosc.,* **37,** 192 (1983).
56. J. R. Cooper and L. T. Taylor, *Appl. Spectrosc.,* **38,** 366 (1984).
57. K. H. Shafer, T. L. Hayes, J. W. Brasch, and R. J. Jakobsen, *Anal. Chem.,* **56,** 237 (1984).
58. P. R. Griffiths, J. A. deHaseth, and L. V. Azarraga, *Anal. Chem.,* **55,** 1361A (1983).
59. L. V. Azarraga, *Appl. Spectrosc.,* **34,** 224 (1980).
60. G. N. Giss and C. L. Wilkins, *Appl. Spectrosc.,* **38,** 17 (1984).

61.   P. R. Griffiths, *Appl. Spectrosc.*, **31**, 284 (1977).
62.   P. Coffey, D. R. Mattson, and J. C. Wright, *Amer. Lab.*, **10**(5), 126 (1978).
63.   J. Å. deHaseth and T. L. Isenhour, *Anal. Chem.*, **49**, 1977 (1977).
64.   B. A. Hohne, G. Hangac, G. W. Small, and T. L. Isenhour, *J. Chromatogr. Sci.*, **19**, 283 (1981).
65.   G. T. Reedy, S. Bourne, and P. T. Cunningham, *Anal. Chem.*, **51**, 1535 (1979).
66.   S. Bourne, G. Reedy, P. Coffey, and D. Mattson, *Amer. Lab.*, **16**(6), 90 (1984).
67.   K. H. Shafer, S. V. Lucas, and R. J. Jakobsen, *J. Chromat. Sci.*, **17**, 464 (1979).
68.   R. W. Crawford, T. Hirschfeld, R. H. Sandborn, and C. M. Wong, *Anal. Chem.*, **54**, 817 (1982).
69.   C. L. Wilkins, G. N. Giss, R. C. White, G. M. Brissey, and E. C. Onyiriuka, *Anal. Chem.*, **54**, 2260 (1982).
70.   D. A. Laude, Jr., G. M. Brissey, C. F. James, R. S. Brown, and C. L. Wilkins, *Anal. Chem.*, **56**, 1163 (1984).
71.   A. L. Smith, "Infrared Spectroscopy," in J. W. Robinson, *Handbook of Spectroscopy, Vol. II.*, CRC Press, Cleveland, 1974, pp. 24–35.
72.   D. W. Vidrine, in J. R. Ferraro and L. J. Basile, Eds., *Fourier Transform Infrared Spectroscopy*, Vol. 2, Academic, New York, 1979, p. 129.
73.   A. J. Rein and P. Wilks, *Am. Lab.*, **14**(10), 152 (1982).
74.   D. W. Vidrine, *J. Chromatogr. Sci.*, **17**, 477 (1979).
75.   R. S. Brown and L. T. Taylor, *Anal. Chem.*, **55**, 1492 (1983).
76.   D. Kuehl and P. R. Griffiths, *J. Chromatogr. Sci.*, **17**, 471 (1979).
77.   M. M. Gomez-Taylor and P. R. Griffiths, *Appl. Spectrosc.*, **31**, 528 (1977).
78.   M. M. Gomez-Taylor, D. Kuehl, and P. R. Griffiths, *Appl. Spectrosc.*, **30**, 447 (1976).
79.   C. J. Percival and P. R. Griffiths, *Anal. Chem.*, **47**, 154 (1975).
80.   M. P. Fuller and P. R. Griffiths, *Appl. Spectrosc.*, **34**, 533 (1980).
81.   P. A. Peaden and M. L. Lee, *J. Liq. Chromatogr.*, **5**, 179 (1982).
82.   K. H. Shafer and P. R. Griffiths, *Anal. Chem.*, **55**, 1939 (1983).
83.   J. C. Fjeldsted and M. L. Lee, *Anal. Chem.*, **56**, 619A, 1984.
84.   R. R. Willey, *Appl. Spectrosc.*, **30**, 593 (1976).
85.   M. P. Fuller and P. R. Griffiths, *Anal. Chem.*, **50**, 1906 (1978).
86.   S. A. Yeboah, S. H. Wang, and P. R. Griffiths, *Appl. Spectrosc.*, **38**, 259 (1984).
87.   R. W. Hannah and R. E. Anacreon, *Appl. Spectrosc.*, **37**, 75 (1983).
88.   P. R. Griffiths, *Appl. Spectrosc.*, **26**, 73 (1972).
89.   P. Baraldi, *Spectrochim. Acta*, **40A**, 81 (1984).
90.   D. Kember, D. H. Chenery, N. Sheppard, and J. Fell, *Spectrochim. Acta*, **35A**, 455 (1979).
91.   D. B. Chase, *Appl. Spectrosc.*, **35**, 77 (1981).
92.   J. L. Lauer and M. E. Peterkin, *J. Lubr. Technol.*, **97**, 145 (1975).
93.   J. L. Lauer and M. E. Peterkin, *J. Lubr. Technol.*, **98**, 230 (1976).
94.   Y. Nagasawa and A. Ishitani, *Appl. Spectrosc.*, **38**, 168 (1984).
95.   T. Hirschfeld, *Appl. Spectrosc.*, **30**, 353 (1976).
96.   P. R. Griffiths and F. Block, *Appl. Spectrosc.*, **27**, 431 (1973).
97.   S. S. T. King, *J. Ag. Food Chem.*, **21**, 526 (1973).

98.   H. R. Garner and H. Packer, *Appl. Spectrosc.*, **22**, 122 (1968).

99.   F. T. Tweed, R. Cameron, J. S. Deak, and P. G. Rogers, *Forensic Sci.*, **4**, 211 (1974).

100. J. U. White, *J. Opt. Soc. Am.*, **32**, 285 (1942).

101. P. L. Hanst, in J. R. Ferraro and L. J. Basile, Eds., *Fourier Transform Infrared Spectroscopy*, Vol. 2, Academic, New York, 1979, p. 79.

102. P. L. Hanst, *Appl. Opt.*, **17**, 1360 (1978).

103. E. C. Tuazon, R. A. Graham, A. M. Winer, R. R. Easton, J. N. Pitts, and P. L. Hanst, *Atmos. Environ.*, **12**, 865 (1978).

104. P. L. Hanst, A. S. Lefohn, and B. W. Gay, Jr., *Appl. Spectrosc.*, **27**, 188 (1973).

105. P. L. Hanst and B. W. Gay, Jr., *Env. Sci. Technol.*, **11**, 1105 (1977).

106. R. A. Nyquist and C. C. Putzig, Dow Chemical Co., Private communication (1984).

107. P. L. Hanst, *Adv. Environ. Sci. Technol.*, **2**, 91 (1971).

108. P. L. Hanst, L. L. Spiller, D. M. Watts, J. W. Spence, and M. F. Miller, *J. Air Poll. Control Assoc.*, **25**, 1220 (1975).

109. S. M. Freund, W. B. Maier, II, R. F. Holland, and W. H. Beattie, *Anal. Chem.*, **50**, 1260 (1978).

110. D. L. Wood, J. P. Luongo, and S. S. DeBala, *Anal. Chem.*, **53**, 1967 (1981).

111. A. Rosencwaig and A. Gersho, *J. Appl. Phys.*, **47**, 64 (1976).

112. D. W. Vidrine, "Photoacoustic Fourier Transform Infrared Spectroscopy of Solids and Liquids," in J. R. Ferraro and L. J. Basile, Eds., *Fourier Transform Infrared Spectroscopy*, Vol. 3, Academic, New York, 1982, p. 125.

113. J. B. Kinney and R. H. Staley, *Anal. Chem.*, **55**, 343 (1983).

114. H. G. Tomkins, *Appl. Spectrosc.*, **30**, 377 (1976).

115. R. G. Greenler, *J. Chem. Phys.*, **44**, 310 (1966).

116. D. L. Allera, "Organic Monolayer Studies using FTIR Reflection Spectroscopy," in A. T. Bell and M. L. Hair, Eds., *Vibrational Spectroscopies for Adsorbed Species* (ACS Symposium Series, Vol. 137), American Chemical Society, Washington, D.C., 1980, p. 37.

117. A. T. Bell and M. L. Hair, Eds., *Vibrational Spectroscopies for Adsorbed Species* (ACS Symposium Series, Vol. 137), American Chemical Society, Washington, D.C., 1980.

118. J. F. Blanke, S. E. Vincent, and J. Overend, *Spectrochim. Acta*, **32A**, 163 (1976).

119. A. E. Dowrey and C. Marcott, *Appl. Spectrosc.*, **36**, 414 (1982).

120. H. J. Humecki and R. Z. Muggli, "Micro Sample Identification with FTIR Spectroscopy," in K. F. J. Heinrich, Ed., *Microbeam analysis—1982*, Proc. 17th Ann. Conf. Microbeam Anal. Soc., San Francisco Press, 1982, p. 243.

121. R. Cournoyer, J. C. Shearer, and D. H. Anderson, *Anal. Chem.*, **49**, 2275 (1977).

122. K. Krishnan and D. Kuehl, Paper No. 759, Pittsburgh Conf. Anal. Chem. Appl. Spectroscopy, March 1984.

123. Analect Instruments, Utica, N.Y., Infrared microscope accessory.

124. R. Cook and M. D. Paterson, *Forensic Sci. Int.*, **12**, 237 (1978).

125. J. R. Ferraro and L. J. Basile, *Appl. Spectrosc.*, **28**, 505 (1974).

126. J. C. Shearer, D. C. Peters, G. Hoepfner, and T. Newton, *Anal. Chem.*, **55**, 874A (1983).

127. J. E. Stewart, *Infrared Phys.*, **7**, 77 (1967).

128. P. D. Willson and T. H. Edwards, *Appl. Spectrosc. Rev.*, **12**, 1 (1976).

129. A. Savitzky and M. J. E. Golay, *Anal. Chem.*, **36**, 1627 (1964).

130. P. R. Griffiths, Private communication (1984).

131. J. L. Koenig, *Appl. Spectrosc.*, **29**, 293 (1975).

132. J. Strassburger and I. T. Smith, *Appl. Spectrosc.*, **33**, 283 (1979).

133. T. Hirschfeld and K. Kizer, *Appl. Spectrosc.*, **29**, 345 (1975).

134. T. Hirschfeld, *Appl. Spectrosc.*, **30**, 550 (1976).

135. T. Hirschfeld, *Appl. Spectrosc.*, **30**, 549 (1976).

136. P. C. Gillette and J. L. Koenig, *Appl. Spectrosc.*, **38**, 334 (1984).

137. P. R. Griffiths, *Appl. Spectrosc.*, **31**, 497 (1977).

138. R. J. Anderson and P. R. Griffiths, *Anal. Chem.*, **50**, 1804 (1978).

139. D. L. Allera, *Appl. Spectrosc.*, **33**, 358 (1979).

140. R. H. Wiens and H. H. Zwick, "Trace Gas Detection by Correlation Spectroscopy," in J. S. Matson, H. B. Mark, Jr., and H. C. MacDonald, Jr., Eds., *Infrared, Correlation and Fourier Transform Spectroscopy*, Dekker, New York, 1977.

141. R. H. Wiens and H. H. Zwick, "Trace Gas Detection by Correlation Spectroscopy," *Comput. Chem. Instrum.*, **7**, 1977, p. 119.

142. D. M. Haaland and R. G. Easterling, *Appl. Spectrosc.*, **34**, 539 (1980).

143. Dept. of Health, Education and Welfare, Public Health Service, *Manual of Analytical Methods: Organic Solvents in Air*, PSCAM127, Center for Disease Control, Cincinnati, OH, 1973.

144. J. Diaz-Rueda, H. J. Sloane, and R. J. Obremski, *Appl. Spectrosc.*, **31**, 298 (1977).

145. K. Golding, *Process Eng.*, **70**, 1 (1974).

146. P. E. Wilkins, *Air Qual. Instrum.*, **2**, 246 (1974).

147. P. A. Wilks, Jr., *Env. Sci. Tech.*, **10**, 1204 (1976).

148. A. W. Mantz, *Appl. Spectrosc.*, **30**, 539 (1976).

149. J. Reid and B. K. Garside, *Opt. Quantum Electron.* **11**, 385 (1979).

150. J. N. Pitts, B. J. Finlayson-Pitts, and A. M. Winer, *Environ. Sci. Technol.*, **11**, 568 (1977).

151. B. M. Golden and E. S. Yeung, *Anal. Chem.*, **47**, 2132 (1975).

152. L. W. Chaney and W. A. McClenny, *Env. Sci. Technol.*, **11**, 1186 (1977).

153. A. S. Zachor, T. Zehnpfennig, and A. T. Stair, Jr., in D. G. Killinger and A. Mooradian, Eds., *Optical and Laser Remote Sensing*, Springer-Verlag, New York, 1983, pp. 81–89.

154. A. S. Zachor, B. Bartchi, and F. P. DelGreco, *Proc. AGARD Conf.*, 1981. AGARD-CP-300, 25/1-25/6.

155. W. F. Herget, in J. R. Ferraro and L. J. Basile, Eds., *Fourier Transform Infrared Spectroscopy*, Vol. 2, Academic, New York, 1979, p. 111.

156. D. I. Sebacher, Rev. Sci. Instrum., **49**, 1520 (1978).

157. R. J. Nordstrom, J. H. Shaw, W. R. Skinner, W. H. Chan, J. G. Calvert, and W. M. Uselman, *Appl. Spectrosc.*, **31**, 224 (1977).

158. D. Rosenthal, *Anal. Chem.*, **54**, 63 (1982).

159. J. Ducreax, R. Boulet, N. Petroff, and J. C. Roussel, *Int. J. Environ. Anal. Chem.*, **12**, 195 (1982).

160. M. M. Gomez-Taylor, D. Kuehl, and P. R. Griffiths, *Int. J. Environ. Anal. Chem.*, **5,** 103 (1978).

161. H. Hellmann, *Fresenius' Z. Anal. Chem.*, **293,** 359 (1978).

162. D. F. Gurka, M. Hiatt, and R. Titus, *Anal. Chem.*, **56,** 1102 (1984).

163. D. R. Battiste, J. P. Butler, J. B. Cross, and M. P. McDaniel, *Anal. Chem.*, **53,** 2232 (1981).

164. M. Ito and R. S. Porter, *J. Appl. Polym. Sci.*, **27,** 4471 (1982).

165. C. A. S. Hill, *J. Appl. Polym. Sci.*, **27,** 3313 (1982).

166. J. A. Gardella, Jr., G. L. Grobe III, W. L. Hopson, and E. M. Eyring, *Anal. Chem.*, **56,** 1169 (1984).

167. F. R. S. Clark and D. J. Moffatt, *Appl. Spectrosc.*, **32,** 547 (1978).

168. D. G. Mead and S. R. Lowry, *Appl. Spectrosc.*, **34,** 167 (1980).

169. D. W. Vidrine, *Anal. Chem.*, **52,** 92 (1980).

170. S. C. Baber, *Thin Solid Films*, **72,** 201 (1980).

171. D. H. Lemon and J. C. Swartz, *J. Mol. Struct.*, **61,** 415 (1980).

172. R. M. Scott and J. N. Ramsey, *Microbeam Analysis—1982*, Proc. 17th Ann. Conf. Microbeam Analysis Soc., San Francisco Press, 1982, p. 239.

173. R. A. Bowling and G. B. Larrabee, *Anal. Chem.*, **55,** 133R (1983).

174. R. S. McDonald, *Anal. Chem.*, **56,** 349R (1984).

175. N. Sheppard, *J. Mol. Struct.*, **80,** 163 (1982).

176. C. L. Angell, "FTIR Spectroscopy in the Study of Catalysts," in J. R. Ferraro and L. J. Basile, Eds., *Fourier Transform Infrared Spectroscopy,* Vol. 3, Academic, New York, 1982, p. 1.

177. H. G. Tompkins and R. G. Greenler, *Surf. Sci.*, **28,** 194 (1971).

178. H. Ishida and J. L. Koenig, *J. Polym. Sci., Polym. Phys. Ed.*, **17,** 615 (1979).

179. H. Ishida, C. Chiang, and J. L. Koenig, *Polymer*, **23,** 251 (1982).

180. S. R. Culler, H. Ishida, and J. L. Koenig, *Appl. Spectrosc.*, **38,** 1 (1984).

181. R. T. Graf, J. L. Koenig, and H. Ishida, *Anal. Chem.*, **56,** 773 (1984).

182. R. T. Yang, M. J. D. Low, G. L. Haller, and J. Fenn, *J. Coll. Interface Sci.*, **44,** 249 (1973).

183. R. Iwamoto and K. Ohta, *Appl. Spectrosc.*, **38,** 359 (1984).

184. R. M. Gendreau and R. J. Jakobsen, *Appl. Spectrosc.*, **32,** 326 (1978).

185. R. M. Gendreau, S. Winters, R. I. Leininger, D. Fink, C. R. Hassler, and R. J. Jakobsen, *Appl. Spectrosc.*, **35,** 353 (1981).

186. F. S. Parker, *Applications of Infrared, Raman, and Resonance Raman Spectroscopy in Biochemistry,* Plenum, New York, 1983.

CHAPTER

4

NUCLEAR MAGNETIC RESONANCE SPECTROSCOPY AND ITS APPLICATION TO TRACE ANALYSIS

DALLAS L. RABENSTEIN

Department of Chemistry
University of California
Riverside, California

THOMAS T. NAKASHIMA

Department of Chemistry
University of Alberta
Edmonton, Alberta, Canada

1. INTRODUCTION

Nuclear magnetic resonance (NMR) spectroscopy offers several important advantages as a tool for chemical analysis, including the absence of quantities analogous to absorption coefficients, the ease with which multicomponent mixtures can be analyzed directly, and its nondestructive nature. Unfortunately, NMR has the disadvantage of inherently low sensitivity relative to many other spectroscopic methods. The sensitivity of NMR has increased dramatically, however, as a result of developments over the past decade, in both NMR techniques and instrumentation, the two most important being the pulse/Fourier transform (P/FT) method of measuring NMR spectra and the use of the large magnetic fields which can be obtained with superconducting solenoids.

With state-of-the-art NMR instrumentation, it now is possible to obtain ^{1}H and natural-abundance ^{13}C NMR spectra from nanomoles and micromoles of material, respectively, which has greatly increased the utility of NMR for chemical analysis. It is the objective of this chapter to consider NMR as a technique for chemical analysis, with emphasis placed on those aspects which limit its inherent sensitivity, and on the various developments in both techniques and instrumentation which have resulted in the dramatic increase in sensitivity of the past decade. Since the P/FT experiment provides the greatest sensitivity, much of the discussion is concerned with P/FT NMR. Also, some emphasis is given to instrumental and experimental methods by which resolution can be increased.

2. BASIC THEORY

2.1. Magnetic Properties of Nuclei

NMR spectroscopy is based on the magnetic properties of atomic nuclei. The nuclei of some isotopes behave as spinning charged particles, and thus possess both angular momentum $\mathbf{p}$ and a magnetic moment $\boldsymbol{\mu}$. The angular momentum is colinear with and proportional to $\boldsymbol{\mu}$:

$$\boldsymbol{\mu} = \gamma \mathbf{p} \tag{1}$$

where γ is the magnetogyric ratio. γ may be positive or negative, depending on whether $\boldsymbol{\mu}$ is parallel or antiparallel to the $\mathbf{p}$ vector. The maximum observable component of $\mathbf{p}$ for a particular nucleus is $I\hbar$, where I is the spin quantum number of the nucleus and $\hbar$ is Planck's constant divided by 2π. I can be 0, ½, 1, ³⁄₂, etc. Since $\boldsymbol{\mu}$ is nonzero only if I is nonzero, the basic requirement for an isotope to be NMR active is that its spin quantum number be nonzero. I is zero for nuclei with even mass and atomic numbers, integral for nuclei with even mass

Table 1. NMR Properties of Selected Isotopes[a]

Isotope	I	Natural Abundance (%)	NMR Frequency in a 4.7-T Field (MHz)	Magnetic Moment μ (Units of μ_N)	Magneto-gyric ratio γ (10^8 rad $T^{-1} s^{-1}$)	Quadrupole Moment (10^{-28} m^2)	Relative Sensitivity at Constant Field		Chem Shift Range (ppm)
							For Equal Numbers of Isotopes	At Natural Isotopic Abundance	
^{1}H	1/2	99.985	200	4.8371	2.67510		1.000	1.000	10
^{2}H	1	0.015	30.7	1.2125	0.41064	2.73×10^{-3}	9.65×10^{-3}	1.45×10^{-6}	10
^{11}B	3/2	80.42	64.2	3.4702	0.85827	3.55×10^{-2}	0.165	0.133	250
^{13}C	1/2	1.108	50.3	1.2162	0.67263		1.59×10^{-2}	1.76×10^{-4}	250
^{14}N	1	99.63	14.4	0.5706	0.19324	1.6×10^{-2}	1.01×10^{-3}	1.00×10^{-3}	900
^{15}N	1/2	0.37	20.3	-0.4901	-0.27107		1.04×10^{-3}	3.85×10^{-6}	900
^{17}O	5/2	0.037	27.1	-2.2398	-0.36266	-2.6×10^{-2}	0.0291	1.08×10^{-5}	700
^{19}F	1/2	100	188	4.5506	2.51665		0.833	0.833	800
^{23}Na	3/2	100	53	2.8610	0.70760	0.12	0.095	0.095	
^{29}Si	1/2	4.70	39.7	-0.9609	-0.53141		7.84×10^{-3}	3.68×10^{-4}	400
^{31}P	1/2	100	81	1.9581	1.0829		0.0663	0.0663	700
^{35}Cl	3/2	75.53	19.6	1.0598	0.26212	-7.89×10^{-2}	4.70×10^{-3}	3.55×10^{-3}	
^{37}Cl	3/2	24.47	16.3	0.8821	0.2182	-6.12×10^{-2}	2.71×10^{-3}	6.64×10^{-4}	

[a] Adapted from ref. 45.

and odd atomic numbers, and half-integral for nuclei with odd mass numbers.

Some of the important NMR-active isotopes, along with useful data for discussing their NMR behavior, are listed in Table 1. It is important to note that $I = 0$ for the most abundant isotopes of carbon (^{12}C), oxygen (^{16}O), and sulfur (^{32}S).

2.2. The NMR Experiment

The quantum mechanical description of NMR states that, when a nucleus with spin quantum number I is placed in a magnetic field $\mathbf{B_0}$, it can exist in $2I + 1$ distinct energy levels, whose energy (relative to that at zero magnetic field) is given by (bold face indicates a vector quantity, and non-bold face a magnitude)

$$E = \frac{-m\mu B_0}{I} \qquad (2)$$

where m is the magnetic quantum number ($m = I, I\text{-}1, \ldots, -I$). Thus, a nucleus of spin ½ can exist in two energy levels with energies of $-\mu B_0$ and $+\mu B_0$. In the NMR experiment, radiation of an energy which exactly matches the difference between adjacent levels stimulates transitions between them. This exact matching of the energy difference and the energy of the radiation is the so-called resonance condition. The frequency of radiation which satisfies this condition for $I = $ ½ nuclei is

$$\nu = \frac{2\mu B_0}{h} \qquad (3)$$

or, in terms of the magnetogyric ratio,

$$\nu = \frac{\gamma B_0}{2\pi} \qquad (4)$$

The resonance frequencies for nuclei in typical magnetic fields used in NMR (1–14 T) are in the megahertz range; the frequencies listed in Table 1 are for a 4.7-T magnetic field.

The two different energy states may be viewed as arising from two orientations of μ realtive to $\mathbf{B_0}$ (Fig. 1), the orientation in the direction of B_0 having the lower energy. Because the nucleus possesses angular momentum, μ precesses around $\mathbf{B_0}$ rather than aligning with it, as would a bar magnet, at a frequency given by Eq. (4). The precession frequency is called the Larmor frequency.

The energy separating adjacent energy levels is small. For example, $\Delta E \simeq$ 0.08 J mol^{-1} for the ^1H nucleus in a magnetic field of 4.7 T. Thus, at equilibrium,

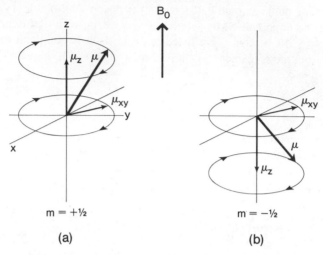

Fig. 1. Vectorial representation of the two allowed orientations and the Larmor precession of the magnetic moment of a spin-½ nucleus in magnetic field B_0. The $m = +½$ state (*a*) is taken to be the α state and $m = -½$ the β state (*b*).

the two levels are nearly equally populated. The ratio of the populations can be predicted by the Boltzmann distribution law to be

$$\frac{N_2}{N_1} = \exp{-2\mu\beta_0/kT} \tag{5}$$

where N_1 and N_2 are the number of nuclei in the lower and higher energy states, k the Boltzmann constant, and T the absolute temperature. Since $2\mu B_0/kT \ll 1$, this ratio can be approximated as

$$\frac{N_2}{N_1} \simeq 1 - \frac{2\mu B_0}{kT} \tag{6}$$

N_2/N_1 is approximately 0.999968 for the ^1H nucleus in a magnetic field of 4.7 T, which corresponds to the population of the lower energy state being larger by 0.0032%.

The probabilities of absorption and stimulated emission of radiation ($m = +½$ to $-½$ and $m = -½$ to $+½$, respectively, for $I = ½$) are equal and much larger than the probability of spontaneous emission at megahertz frequencies. Since the $m = +½$ level is slightly more populated, there is an excess of absorption transitions and thus a net absorption of energy. Since only the net absorption is detected, the signal strength is determined by the very small excess of nuclei in the lower energy state, e.g., only 0.0016% of the ^1H nuclei in a sample contribute to its ^1H-NMR spectrum measured at 4.7 T, and thus the sensitivity of NMR is inherently low.

In a more classical picture of NMR, the nearly equal populations of the two allowed orientations are represented as in Fig. 2a. The precession of each magnetic moment can be resolved into a static component along the z axis and a rotating component in the xy plane (Fig. 1). Since the magnetic moments are randomly precessing around $\mathbf{B}_0$ (Fig. 2a), there is no net macroscopic magnetization in the xy-plane from either orientation. Each orientation has a net magnetization along the z axis, and since the lower energy level is slightly more populated, there is a nonzero macroscopic magnetization M_z in the direction of $\mathbf{B}_0$ (Fig. 2b).

The NMR spectrometer is designed so that it will detect a nonzero macroscopic magnetization in the xy plane, M_{xy}. To create M_{xy}, the nuclei are subjected to a second magnetic field, $\mathbf{B}_1$, which is rotating in a plane perpendicular to $\mathbf{B}_0$. When $\mathbf{B}_1$ rotates at precisely the Larmor frequency, the nuclear magnetic moments experience an effective field, $\mathbf{B}_{eff}$, comprised of $\mathbf{B}_0$ and $\mathbf{B}_1$, which tips the nuclear magnetic moments away from their equilibrium positions towards the xy plane. The nonzero M_{xy} rotates around $\mathbf{B}_0$ at the Larmor frequency and is detected as the NMR signal by the voltage it induces in a receiver coil. Relative to $\mathbf{B}_1$ (which is directed along the x axis), M_{xy} has two components 90° out of phase with each other, one of which is perpendicular to the rotating field $\mathbf{B}_1$ and the other of

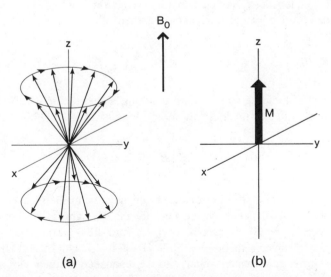

Fig. 2. (a) Vectorial representation of the magnetic moments from an ensemble of spin-½ nuclei in $\mathbf{B}_0$. (b) Macroscopic magnetization from the ensemble of spin-½ nuclei.

which is in phase with $\mathbf{B}_1$. That component which is perpendicular to $\mathbf{B}_1$ gives a pure absorption signal, while the other component gives a pure dispersion signal. The two different signals can be detected by phase-sensitive detection of a selected component of M_{xy}.

Experimentally, $\mathbf{B}_1$ is produced by passing a current of the appropriate frequency through a coil whose axis is perpendicular to $\mathbf{B}_0$. This axis is generally taken to be along the x axis. The coil generates a linearly oscillating, polarized magnetic field, which can be represented as two fields rotating in opposite directions. The field which is rotating in the same sense as the nuclear precession is $\mathbf{B}_1$ in the NMR experiment.

2.3. NMR Parameters

Some of the important NMR parameters are chemical shifts, spin–spin coupling constants, relaxation times, intensities, and widths of resonance lines.

2.3.1. Chemical Shifts

Equation (4) predicts that the resonance frequency for a particular nucleus depends only on its γ and $\mathbf{B}_0$. In fact, the actual resonance frequency also depends on the chemical environment of the nucleus because $\mathbf{B}_0$ produces a secondary magnetic field of a magnitude proportional to $\mathbf{B}_0$ by inducing motion in the electron cloud within the molecule. Thus, the actual magnetic field experienced by the nucleus, $\mathbf{B}_{local}$, depends to some extent on the electron density around the nucleus, which is determined by its chemical environment. $\mathbf{B}_{local}$ is less than $\mathbf{B}_0$ according to the relation

$$\mathbf{B}_{local} = \mathbf{B}_0 - \sigma\mathbf{B}_0 = \mathbf{B}_0 (1\text{-}\sigma) \tag{7}$$

where σ is the shielding or screening constant. Thus, the actual resonance frequency for a particular nucleus in a molecule is given by

$$\nu = \frac{\gamma}{2\pi} B_0 (1 - \sigma) \tag{8}$$

The magnitude of this shift in resonance frequency due to chemical environment, the chemical shift, depends on the nucleus and $\mathbf{B}_0$. Thus, ^{1}H chemical shifts cover a range of ~600 Hz at a $\mathbf{B}_0$ of 1.4 T and 4000 Hz at 9.4 T, while ^{13}C chemical shifts cover a range of ~3800 Hz at 1.4-T and 25,000 Hz at 9.4-T. Chemical shifts are measured relative to the resonance frequency of a reference compound. Specific reference compounds are discussed in Section 7.1.4. It has become general practice to describe nuclei as being shielded or deshielded relative to a reference resonance if their resonance occurs at lower or higher frequency (at constant B_0).

Chemical shifts measured relative to a reference compound are dependent on the magnitude of B_0 used in the measurement. Thus, a chemical shift of 326 Hz in an ^{1}H-NMR spectrum measured with a 1.4-T (60-MHz) spectrometer would be 1086 Hz in a spectrum measured with a 4.7-T (200-MHz) spectrometer. To eliminate the dependence on B_0, chemical shifts are usually reported as dimensionless quantities, defined as

$$\delta = \frac{\nu_{obs} - \nu_{ref}}{\nu_{spect}} \times 10^6 \tag{9}$$

where $\nu_{obs} - \nu_{ref}$ is the difference (in Hz) between the frequency observed for the resonance of interest and the reference resonance, ν_{spect} is the nominal frequency (in Hz) of the spectrometer, e.g., 200×10^6 Hz for ^{1}H-NMR on a 4.7-T spectrometer, and δ is the chemical shift in units of parts per million (ppm). Thus, the chemical shift differences of 326 Hz at 60 MHz and 1086 Hz at 200 MHz both correspond to 5.43 ppm. According to Eq. (9), nuclei which are deshielded relative to the reference will have positive δ values. NMR spectra are normally plotted with δ increasing from right to left, i.e., with shielding increasing from left to right.

2.3.2. Spin–Spin Coupling Constants

The above discussion of chemical shifts would suggest that the ^{1}H-NMR spectrum for a molecule having hydrogen atoms in n chemically different environments would consist of n resonances. This is generally not the case; rather, some of the resonances are usually split into multiplets. There are examples of both singlets and multiplets in Fig. 3. The multiplets arise from through-bond spin–spin coupling interactions, which split the energy levels of neighboring nuclear magnetic moments.

2.3.2.1. Spin–Spin Coupling.   The multiplet patterns for the various chemically equivalent groups of nuclei in a spin system depend on (1) the number of nuclei in each of the different groups, (2) $\Delta\nu/J$ where $\Delta\nu$ is the difference in the chemical shift of the resonances for the coupled groups, and (3) whether the nuclei in each group are magnetically equivalent. Nuclei are magnetically equivalent if they have the same chemical shift and couple identically to all the nuclei in each chemically equivalent group. Magnetic equivalence may result from rapid rotation, e.g., the three methyl protons of $—CH_2CH_3$ are magnetically

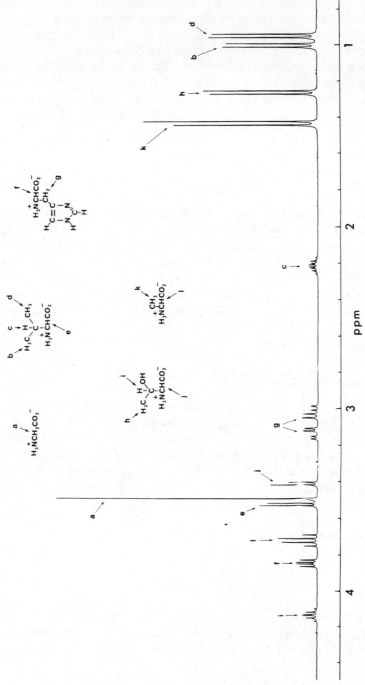

Fig. 3. A portion of the ^{1}H-NMR spectrum (360 MHz) of a D$_2$O solution containing glycine, alanine, valine, threonine, and histidine.

equivalent as a result of rapid rotation around the C—C bond. The H_3 and H_6 protons of o-dichlorobenzene are magnetically nonequivalent because, although they are chemically equivalent, they couple differently to H_4 and H_5.

When $\Delta v/J$ for groups of spin-coupled, magnetically equivalent, $I = \frac{1}{2}$ nuclei is large, the number of lines in multiplet patterns and their intensities can be predicted by the following rules:

1. A resonance is split into $n + 1$ lines, where n is the number of magnetically equivalent nuclei which cause the splitting (the ^1H nuclei of CH, CH_2, and CH_3 groups will split neighboring groups of nuclei into doublets, triplets, and quartets).

2. The relative intensities of lines in a multiplet are given by the coefficients of the binominal expansion (1:1 for a doublet, 1:2:1 for a triplet, 1:3:3:1 for a quartet).

3. The separation in Hz between adjacent lines in a multiplet pattern is the coupling constant J.

4. The multiplet patterns are independent of the signs of the J's.

5. Spin–spin coupling between magnetically equivalent nuclei does not appear in the spectrum.

Nuclei which satisfy the conditions of $\Delta v/J$ being large (greater than ~ 7) and magnetic equivalence are said to be weakly coupled and the splitting is first order. If Δv is comparable to J, or if the nuclei are chemically but not magnetically equivalent, the system is said to be strongly coupled and the resonance patterns, which will be systems of lines having little or no regularity of spacing or intensity, are second order. More specifically, as Δv approaches J, the intensities of lines in a multiplet pattern become skewed toward the resonance position of the nuclei which cause the splitting, and the position of the lines no longer gives directly the chemical shift or the coupling constant. The chemical shifts and coupling constants can be extracted from such spectra by quantum mechanical calculations, and there are several standard computer programs which treat spin systems containing up to 6–8 nuclei. The use of these programs, which generally are supplied by instrument companies as part of their software package, falls into the realm of spectral analysis (1–4), which is beyond the scope of this chapter.

It is important to note that, since Δv is directly proportional to B_0 whereas J is not, a multiplet pattern becomes more first order as B_0 increases. Thus, spectrum analysis becomes easier the larger B_0 because the separation (in Hz) between multiplet patterns increases and the multiplet patterns become more first order. These effects are illustrated by the spectra in Fig. 4 for a mixture of amino acids in D_2O solution. The 1.5–2.5 and 3.5–4 ppm regions of the 100-MHz spectrum contain a number of overlapping resonances, which increasingly spread apart at 200 and 360 MHz.

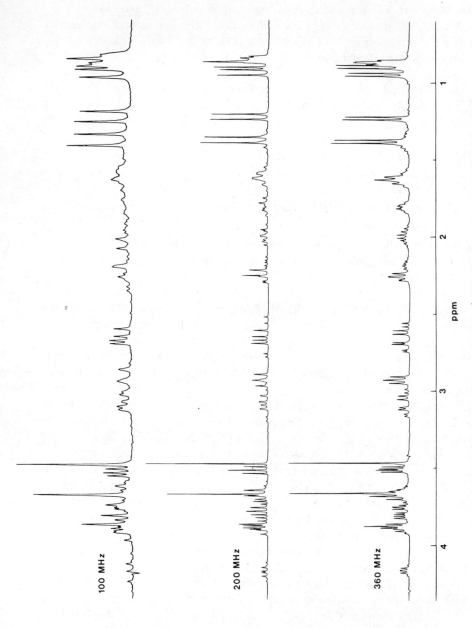

Fig. 4. Portions of the 100, 200, and 360 MHz ^{1}H-NMR spectra of a mixture of amino acids in D$_2$O solution.

The nomenclature generally used to describe spin systems is:

1. weakly coupled groups of chemically and magnetically equivalent nuclei are represented by letters from different parts of the alphabet, and the number of nuclei in each group is denoted by a subscript (e.g., A_3X_2 and AMX).

2. strongly coupled chemically and magnetically equivalent groups of nuclei are represented by letters from the same part of the alphabet (A_3B_2).

3. systems with both weak and strong coupling are represented by a combination of the above (ABX).

4. magnetic nonequivalence in a chemically equivalent group of nuclei is represented by using the same letter for each of the nuclei in the group to indicate chemical equivalence and primes to indicate magnetic nonequivalence ($AA'A''B$ and $AA'BB'$).

Spin–spin coupling to nuclei with a spin greater than $\frac{1}{2}$ usually is not observed because of the short relaxation times for these nuclei. For example, coupling of ^1H to chlorine ($I = \frac{3}{2}$ is not observed and coupling of $_1H$ to nitrogen ($I = 1$) is seldom observed.

2.3.2.2. Spin Decoupling.   Multiplet patterns can be simplified by irradiating the sample with a second magnetic field B_2 perpendicular to B_0 and oscillating at the resonance frequency of the nucleus which causes the splitting. If the X transitions of an AX spin system are irradiated with a sufficiently large B_2, the magnetic moments of X are no longer quantized along B_0, but rather along an axis perpendicular to B_0. Since A and X are quantized in directions perpendicular to one another, the spin–spin coupling interaction vanishes and no spin coupling from X is observed on the A resonance.

Spin decoupling is illustrated by the spectra in Fig. 5, which were obtained for the same mixture of amino acids as in Fig. 3. Irradiation with B_2 at 4.139 ppm (Fig. 5A) causes the doublets at 3.410 and 1.258 ppm to collapse to singlets, indicating these multiplets to be connected by spin–spin coupling. With this information and the nature of the multiplets, they can be assigned to the carbon-bonded protons of threonine (Fig. 3). In the same way, the connectivities established in Fig. 5B and 5C identify the resonances for valine and alanine, respectively. Spin decoupling is an important aid for the assignment of resonances in spectra of large molecules and complex mixtures.

Depending on whether the irradiated and observed nuclei are of the same type, the decoupling is said to be homonuclear or heteronuclear. The usual notation is to represent the irradiated nucleus within brackets, e.g., ^1H-NMR with ^1H decoupling would be represented as $^1H\{^1H\}$ and ^{13}C-NMR with ^1H decoupling as $^{13}C\{^1H\}$.

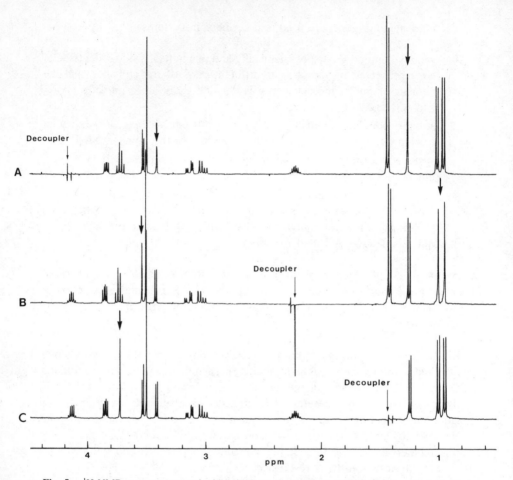

Fig. 5. ¹H-NMR spectra measured with spin decoupling at the indicated positions. The sample is the same mixture of amino acids as in Fig. 3. The connectivities, and thus the assignment, of resonances can be established by these results.

The major applications of spin decoupling are for assignment purposes and for removing splitting by ¹H from NMR spectra of other nuclei, e.g., ¹³C and ¹⁵N spectra. In the first case, the decoupling is selective since a small frequency range is irradiated; in the second case, broadband decoupling is used to remove all ¹H coupling. Selective homonuclear spin decoupling works best when the spin system is first order. Any resonance which is close to the B_2 frequency will be affected by B_2 and its intensity may be altered. Also, in homonuclear spin decoupling, resonance positions will shift due to the B_2 field. When $(\gamma B_2)^2 \ll (\nu_A - \nu_X)^2$, the shift is given by (5)

$$\nu_{obs} = \nu_A + \frac{(\not\!\gamma B_2)^2}{2(\nu_A - \nu_X)} \qquad (10)$$

where $\not\!\gamma = \gamma/2\pi$, ν_A is the frequency of resonance A, and ν_X the frequency of B_2. This shift in resonance frequency, known as the Bloch–Siegert shift, occurs because the effective fields that the various nuclei experience are altered by the $\mathbf{B}_2$ field. Resonances are shifted away from the irradiation frequency, the shift being larger the closer the resonance is to the irradiation frequency. Thus, care must be taken when obtaining chemical shifts from decoupled spectra. For example, the chemical shift of the glycine resonance is 3.495 ppm in Fig. 3. In Fig. 5, it shifts to 3.488, 3.502, and 3.496 ppm with the decoupler set at 4.139, 2.218, and 1.431 ppm, respectively. Bloch–Siegert shifts are not observed in heteronuclear decoupling because of the large frequency differences. Bloch–Siegert shifts provide a convenient experimental method for determining the magnitude of $\mathbf{B}_2$.

Intensities of resonances in first-order spectra can be affected by spin decoupling even when the $\mathbf{B}_2$ field does not perturb the observed resonance, as can the intensities of resonances which are not spin-coupled to the irradiated nucleus. This is known as the nuclear Overhauser effect (NOE) (6) and is discussed in the section on resonance intensities.

2.3.3. Nuclear Relaxation

When a spin system is placed in a magnetic field, it exchanges energy with its surroundings or lattice to reach the equilibrium distribution predicted by the Boltzmann equation. This is a nonradiative process, characterized by the spin–lattice relaxation time T_1.

At equilibrium, the number of spin ½ nuclei in each of the 2 possible energy levels remains constant. However, it is a dynamic equilibrium, since the mechanisms by which spin–lattice relaxation occurs are still operative, and thus nuclei are continually being transferred between the two levels. If W_u is the probability that a nucleus will undergo a transition from the lower to the upper energy level and W_d the probability for a downward transition, this dynamic equilibrium can be represented as

$$W_u N_1 = W_d N_2 \qquad (11)$$

where N_1 and N_2 are the number of nuclei in the lower and higher energy levels [Eq. (6)]. Since $N_1 > N_2$, the probabilities of the upward and downward nonradiative transitions are not equal.

As discussed in Section 2.2, there is a net macroscopic magnetization in the z direction, M_z, but M_x and M_y are zero at equilibrium. When the resonance condition is reached, M_z is decreased from its equilibrium value and M_{xy} is no longer zero. If, after resonance is established, the magnitude of $\mathbf{B}_0$ or the frequency of $\mathbf{B}_1$ is changed so that the resonance condition is no longer satisfied, M_z will return to its equilibrium value at a rate determined by its spin–lattice or longitudinal relaxation time T_1, and M_{xy} will return to 0 at a rate determined by the spin–spin or transverse relaxation time T_2. Changes in M_z result from energy transfer between the spin system and the lattice, and thus spin–lattice relaxation is an enthalpy effect. On the other hand, spin–spin relaxation can occur with no change in the total energy of the spin system by a mutual exchange of spin states in the spin system, and thus is regarded as an entropy effect.

Both spin–lattice and spin–spin relaxation processes result from time-dependent magnetic or electric fields generated in the lattice by molecular motion. A nuclear spin may experience fluctuations in its local magnetic field as a result of fluctuating magnetic fields from electron or nuclear spins which are tumbling or rotating nearby, or as a result of changes in its chemical shielding caused by rotation of the molecule. If these fluctuations can couple directly with the nuclear spin and if they are of the proper frequency, they will be effective at causing relaxation. Fluctuations at frequencies of 0, ω_1 and $2\omega_1$, where ω_1 is the Larmor frequency in rad/s, are effective in T_2 relaxation processes, whereas T_1 relaxation is caused only by fluctuations of ω_1 and $2\omega_1$ (7). Consequently, T_2 is always shorter than or equal to T_1.

The important mechanisms for relaxation are nuclear–nuclear and electron–nuclear dipole–dipole interactions, electric quadrupole interaction, chemical shift anistropy, spin rotation, and scalar coupling. The first three will be discussed briefly; more thorough treatments are given in several excellent texts (1–3,6,7).

2.3.3.1. Nuclear Dipole–Dipole Interaction.

Consider a nucleus of $I = \frac{1}{2}$ which experiences a magnetic field from other similar nuclear magnetic dipoles at a distance r. Tumbling of molecules causes the magnetic field to fluctuate, giving rise to a relaxation mechanism. The contributions of this mechanism to the spin–lattice and spin–spin relaxation rates are (8)

$$\frac{1}{T_1} = \frac{3\gamma^4\hbar_2}{10\,r^6}\left[\frac{\tau_c}{1\,+\,\omega_1{}^2\tau_c{}^2} \,+\, \frac{4\tau_c}{1\,+\,4\omega_1{}^2\tau_c{}^2}\right] \qquad (12)$$

and

$$\frac{1}{T_2} = \frac{3\gamma^4\hbar^2}{10\,r^6}\left[1.5\,\tau_c \,+\, \frac{2.5\tau_c}{1\,+\,\omega_1{}^2\tau_c{}^2} \,+\, \frac{\tau_c}{1\,+\,4\omega_1{}^2\tau_c{}^2}\right] \qquad (13)$$

where ω_1 is the Larmor frequency in rad/s and τ_c is the rotational correlation time. When $\omega_1\tau_c \ll 1$, as for example in nonviscous solutions, we have the so-called extreme narrowing condition and $T_1 = T_2$. At long τ_c, as for example with a protein, $T_2 \ll T_1$. The equations indicate a dependence of T_1 on the frequency of the spectrometer when the extreme narrowing condition is not met. In ^{1}H-NMR, the dipole–dipole interaction generally provides the dominant relaxation mechanism.

2.3.3.2. Electron–Nuclear Dipole–Dipole Interaction.

An unpaired electron in a paramagnetic substance provides a relaxation mechanism similar to the nuclear–nuclear dipole interaction but with the nuclear magnetic moment replaced by the electron magnetic moment. This is a very efficient mechanism for nuclear relaxation, because the electron magnetic moment is about 1000 times larger than nuclear magnetic moments. Consequently, traces of paramagnetic impurity, e.g., Cu^{2+} or Mn^{2+}, can cause large decreases in relaxation times with corresponding increases in line widths. As discussed in Section 7, this forms the basis for relaxation and broadening reagents. Traces of molecular oxygen can also affect relaxation times and resonance line widths.

2.3.3.3. Electric Quadrupole Interaction.

Nuclei with $I > \frac{1}{2}$ possess an electric quadrupole moment which can interact with electric field gradients near the nucleus. Molecular motion causes a fluctuation in these electric field gradients which provides another mechanism for relaxation for $I > \frac{1}{2}$ nuclei. In the extreme narrowing limit, the relaxation rates due to quadrupolar interactions are given by (7)

$$\frac{1}{T_1} = \frac{1}{T_2} = \frac{3}{40} \frac{2I+3}{I^2(2I-1)} \left(1 + \frac{\eta^2}{3}\right) \left(\frac{e^2Qq}{\hbar}\right)^2 \tau_c \qquad (14)$$

where η represents an asymmetry parameter and $e^2Qq/\hbar$ represents the quadrupolar coupling. In highly symmetric molecules, for example in ammonium ions, the ^{14}N-NMR signal is very sharp and splittings due to proton couplings are observed. In asymmetric molecules, as in ammonia, the quadrupolar relaxation mechanism is extremely important. Consequently, relaxation is rapid and resonances are broad.

2.3.3.4. Magnetic Field Inhomogeneity.

M_{xy} is a resultant from the magnetic moments of an ensemble of nuclei which are spread throughout the sample. If nuclei in different regions of the sample experience a different $\mathbf{B}_0$, i.e., if $\mathbf{B}_0$ is not perfectly homogeneous, they will precess with slightly different Larmor frequencies and M_{xy} will tend to fan out. Since the total rate of decay is caused by $\mathbf{B}_0$ inhomogeneity and natural spin–spin relaxation, the apparent spin–spin relaxation rate, $1/T_2^*$, is given by

$$\frac{1}{T_2^*} = \pi \Delta W_{1/2} + \frac{1}{T_2} \tag{15}$$

where $\Delta W_{1/2}$ is the increase in resonance width (in hertz) at half height due to magnetic field inhomogeneity. Magnetic field inhomogeneity is generally the greatest contributor to the time dependence of M_{xy}.

2.3.4. Resonance Intensities

2.3.4.1. Factors Affecting the Intensity.   The resonance intensity, that is, the area under the resonance peak, is directly proportional to the number of nuclei giving the resonance. An important property of NMR is that the proportionality constant only depends on the nucleus and not on its chemical environment if the experiment is performed properly (a sufficiently small B_1 in continuous wave NMR and a sufficiently slow repetition rate in P/FT NMR). The proportionality constant is different for ^1H and ^{13}C resonances, but is the same for all ^1H resonances in a spectrum, and thus the relative intensities of the resonances in the ^1H spectrum correspond directly to the relative numbers of nuclei. This property has important consequences in the use of NMR as a technique for chemical analysis. For example, pure samples are not required to calibrate the instrument; rather, a pure sample of another compound can be used. This is discussed further in Section 7.

The inherent NMR sensitivity of a nucleus depends upon a number of factors. As discussed earlier, an NMR signal is detected for an $I = \frac{1}{2}$ nucleus because there is a very small excess of nuclei in the lower energy state. The magnitude of the signal is limited by the magnitude of this population difference, which in turn depends on the μ of the nucleus, B_0, and the temperature [Eq. (6)]. The actual dependence of intrinsic signal intensity, S, on μ and B_0 is given by (1):

$$S \propto \frac{I + 1}{I^2} \mu^3 B_0^{\,2} \tag{16}$$

According to this equation, the peak height increases as $B_0^{\,2}$. However, the r.m.s. noise of the spectrometer increases as $B_0^{\,1/2}$, so that the overall signal-to-noise ratio increases as $B_0^{\,3/2}$.

The relative inherent sensitivities of various nuclei as predicted by Eq. (16) are compared in column 7 of Table 1. The inherent sensitivity of ^{13}C relative to ^1H, for equal numbers of ^1H and ^{13}C nuclei at constant B_0, is 0.016. The actual relative sensitivity of ^{13}C, and of many other nuclei, is considerably below that listed in column 7 due to their low natural abundance (given in column 4). Their relative sensitivities at natural abundance levels are compared in column 8. The low natural abundance coupled with small nuclear magnetic moments limits the usefulness of many magnetically active nuclei for trace analysis. Signal averaging is generally necessary, and thus the P/FT method is used almost exclusively in NMR measurements on low-sensitivity nuclei. Recently, several ingenious multiple-pulse schemes have been devised with which resonance intensities for isotopically dilute, low-sensitivity nuclei can be increased by transferring magnetization from protons to these nuclei. These methods will be discussed in the section on P/FT NMR. Resonance intensities for nuclei such as ^{13}C and ^{15}N can also be affected if ^1H-spin decoupling is used due to the nuclear Overhauser effect.

2.3.4.2. The Nuclear Overhauser Effect.

When spin decoupling, one often observes changes in the intensities of resonances, even though they are not spin coupled to the resonance irradiated. These changes are due to the nuclear Overhauser effect (NOE), which is described by

$$M_A^X = (1 + \eta_A^X)M_A^0 \qquad (17)$$

where M_A^X and M_A^0 represent the maximum macroscopic magnetization of A which can be rotated into the xy plane (i.e., M_z) when X is saturated and at equilibrium, respectively, and η_A^X is the NOEF (nuclear Overhauser effect factor) on A when X is irradiated. The NOE arises when changes in the energy level populations of the X spin affect those of the A spin. This effect is intimately associated with their respective relaxation processes.

Consider the energy level diagram in Fig. 6 for the AX spin system. A and X may be weakly coupled or there may be no coupling; the important criterion for the NOE to change the intensity of A when X is irradiated is that the relaxation

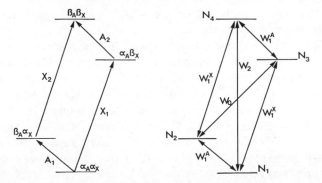

Fig. 6. Energy level diagram for the AX spin system showing (*left*) A and X spin states and transitions and (*right*) energy level populations (*N*) and transition probabilities (*W*).

of A and X be by mutual dipole–dipole interaction. W_0, W_1, and W_2 in Fig. 6 represent, respectively, the probabilities for zero ($\alpha_A\beta_X \leftrightarrow \beta_A\alpha_X$), single, and double ($\alpha_A\alpha_X \leftrightarrow \beta_A\beta_X$) quantum transitions due to spin–lattice relaxation.

In the region of extreme narrowing, the relative transition probabilities for dipolar relaxation, $W_0:W_1{}^A:W_2$, are 2:3:12. In the absence of any irradiation at the X frequency, the equilibrium magnetization of the A nucleus (Fig. 6) is proportional to $(N_1 - N_2) + (N_3 - N_4)$. When the X transitions are irradiated with a sufficiently large B_2, $N_1 = N_3$ and $N_2 = N_4$. Relaxation still operates, but since W_2 is so much larger than W_0 and $W_1{}^A$, the W_2 pathway dominates with a resulting increase in N_1. However, irradiation of the X frequency transfers some of these nuclei to N_3 so that, at the new, non-Boltzmann steady-state condition, there is a net gain in the populations of levels 1 and 3. The result is that $N_1 - N_2$ and $N_3 - N_4$ are increased, and thus the intensity of the A resonance is increased.

Quantitatively, this increase is given as the NOEF. If the dipole–dipole mechanism is dominant over all other relaxation mechanisms:

$$\eta_A{}^X = \frac{\gamma_X}{2\gamma_A} \tag{18}$$

This equation gives the maximum value for the NOE. When other mechanisms contribute to the relaxation, η_A^X is given by

$$\eta_A^X = \frac{\rho_{AX}}{\rho_{AX} + \rho_A}\left(\frac{\gamma_X}{2\gamma_A}\right) \tag{19}$$

where ρ_{AX} represents the dipolar relaxation rate and ρ_A the relaxation rate by other mechanisms. If ρ_A is appreciable relative to ρ_{AX}, internuclear distance information can be obtained (6).

According to Eq. (18), the maximum NOEF for the homonuclear case is 0.5; for the heteronuclear case, it depends on the nuclei and which is irradiated. For the $^{13}C\{^1H\}$ experiment, the maximum NOEF is 1.988, whereas for the $^1H\{^{13}C\}$ experiment, it is 0.126. It also is important to note that the NOE can cause a reduction in signal intensity if one γ is negative; e.g., for the $^{15}N\{_1H\}$ experiment the maximum NOEF is -4.935. If the maximum NOE is achieved in this experiment, the ^{15}N resonance will be inverted and of a greater intensity. However, if less than the maximum NOE is realized, the ^{15}N resonance can be reduced in intensity or even zero.

Carbon-13 spectra are often measured with broadband (BB) ^1H decoupling to simplify the spectrum to a single resonance for each carbon. The intensities of ^{13}C resonances measured with BB ^1H decoupling are increased by the NOE, with the maximum intensity of a resonance, I_{max}, given by

$$I_{max} = (1 + \eta_C^H) I_0 = 2.988 \, I_0 \tag{20}$$

where I_0 is the intensity in the absence of any NOE. Not all resonances will experience a maximum NOE, and thus the relative intensities in the BB ^1H-decoupled ^{13}C spectrum may not correspond directly to the relative numbers of ^{13}C nuclei. This is particularly true if some carbons do not have directly bonded protons. This complicates the use of ^{13}C NMR for quantitative chemical analysis. The NOE can be eliminated by gated decoupling schemes or with relaxation reagents, as discussed later; however, the signal intensities are considerably reduced when the NOE is eliminated.

2.3.5. Resonance Lineshapes

The NMR absorption signal has a Lorentzian line shape, with a width at half maximum intensity, $W_{1/2}$, given by

$$W_{1/2} = \frac{1}{\pi T_2} \tag{21}$$

Since it generally is impossible to obtain a perfectly homogeneous B_0 throughout the sample, the observed $W_{1/2}$ will generally be larger according to

$$W_{1/2} = \frac{1}{\pi T_2^*} \tag{22}$$

where T_2^* is the apparent T_2 [Eq. (15)].

The T_2^* is also shortened if the nuclei periodically change their Larmor frequency, e.g., through chemical exchange. The effect of chemical exchange on the resonance lineshape depends on the rate of chemical exchange. At rates of exchange which are slow on the NMR time scale—i.e., $\tau_1, \tau_2 \gg (\nu_1 - \nu_2)^{-1}$, where τ_1 and τ_2 are the average lifetimes in sites 1 and 2 and ν_1 and ν_2 the chemical shifts of the nucleus in sites 1 and 2—separate resonances are observed for nuclei in the two different sites. As τ_1 and τ_2 decrease, the resonances broaden and then coalesce into a single, exchange-averaged resonance at an averaged chemical shift, from which average lifetimes can be estimated. NMR has been used extensively to study chemical exchange processes, many of which would be difficult or impossible to study by other kinetic methods. (See, for example, reference 9.)

3. CONTINUOUS WAVE NMR

In continuous wave (CW) NMR, the resonance condition is met successively by slowing varying either B_0 (field sweep mode) or v (frequency-sweep mode) while holding the other constant. In recent years, the CW technique has given way to the P/FT method. However, the CW technique will be described briefly since there still are many CW spectrometers in use.

3.1. Modes of Operation

In the frequency-sweep mode, the frequency of B_1 is slowly varied while B_0 is held constant. The recorder arm x drive is tied directly to a voltage controlled oscillator whose frequency output is determined by the recorder arm position. Movement of the recorder arm from left to right generally corresponds to decreasing the frequency, i.e., increased shielding. When the resonance condition for a particular nucleus is satisfied, the B_1 field tips the macroscopic magnetization for these nuclei slightly from the z axis, which results in a component in the xy plane. This component in turn induces an alternating EMF in the receiver coil, which, after amplification, causes a y displacement of the recorder pen. The frequency is swept through the spectral region of interest, which depends on the chemical-shift differences of the nuclei in the sample and the nucleus itself. A typical range for ^{1}H NMR at 2.3 T would be 1000 Hz.

In the field-sweep mode, the B_1 frequency is held constant while B_0 is slowly varied. The recorder arm x drive is tied directly to B_0, whose magnitude is determined by the recorder arm position. Movement of the recorder arm from left to right generally corresponds to increasing B_0, i.e., to increased nuclear shielding.

The B_1 power level and the scanning rate influence the shape of the NMR signal. Too high a power level causes saturation of the nuclear spins and a reduction in signal intensity, as discussed in 3.2. The B_1 power level at which saturation occurs depends on the scanning rate. The slower the scanning rate, the lower the power must be to avoid saturation. However, the slower the scanning rate, the better the resolution attainable. The B_1 power level is usually adjusted to just below saturation for the scanning rate which gives acceptable resolution. Typically, for a 10-ppm (1000-Hz) spectral width in ^{1}H NMR, a scan time of 500 s is used. As the scanning rate increases, peak shapes become skewed in the direction of the scan, the resolution becomes poorer, and the accuracy of peak positions decreases. These distortions in resonance lineshapes can be corrected for mathematically. This is done in the CW technique of rapid scan/ Fourier Transform NMR (10) which is discussed in Section 3.3.

3.2. Resonance Intensities and Saturation

As discussed in Section 2.2, the transition probabilities for absorption and stimulated emission of radiation are equal ($W_a = W_e = W$) and depend on the intensity of radiation at the frequency which causes the transitions. Since there is an excess of nuclei in the lower energy level for spin ½ nuclei ($N_{eq} = N_1 - N_2$), there is a net absorption of energy, the actual amount depending on the B_1 power level. It follows, however, that with time the population difference and thus the signal strength will go to zero since there is an excess of upward transitions. Whether or not this happens depends on the B_1 power level because, while the excess of upward radiative transitions are tending to equalize the populations of the energy levels, the nonradiative spin–lattice relaxation processes are tending to restore the spin system to equilibrium. As long as B_1 is small enough that spin–lattice relaxation processes can keep the energy level populations at equilibrium, the steady-state population difference N_{ss} is equal to N_{eq} and the signal strength will increase as B_1 is increased. At some B_1, however, spin–lattice relaxation can no longer maintain $N_{ss} = N_{eq}$ and the signal is saturated. At sufficiently large B_1, the signal virtually disappears due to saturation. The dependence of N_{ss}, and thus the signal strength, on B_1, T_1, and T_2 is given by (1)

$$N_{ss} = \frac{N_{eq}}{1 + \gamma^2 B_1^2 T_1 T_2} \tag{23}$$

The signal begins to saturate when B_1 is large enough that $\gamma^2 B_1^2 T_1 T_2$ is no longer negligible relative to 1, and thus, in CW NMR, signal intensity cannot be increased indefinitely by increasing B_1. This equation also indicates that for relative intensities to be equal to relative numbers of nuclei, B_1 has to be below the level that would cause saturation of the nucleus having the longest T_1 and T_2.

The areas of resonances in CW NMR spectra are generally determined by electronic integration of the signal as the spectrum is scanned. The accuracy of the integral is generally limited by noise in the spectrum. Under ideal conditions, the integrals are accurate to 1–2%. Other factors such as spinning side bands and satellite resonances, e.g., from coupling to ^{13}C in ^1H NMR, also affect the accuracy of quantitative analysis by CW NMR (11).

3.3. Sensitivity Enhancement by Signal Averaging

The concentration levels necessary for CW NMR are relatively high, e.g., the minimum concentration from which an ^1H NMR spectrum can be measured by a single scan through the spectrum is in the range of 0.01 to 0.1 M. From these limits and the data in Table 1, it is clear that it is practical to do only ^{1}H, ^{19}F, and ^{31}P NMR by the CW method.

The detection limits are governed by electronic noise. The signal strength can be increased relative to the noise level by signal averaging, i.e., repetitively sweeping through the spectrum and adding successive spectra coherently in computer memory. Signal strength will add as N, the number of scans, whereas noise will add as $\sqrt{N}$. Thus, the signal-to-noise ratio (S/N) will improve as $\sqrt{N}$. However, this is a very inefficient method of enhancing sensitivity because of the slow scan rates that must be used to prevent distortions in resonance line shape, e.g, a 500 s scan time for a 1000 Hz spectral width. To increase the S/N by 10 using these conditions would require 14 hours, which obviously is impractical. The rapid scan/Fourier transform (RS/FT) technique was developed to reduce the time required for signal averaging in CW NMR.

In the RS/FT method (10), a very fast scan rate is used, e.g., sweep times of just a few seconds. Distortions in line shapes due to fast passage conditions are removed by applying mathematical operations which cross-correlate the observed response with that for a single line scanned under the same conditions. The procedure involves first collecting the rapid scan spectrum in computer memory. The spectrum is subjected to a Fourier transformation, and then multiplied by an exponential function of time, the result of which is a spectrum equivalent to the free induction decay obtained in the P/FT method. Inverse Fourier transformation yields the slow passage spectrum.

Sweep times used in RS/FT are comparable to the times used to collect the free induction decay in P/FT NMR. Thus, when sensitivity enhancement by signal averaging is considered, the efficiency of RS/FT NMR approaches that of P/FT NMR. Although not widely used, RS/FT NMR offers several advantages over P/FT NMR, the most important being the absence of dynamic range problems when the solvent gives a large resonance (see Section 7.7). The solvent resonance can be avoided by simply limiting the spectral region which is scanned to one side or the other of the solvent resonance. Another advantage is that the sweep can be confined to any portion of the spectrum without foldover effects.

4. PULSE/FOURIER TRANSFORM NMR

In P/FT NMR, a high-power rf pulse is applied to the sample to produce a net transverse magnetization (M_{xy}). This is detected as a transient response, the free induction decay (FID) in the time domain. The FID is sampled and stored in a data-acquisition system, where it is Fourier transformed to give the frequency domain spectrum. This is the so-called single-pulse experiment, that is, a single rf pulse is applied, followed by acquisition of the FID. If the S/N is low, signal averaging can be done by repeated application of the single-pulse sequence, with coherent addition of the FIDs in the computer. As compared to CW NMR, P/ FT NMR has a tremendous time advantage when sensitivity enhancement is necessary because the time per FID is much less than the time per scan under slow passage conditions.

State-of-the-art pulse spectrometers also have the capability of performing multiple pulse experiments, i.e., experiments in which a series of carefully timed rf pulses are applied to the sample prior to acquisition of the FID. A variety of multiple-pulse sequences have been described, including sequences for measuring spin–lattice and spin–spin relaxation times, resolution enhancement, sensitivity enhancement, and two-dimensional experiments for increasing resolution in ^{1}H-NMR spectra and for correlating resonances between ^{1}H- and ^{13}C-NMR spectra. Some of these experiments provide information not obtainable by CW NMR, and are now part of the standard arsenal of techniques of the modern NMR spectroscopist. In this section, we describe in detail the single-pulse experiment, with emphasis on experimental aspects of importance for its application to trace analysis, and then we describe some multiple-pulse experiments. For a more detailed treatment of P/FT NMR, see references 8, 12, and 13.

4.1. Magnetization in the Rotating Frame

When a $\mathbf{B}_1$ pulse is applied, the individual nuclear magnetic moments follow a complex path in the laboratory frame of reference (a system of fixed Cartesian coordinates x, y, z), consisting of a rapid precession around $\mathbf{B}_0$ and a much slower precession around $\mathbf{B}_1$. The path is much simpler, however, if the motion is viewed in a coordinate system which rotates around $\mathbf{B}_0$ at the Larmor frequency. In this rotating frame of reference (a system of rotating Cartesian coordinates x', y', z'), each individual $\boldsymbol{\mu}$ appears to be stationary (Fig. 7a as compared to Fig. 2a). Thus, when a $\mathbf{B}_1$ pulse which is rotating around $\mathbf{B}_0$ at the Larmor frequency is applied along the x' axis, $\mathbf{B}_1$ appears to be stationary in the rotating frame and the individual nuclear magnetic moments, and thus the macroscopic magnetization, appear to precess around $\mathbf{B}_1$ only (Fig. 7). In this discussion, the frequency at which the reference frame rotates will be equal to the frequency of $\mathbf{B}_1$.

The motion of $\mathbf{M}$ in the rotating frame can be described by

$$\left(\frac{d\mathbf{M}}{dt}\right)_{\text{rot}} = \gamma\mathbf{M} \times \mathbf{B}_{\text{eff}} \tag{24}$$

where

$$\mathbf{B}_{\text{eff}} = \mathbf{B}_0 + \frac{\omega}{\gamma} \tag{25}$$

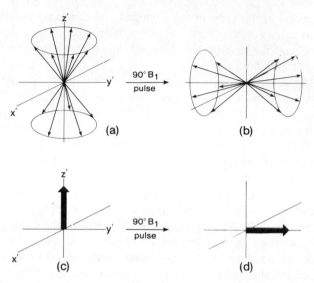

Fig. 7. Schematic representation of the effect of a $\mathbf{B}_1$ pulse on the nuclei of an $I = \frac{1}{2}$ spin system. (a) An ensemble of $I = \frac{1}{2}$ nuclei as viewed in the rotating reference frame. (b) The ensemble of nuclei as viewed in the rotating frame after being rotated through 90° around the x' axis by a $\mathbf{B}_1$ pulse applied along x'. (c) The macroscopic magnetization from the ensemble of nuclei. (d) The macroscopic magnetization in the rotating frame after the $\mathbf{B}_1$ pulse.

and $\boldsymbol{\omega}$ is the angular velocity of the rotating frame. ω/γ has units of magnetic field and can be regarded as a "fictitious field," $\mathbf{B}_f$, i.e., a field which accounts for the changed precessional frequency of the nuclei when referred to the rotating reference frame. $\mathbf{B}_f$ exists only as a result of the relative motion of our two coordinate systems. For $\omega = 0$, $\mathbf{B}_f = 0$ and $\mathbf{B}_{eff} = \mathbf{B}_0$, while for $\omega = -\gamma \mathbf{B}_0$, i.e., the reference frame is rotating at the Larmor frequency, $\mathbf{B}_{eff} = 0$, and the nuclear magnetic moments appear to be stationary.

Now, if a $\mathbf{B}_1$ which is rotating around $\mathbf{B}_0$ in the laboratory frame at ω rad s^{-1} is applied, then in a frame rotating at $\boldsymbol{\omega}$

$$\mathbf{B}_{eff} = \mathbf{B}_0 + \omega/\gamma + \mathbf{B}_1 \tag{26}$$

The frequency of $\mathbf{B}_1$ is often called the carrier frequency. When the carrier frequency equals the Larmor frequency, $\mathbf{B}_f$ cancels $\mathbf{B}_0$ and $\mathbf{B}_{eff} = \mathbf{B}_1$, i.e., the motion of $\mathbf{M}$ in the rotating frame is determined solely by $\mathbf{B}_1$. If $\mathbf{B}_1$ is along the x' axis, $\mathbf{M}$ precesses around the x' axis through an angle α (radians) given by

$$\alpha = \gamma B_1 t_p \tag{27}$$

where t_p is the length of the carrier pulse (pulse width) in seconds. If t_p is of a length such that M rotates through $90°$, $M_{y'} = M_0$, the magnetization along z' before the pulse, and the pulse is said to be a $90°$ or $\pi/2$ pulse (as depicted, e.g., in Fig. 7). Immediately after the pulse, $M_{z'}$ is zero.

Rarely is the resonance condition satisfied, since samples of interest generally give rise to a number of resonances due to chemical-shift effects and spin–spin coupling. In this situation, $\mathbf{B}_{eff}$ changes to

$$|\mathbf{B}_{eff}| = \frac{1}{\gamma}[(\omega_i - \omega)^2 + (\gamma B_1)^2]^{1/2} \tag{28}$$

where ω_i and ω represent the ith frequency in the spectrum and the frequency of the rotating frame, respectively. Thus, in the rotating frame, $\mathbf{M}_i$ precesses around $\mathbf{B}_{eff}$ at a frequency of $[(\omega_i - \omega)^2 + (\gamma B_1)^2]^{1/2}$. In the P/FT experiment, the pulse should cause all the nuclei of interest to precess through the same angle in the $y'z'$ plane, i.e., $\mathbf{B}_{eff}$ should be frequency independent in the spectral region of interest. Equation (28) indicates that for this to be the case, B_1 must be chosen so that

$$\gamma B_1 \gg (\omega_i - \omega) = 2\pi\Delta \tag{29}$$

where Δ represents the entire range of chemical shifts in hertz. When this is the case $\mathbf{B}_{eff} \sim \mathbf{B}_1$ and all nuclei in the spectral region Δ will be rotated through an angle given by Eq. (27). Substitution of Eq. (29) into Eq. (27) gives the result

$$t_p \ll \frac{1}{4\Delta} \tag{30}$$

for a $90°$ pulse.

After the pulse, the spin system begins immediately to relax back to equilibrium by T_1 and T_2 processes, as discussed previously. The longitudinal magnetization, $M_{z'}$, returns to equilibrium by T_1 relaxation, as described by

$$M_{z'} = M_0(1 - \exp(-t/T_1)) \tag{31}$$

where $M_{z'}$ is the longitudinal magnetization at time t following the pulse. The transverse magnetization, $M_{x'y'}$, decreases to zero by T_2 relaxation at a rate described by

$$M_{x'y'} = M_0\exp(-t/T_2^*) \tag{32}$$

where $M_{x'y'}$ is the transverse magnetization at time t after the pulse. Relaxation of the transverse magnetization can be pictured as a dephasing or spreading out of the individual nuclear magnetic moments in the $x'y'$ plane, due both to natural processes which cause T_2 relaxation and to inhomogeneity of the magnetic field. Chemically equivalent nuclei in different regions of the sample experience slightly different values of $\mathbf{B}_0$, and thus some precess faster than their average precession frequency, while others precess more slowly. Thus, $T_2{}^* \leqslant T_1$, and the transverse magnetization generally decays to zero before the longitudinal magnetization has returned to equilibrium.

4.2. The Single-Pulse Experiment

4.2.1. A Single 90° Pulse

In the single-pulse experiment, the $\mathbf{B}_1$ pulse is followed immediately by acquisition of the free induction decay. It is convenient to begin by considering this experiment when the pulse (flip angle) is 90°. As discussed above, for all nuclei in the spectral region of interest (Δ) to be rotated through 90° requires that the pulse be of a very short duration [Eq. (30)]. The required pulse widths depend on the width of the spectral region, which in turn depends on the nucleus and $\mathbf{B}_0$. For ^{1}H-NMR, Δ is typically 1000 Hz when $\mathbf{B}_0$ is 2.35 T and 4000 Hz at 9.4 T, which requires $t_p \ll 2.5 \times 10^{-4}$ s and 6.25×10^{-5} s, respectively. For ^{13}C-NMR, Δ is typically 6000 Hz when $\mathbf{B}_0$ is 2.35 T and 25,000 Hz at 9.4 T, which corresponds to $t_p \ll 4 \times 10^{-5}$ s and 1×10^{-5} s, respectively.

Following the 90° pulse, the transverse magnetization precesses *freely* in the $x'y'$ plane, and *induces* an alternating EMF in the receiver coil which *decays* with time due to $T_2{}^*$ processes. The signal detected as a function of time is the free induction decay (FID). The FID is detected with reference to the carrier frequency, and thus its intensity shows a frequency dependence superimposed on the exponential $T_2{}^*$decay. If phase-sensitive detection is used to detect $M_{y'}$, the FID for resonance A will have a time dependence described by

$$M_{y'} = M_A^0 \cos((\omega_A - \omega)t) \exp(-t/T_2{}^*) \tag{33}$$

where ω_A and ω are the resonance frequency and the carrier frequency, and M_A^0 is the equilibrium longitudinal magnetization for A. If the carrier frequency is set to the resonance frequency, the FID is a simple exponential decay. If the carrier is set to one end of the spectral region of interest, a FID of the type shown in Fig. 8A is obtained. The resonance frequency can be determined from the carrier frequency and the frequency of the intensity variation. Generally, spectra contain more than one resonance, in which case the observed FID will be a superposition of the individual FIDs. The FID shown in Fig. 8B was obtained

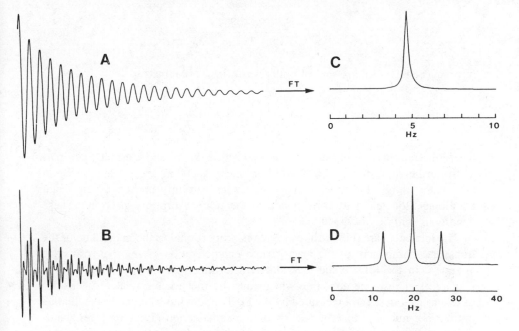

Fig. 8. Free induction decay (*A*) and (*B*) and frequency domain spectra (*C*) and (*D*) of a singlet and a triplet, respectively.

for three resonances, none of which is at the carrier frequency. When the FID contains more than two or three frequency components, it is difficult to extract the resonance frequencies or intensities by inspection. Thus, Fourier transformation is used to decompose the complex time domain function into its frequency components (spectra C and D in Fig. 8).

The shape of an NMR resonance is given by (1)

$$g(\omega) = \frac{T_2}{1 + T_2{}^2(\omega - \omega_i)^2} \tag{34}$$

which describes a Lorentzian curve. An exponentially decaying signal and a Lorentzian curve form a Fourier transform pair, and thus (8)

$$g(\omega) = \int_0^\infty e^{-t/T_2{}^*} \cos(\omega - \omega_i)t \, dt \tag{35}$$

The cosine transform yields the absorption signal [Eq. (34)] and the sine transform yields the dispersion signal

$$g(\omega) = \frac{(\omega - \omega_i)T_2^{2}}{1 + T_2^{2}(\omega - \omega_i)^2} \tag{36}$$

In practice, the complex sum of both transforms is obtained:

$$g(\omega) = \frac{T_2}{1 + T_2^{2}(\omega - \omega_i)^2} + i\frac{(\omega - \omega_i)T_2^{2}}{1 + T_2^{2}(\omega - \omega_i)^2} \tag{37}$$

where the real part corresponds to the absorption signal and the imaginary part to the dispersion signal. The absorption and dispersion spectra are stored in different regions of computer memory. Under carefully chosen experimental conditions, the real part corresponds to the true absorption spectrum. These conditions will be discussed later.

The intensity (area) of each resonance is proportional to the magnitude of the transverse component of the macroscopic magnetization giving that resonance at time 0 in the FID, while its width is determined by the rate of T_2^{*} decay [Eq. (22)]. Thus, one condition which must be met for the relative intensities of resonances in spectra measured by the P/FT method to be equal to the relative numbers of nuclei is that the macroscopic magnetization giving each resonance must be rotated through the same angle, in this case 90°. If the pulse width does not meet the condition of Eq. (30), resonance intensities will decrease continuously with increasing separation from the carrier frequency. Thus, Eq. (30) must be met if quantitative analysis is to be done by comparing the relative intensities of resonances. If Eq. (30) is not met, this source of error can be corrected for with a calibration curve prepared experimentally by measuring the intensity of a reference resonance as a function of the carrier frequency (14). The magnitude of the error can also be reduced by using smaller flip angles.

4.2.2. Acquisition of the FID

The FID is obtained in digital form, i.e., it is sampled point by point at discrete times (Fig. 9). At each point, a dc signal is obtained which is stored in a computer memory address. The sampling is continued until a signal has been stored in all the specified memory addresses. The time between successive points is called the dwell time.

To properly represent a cosine or sine wave in digital form, it must be sampled at least twice per cycle (Fig. 9), i.e., the highest frequency, the Nyquist frequency, which is uniquely defined is one half the sampling rate. Thus, if the carrier is positioned at one end of a spectral region of width Δ, the maximum frequency in the FID will be Δ. The minimum digitizing rate to properly represent FID's for this spectral width is 2Δ, and the dwell time is $1/(2\Delta)$. Figure 9 shows that cosine waves at frequencies lower than Δ ($=\nu$) will be sampled more than

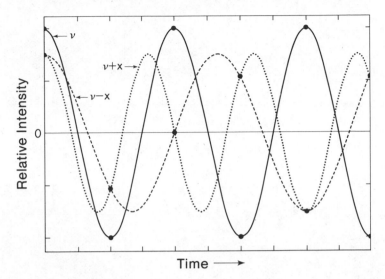

Fig. 9.   Digitization of cosine waves of frequencies v, $v - x$, and $v + x$ at a frequency of $2v$. In digital form, the cosine wave of frequency $v + x$ is indistinguishable from that of frequency $v - x$. The dark circles represent sampled points along the FID.

twice per cycle; however, higher frequencies will not. Rather, a cosine wave of frequency $v + x$ is indistinguishable from one of frequency $v - x$ when the sampling rate is $2v$ (Fig. 9). Thus, a resonance at a frequency of $v + x$ Hz will be detected as a resonance of frequency $v - x$, i.e., it will be folded back or *aliased* to the lower frequency. This effect is illustrated in Fig. 10B. Because noise at frequencies greater than v will also be folded back, a larger spectral width is often used, along with electronic filtering.

The FID is generally sampled at 2^n points, e.g., 4096 (4K), 8192 (8K), 16,384 (16K) or 32,768 (32K) points. The acquisition time is determined by the dwell time, which depends on the spectral width, and the number of points sampled. Thus, the acquisition time for the FID when $\Delta = 2000$ Hz spectral width sampled at 8K points is 2.048 s. Halving the spectral width doubles the acquisition time, as does doubling the number of points. To properly define peaks in the frequency-domain spectrum, the FID should be sampled for at least $3T_2^*$. Thus, the FID's for resonances having $W_{1/2}$ of 0.1, 0.5, and 5 Hz would have to be sampled for at least 9.5, 1.9, and 0.19 s, respectively.

Fourier transformation of the FID yields both absorption and dispersion spectra [Eq. (37)], which together occupy the same amount of computer memory as did the FID (Fig. 11). Thus, if the FID from a 1000-Hz spectral width is sampled at 8K points, the absorption spectrum obtained by Fourier transformation occupies 4K memory addresses, corresponding to 0.244 Hz/point. The narrowest $W_{1/2}$ which will be observed for resonances at this digital resolution is 0.488 Hz.

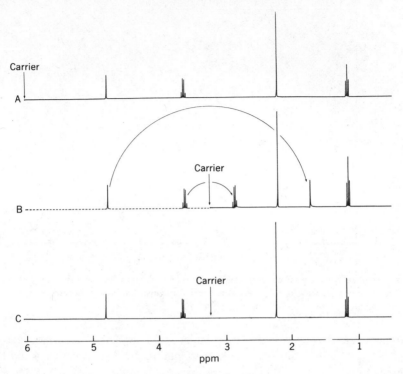

Fig. 10. Frequency domain spectra obtained by Fourier transformation of the FID collected (*A*) by single phase detection (SPD) with the carrier positioned at the left end of the spectrum, (*B*) by SPD with the carrier positioned in the middle of the spectrum, and (*C*) by quadrature phase detection (QPD) with the carrier positioned in the middle of the spectrum.

To obtain the very high resolution of which state-of-the-art spectrometers are capable, e.g., $W_{1/2} < 0.05$ Hz, the spectral width and the size of the FID must be carefully chosen.

To obtain accurate resonance intensities for quantitative analysis, the upper half of a resonance must be defined by at least 5 memory addresses (13) (Fig. 12). Thus, for a resonance having a $W_{1/2}$ of 0.40 Hz, the digital resolution must be < 0.08 Hz/point. If the spectral width is 1000 Hz, the FID must be sampled at 25,000 points. The closest larger value of 2^n is 32K, for which the acquisition time would be 16.384 s. However, since T_2^* for the resonance is 0.8 s, only noise is being sampled during a large fraction of this time; e.g., at $t = 3T_2^*$, $4T_2^*$, and $5T_2^*$, where $t =$ time after the start of sampling of the FID, the intensity of the FID is reduced to 0.050, 0.018, and 0.0067 of its value at $t = 0$. Thus, the FID can be sampled for a shorter time (~3–4 times the longest T_2^* in the spectrum), and then its size increased by zero filling, i.e., adding zeroed computer memory up to the size required for adequate digital resolution, since

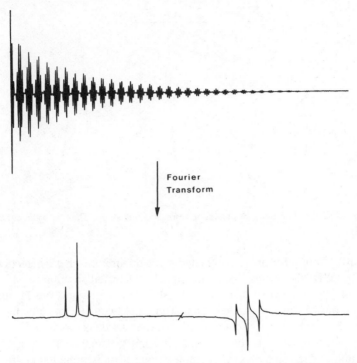

Fig. 11. The absorption and dispersion spectra obtained by Fourier transformation of a FID. The FID was sampled at 1024 points. The absorption and dispersion spectra each are defined by 512 points. The spectral width is 100 Hz.

only noise would have been collected in these memory locations. For example, $3T_2^*$ for $W_{1/2}$ of 0.4 Hz is 2.4 s. The acquisition times for sampling the FID for a 1000-Hz spectral width at 4K and 8K points are 2.048 and 4.096 s, respectively. Thus, an 8K FID is sufficient for a 1000-Hz spectral width when the maximum T_2^* is 0.8 s, and zero filling to 32K provides sufficient resolution to properly define the peak shape for quantitative analysis. The significance of this will become more apparent when we discuss sensitivity enhancement by signal averaging. Quantitative accuracy can also be improved by artificially broadening the resonances by multiplication of the FID by an apodization function (Section 6.5.2).

In the above discussion, the carrier was set to one end of the spectral region of interest, because with single-phase detection (SPD), i.e., detection of a single component of the transverse magnetization ($M_{y'}$), the digital representation of the FID for a nucleus precessing at a frequency of x Hz faster than the carrier is identical to that for one precessing at a frequency of $-x$ Hz. Thus, if the carrier were positioned in the middle of the spectral region, resonances of frequency $-x$ Hz would be aliased into the $+x$ spectral region, and vice versa

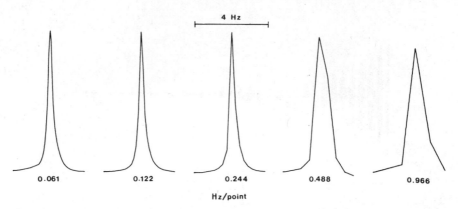

Fig. 12.   The effect of digital resolution on resonance line shape. The true width at half height is 0.24 Hz.

(Fig. 10). This problem is eliminated with quadrature phase detection (QPD) (15). In QPD, the nuclear induction signal is detected simultaneously with two identical phase-sensitive detectors, whose reference signals are in quadrature, i.e., 90° out of phase with each other. Thus, the two detectors are able to detect $M_{x'}$ and $M_{y'}$ simultaneously. The output of the two detectors is stored separately in computer memory as the real and imaginary components of the FID. The spectrum is obtained by doing a complex Fourier transformation on the complex time domain signal, yielding, again, both absorption and dispersion spectra (Fig. 11). Since $M_{x'}$ and $M_{y'}$ are detected simultaneously, the complex time domain signal contains sufficient information to distinguish resonances which are higher in frequency than the carrier from those lower in frequency (Fig. 13). Thus, with QPD, the carrier can be positioned in the center of the spectral region (Fig. 10C). The two components of the signal from a single receiver coil are detected either alternately or simultaneously by two phase-sensitive detectors which are 90° out of phase relative to each other. Alternate signals are stored in different regions of the computer, e.g., in alternate memory addresses.

QPD offers several other advantages over SPD. With SPD, half of the transmitter power is wasted. With QPD, the pulse has to cover only half as large a spectral range on each side of the carrier, i.e., for QPD, t_p must satisfy the condition

$$t_p \ll \frac{1}{2\Delta} \tag{38}$$

rather than Eq. (30). The required transmitter power is reduced by 4 since $\mathbf{B}_1$ is proportional to the square root of the transmitter power. An important advantage of QPD for trace analysis is that, compared to SPD, there is a gain of

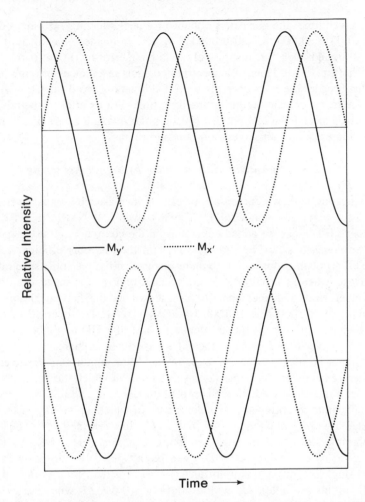

Fig. 13. $M_{x'}$ and $M_{y'}$ for two nuclei whose Larmor frequencies are greater than (*top*) and less than (*bottom*) the carrier frequency by the same amount.

$\sqrt{2}$ in S/N since with SPD the noise in the spectral region covered by Δ to the other side of the carrier is aliased into the spectral region of interest.

Another experimental factor of importance for quantitative analysis is that the start of data acquisition is generally delayed by one or two dwell times to allow the receiver to recover from overloads due to the $\mathbf{B}_1$ pulse. If the T_2^* of a resonance is very short, the intensity of the resonance in the frequency spectrum will be less than its true intensity since the observed intensity is determined by the magnitude of the transverse magnetization at the first point sampled. To illustrate, if the dwell time is 5.0×10^{-4} s, and if data acquisition is delayed by one dwell time, the intensities obtained for nuclei having T_2^* values of 3.18, 0.318, 0.0318, 0.00318, and 0.000318 s, i.e., resonances having $W_{1/2}$ of 0.1, 1, 10, 100, and 1000 Hz, respectively, will be in error by 0.016, 0.16, 1.6, 15, and 79% [Eq. (32)]. Thus, if a spectrum contains resonances of quite different $W_{1/2}$, the broader the resonance the more likely its observed intensity will be in error due to T_2^* relaxation during the dead time. For example, integrals for the broad resonances from NH protons (quadrupolar relaxation) are often low compared to integrals for sharp, carbon-bonded protons.

4.2.3. *Sensitivity Enhancement by Signal Averaging*

The efficiency with which sensitivity can be increased by signal averaging is one of the main advantages of P/FT NMR over CW NMR. In P/FT NMR, successive FID's can be added coherently in the computer, giving a gain in S/N proportional to $\sqrt{N}$ (Fig. 14). The P/FT method is more efficient because a single FID contains information about resonances across the entire spectral region of interest, whereas in CW NMR the spectral region is scanned from one end to the other. Ernst and Anderson (16) have shown that the theoretical time savings is $\Delta/W_{1/2}$, e.g., 1000 if Δ is 1000 Hz and $W_{1/2}$ is 1 Hz. In practice, the time saving is generally somewhat less than this, e.g., the FID might be acquired for $\sim 5T_2^*$ (~ 1.6 s in this case) in the P/FT experiment, whereas a scan time of 500 s might be used in the CW experiment. More important, however, when considering the theoretical time saving, is the rate at which the pulse/acquisition sequence can be repeated, i.e., the repetition rate.

To illustrate the problem, consider a case in which $T_1 = 5T_2$. If the FID is collected for $5T_1$ following a 90° pulse, $M_{z'}$ has returned to 99.33% of its equilibrium value (M_0) when the next pulse is applied (Fig. 15A). Thus, the S/N obtained by co-adding 100 FID's will be 0.9933 $\sqrt{100}$ times that obtained from a single 90° pulse at equilibrium. However, if the FID is collected for $1T_1$ ($= 5T_2$ in this example), $M_{z'}$ will return to only 0.632 M_0 when the second and

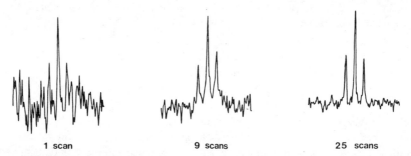

1 scan	9 scans	25 scans

Fig. 14. The effect of adding coherently successive FID's on the signal-to-noise ratio in the frequency domain spectrum. The spectra were obtained by coadding 1, 9, and 25 FID's. The sensitivity enhancement was approximately 3 and 5.5 for the latter two.

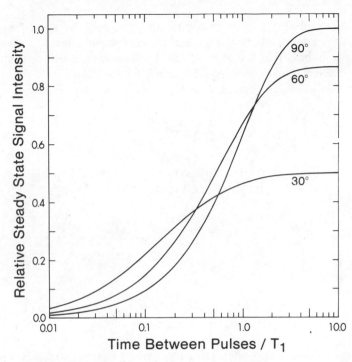

Fig. 15. Relative steady state signal intensities obtained from repetitive 30°, 60°, and 90° pulses as a function of the time between pulses (in units of T_1). Reprinted with permission from reference 89.

successive pulses are applied and the increase in S/N obtained by co-adding 100 FID's will be a factor of 0.632 $\sqrt{100}$. Thus, if $T_2 < T_1$ and the time between pulses is $5T_2$, the S/N of the spectrum obtained by co-adding N FID's will be less than $\sqrt{N}$ times that of the spectrum obtained from a single 90° pulse. If the nuclei have different T_1 values, as would almost always be the case, the extent to which the longitudinal magnetization is less than M_0 when the pulses are applied, and thus the extent to which the resonance intensities differ from $(\sqrt{N}) \times$ (intensity from a single 90° pulse), will be different for different nuclei. This is an important source of error when doing quantitative analysis by P/FT NMR.

Generally, for quantitative work the time between pulses is set to at least 5 times the longest T_1, which further decreases the efficiency of signal averaging by P/FT NMR. If the nucleus in the above example (Δ = 1000 Hz and a $W_{1/2}$ of 1 Hz) had a T_1 of 1 s, the FID should be acquired for at least ~1.6 s followed by a delay of another 3.4 s for T_1 relaxation before applying the next pulse. Thus, rather than doing 500 pulse/acquisition sequences in 500 s, we can only do 100.

If sensitivity enhancement is a main consideration, a 90° flip angle and a repetition rate of $1/5T_1$ are not the optimum conditions for a fixed amount of time since a large fraction of the time is spent waiting for T_1 relaxation. A different combination of flip angle α and repetition time t_r will generally result in more efficient signal averaging. If $t_r \geq 5T_2$ but less than $\sim 5T_1$, a steady state is reached in which, at the end of each pulse interval, $M_{x'y'} \approx 0$ and

$$M_{z'} = \frac{1 - E}{1 - E \cos \alpha} M_0 \qquad (39)$$

where $E = \exp(-t_r/T_1)$. The components of magnetization along z' and y' immediately after the next pulse reach steady state values of

$$M_{z'} = \frac{(1 - E) \cos \alpha}{1 - E \cos \alpha} M_0 \qquad (40)$$

and

$$M_{y'} = \frac{(1 - E) \sin \alpha}{1 - E \cos \alpha} M_0 \qquad (41)$$

The relative steady-state signal strengths obtained from repetitive 30°, 60°, and 90° pulses are plotted as a function of t_r in Fig. 15. An important result in Fig. 15 is that the largest steady-state signal is not necessarily obtained with a 90° pulse. The smaller α, the larger $M_{z'}$ immediately after the pulse which, depending on α and t_r, can more than compensate for the sin α dependence of the signal strength. For example, the relative steady-state signal intensity from 60° pulses is larger than that from 90° pulses when $t_r < 1.31T_1$. Thus, when optimizing pulse conditions for maximum sensitivity enhancement in a given total experimental time, not only is it better to use the minimum t_r allowed by T_2, it is also better to use the α at which the steady-state signal intensity is largest for that t_r. The relative S/N of signals S_1 and S_2 measured with α_1, t_{r1}, and α_2, t_{r2} in the same total experimental time is given by

$$R = \frac{S_1}{S_2} = \frac{(1 - E_1)(1 - E_2 \cos \alpha_2) \sin \alpha_1 \sqrt{t_{r2}}}{(1 - E_2)(1 - E_1 \cos \alpha_1) \sin \alpha_2 \sqrt{t_{r1}}} \qquad (42)$$

where $E_1 = \exp(-t_{r1}/T_1)$ and $E_2 = \exp(-t_{r2}/T_1)$. The S/N obtained in a constant total experimental time with a range of flip angles and repetition times is plotted relative to that obtained with $\alpha = 90°$ and $t_r = 5T_1$ in Fig. 16.

For a particular t_r, the optimum flip angle (the so-called Ernst angle) is given by (16)

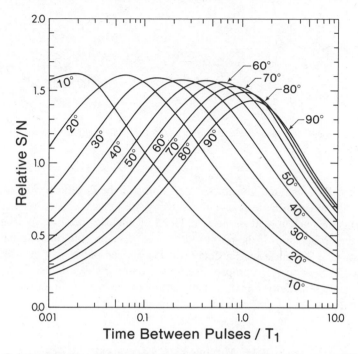

Fig. 16.   Relative S/N ratios obtained in the same total experimental time for a range of flip angles as a function of the time between pulses (in units of T_1). The S/N is relative to that obtained with $\alpha = 90°$ and $t_r = 5T_1$ in the same experimental time.

$$\cos \alpha = \exp(-t_r/T_1) \tag{43}$$

Once the pulse interval has been selected, the optimum flip angle can be calculated with this equation. T_1 values for ^{1}H nuclei are typically in the range 0.5–10 s, decreasing as molecular size increases. They depend on molecular size and solvent, e.g., they are generally longer in deuterated solvents and low-viscosity solutions, T_1 values for ^{13}C can vary over a wide range within the same molecule (17), depending on the number of directly bonded ^{1}H atoms (Fig. 17). If the pulse interval is taken to be the acquisition time, then the optimum pulse angle depends on the digital resolution required. For ^{1}H NMR, optimum pulse angles are typically 90°–35°, while for ^{13}C NMR they can be even smaller, depending on T_1 and the resolution required. It is essential to have estimates of T_1 values to optimize sensitivity in NMR measurements at trace levels. The measurement of T_1 is discussed in Section 4.3.1.

When using a pulse interval equal to t_{acq} and a pulse angle calculated with Eq. (43), the longitudinal magnetization generally does not return to its equilibrium value between pulses and thus intensities of resonances in the frequency

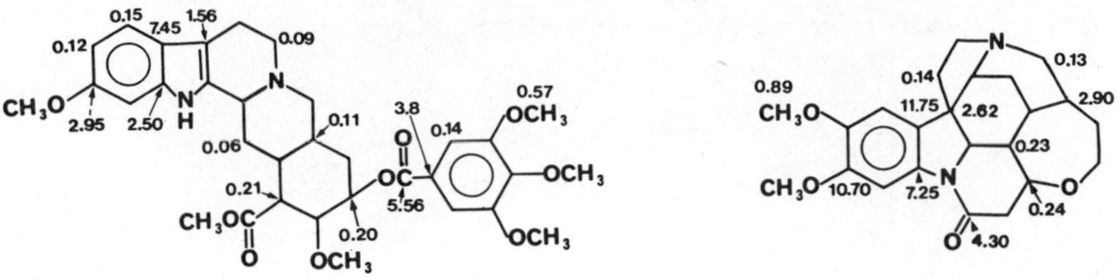

Fig. 17. Carbon-13 spin–lattice relaxation times (s) for reserpine (*left*) and brucine (*right*). From reference 17.

spectrum will be distorted, the amount depending on the T_1 value for the particular nucleus. If the Ernst angle is calculated using the longest T_1, the resonance for that particular nucleus will be most reduced in intensity because of incomplete longitudinal relaxation. If the purpose of the experiment is to obtain quantitative information, a 90° pulse with a delay of $5T_1$ between pulses can be used or, as discussed in Section 7.5, compromise pulse conditions can be used which give the optimum sensitivity enhancement within the maximum tolerable intensity error limits.

4.3. Multiple-Pulse Experiments

4.3.1. Measurement of Spin–Lattice Relaxation Times

The inversion-recovery pulse sequence (18) is the most widely used method for measuring T_1. In this method, the magnetization is inverted with a 180° (π) pulse [Eq. (27)] and then sampled as it recovers to M_0 (Fig. 18A). If $t_p \ll T_1$, the magnetization lies along $-z'$ immediately after the pulse ($\tau_1 = 0$), and is of magnitude $-M_0$. As times passes, $M_{z'}$ relaxes to equilibrium by T_1 processes. The magnitude of the magnetization at time τ_1, $M_{z'\tau_1}$, is determined by rotating it into the $x'y'$ plane with a 90° observation pulse. The total sequence is $180° - \tau_1 - 90° - $ acquisition $-T$, where $T \simeq 5T_1$ (Fig. 18B). As τ_1 is increased, $M_{z'}$ decreases, passes through zero, and then increases along the z' axis. Fourier transformation of the FID's obtained with increasing values of τ_1 yield spectra in which the resonances are of negative, zero, and positive intensity (Fig. 18C), depending on τ_1 and T_1, as follows:

$$I_{\tau_1} = I_\infty(1 - 2\exp(-\tau_1/T_1))\qquad(44)$$

where I_{τ_1} and I_∞ are the intensities at time τ_1 and at very long τ_1, respectively. This equation is usually used in the form

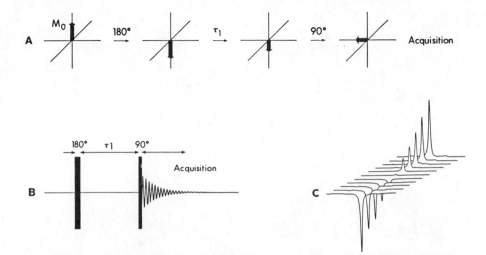

Fig. 18. Schematic representation of (A) macroscopic magnetization during the inversion–recovery pulse sequence (B). (C) shows a typical series of inversion–recovery spectra obtained as a function of τ_1.

$$\ln(I_\infty - I_{\tau_1}) = \ln I_\infty - \tau_1/T_1 \qquad (45)$$

A plot of $\ln(I_\infty - I_{\tau_1})$ versus τ_1 yields a straight line with slope of $-1/T_1$. For an estimate of T_1, the null method may be used; when $I_{\tau_1} = 0$, $\tau_1 = 0.69T_1$.

4.3.2. Measurement of Spin–Spin Relaxation Times

The value of T_2 may be determined from $W_{1/2}$ ($= 1/\pi T_2$) if broadening due to magnetic field inhomogeneity is negligible. Such is the case for broad resonances from quadrupolar nuclei (Section 2.3.3) or from spin-½ nuclei in the presence of paramagnetic impurities (Section 2.3.3). If magnetic field inhomogeneity contributes significantly to the observed $W_{1/2}$, T_2 can be determined with P/FT NMR methods.

The basic problem in measuring T_2 values by P/FT NMR is to eliminate the effect of B_0 inhomogeneity on the rate of decay of $M_{x'y'}$. This is accomplished with a 180° refocusing pulse. The spin–echo sequence is the simplest sequence which uses a refocusing pulse: 90°–τ_2–180°–τ_2–acquisition (Fig. 19B) (19). The behavior of the magnetization during the pulse sequence is shown schematically in Fig. 19A. The 90° pulse along x' rotates the magnetization into the $x'y'$ plane. It immediately begins to decay by spin–spin relaxation and to dephase due to B_0 inhomogeneity. At time τ_2, a 180° pulse is applied along x' to rotate the individual nuclear magnetic moments through 180° around the x' axis. If diffusion

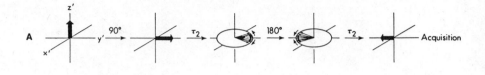

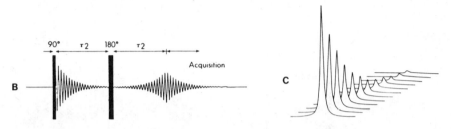

Fig. 19. Schematic representation of (A) macroscopic magnetization during the spin–echo pulse sequence (B). (C) shows typical spectra obtained by the spin–echo pulse sequence as a function of increasing τ_2.

is negligible, those nuclei that precessed faster (or slower) than the average before the 180° pulse continue to do so after the pulse, giving rise to refocused magnetization and an echo at time $2\tau_2$ after the 90° pulse. The echo maximum is of an intensity determined by T_2:

$$I_{2\tau_2} = I_0 \exp(-2\tau_2/T_2) \tag{46}$$

For multiline spectra, the second half of the echo is collected as the FID, which is processed as before. The decay of resonance intensity as τ_2 is increased is shown in Fig. 19C. T_2 values can be obtained from the dependence of the intensity on τ_2.

If nuclei diffuse to regions where B_0 is different during the two delay intervals, refocusing will not be complete and the intensity at time $2\tau_2$ is given by (8)

$$I_{2\tau_2} \ \alpha \ exp\left(-\frac{2\tau_2}{T_2} - \frac{2}{3}\gamma^2 G^3 D\tau_2^3\right) \tag{47}$$

where G is the spatial magnetic field gradient and D the diffusion coefficient.

Resonances which are split by homonuclear coupling are phase modulated (20) in spectra obtained by Fourier transformation of the second half of a spin echo (21), which complicates the measurement of T_2 values. The modulation arises because the 180° pulse is a nonselective pulse (i.e., $\gamma B_1 \gg 2\pi\Delta$) and thus it inverts the spin populations of all the nuclei in the spin system. Consider,

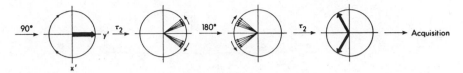

Fig. 20. The behavior of the two components of the A doublet of an AX spin system during the spin–echo pulse sequence. The spin system is homonuclear and the 180° pulse affects both the A and X nuclei.

for example, two coupled spin-½ nuclei in an AX spin system. The 90° pulse places the magnetization for both A and X along y'. The two components of the A (and X) magnetization due to spin coupling separate into a fast and slow component, diverging in the $x'y'$ plane at a rate determined by J_{AX}, as shown in Fig. 20. The magnetization in each component fans out due to B_0 inhomogeneity. At time τ_2, the 180° pulse rotates the two components of the A (and X) magnetization into their mirror images with respect to the $x'z'$ plane. This has the effect of interchanging the spin states of both the A and X nuclei, with the result that fast and slow components are interchanged (Fig. 21). Those A nuclei that were associated with the higher frequency A transition before the 180° pulse are associated with the lower frequency A transition after the pulse, and vice versa. Thus, in the time interval after the 180° pulse, the two A (and X) components continue to diverge. At time $2\tau_2$, the magnetization within each component has refocused, but the two components of each doublet are out of phase with one another by $2(360)\tau_2 J_{AX}$ degrees. They also are out of phase with y' and thus the resonances are out of phase in the frequency domain spectrum. Spin–echo spectra are shown as a function of τ_2 in Fig. 22 for a doublet, triplet, and quartet. The phase modulation of triplets and quartets can be explained with similar arguments. The center line of the triplet is not phase modulated, but the

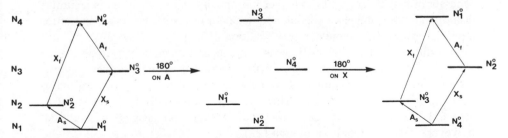

Fig. 21. Energy level diagram for an AX spin system during the spin–echo pulse sequence. A_s and X_s represent the lower frequency A and X transitions. N_1, N_2, etc., represent the populations of levels 1, 2, etc. At equilibrium these are $N_1{}^0$, $N_2{}^0$ . . . . The 180° pulse on A inverts A spin populations (*center*), and then the 180° pulse on X inverts the X spin populations (*right*). (In the spin–echo experiment, these occur simultaneously.) Those nuclei which were associated with the slow transition A_s are associated with the faster transition A_f after the 180° pulses.

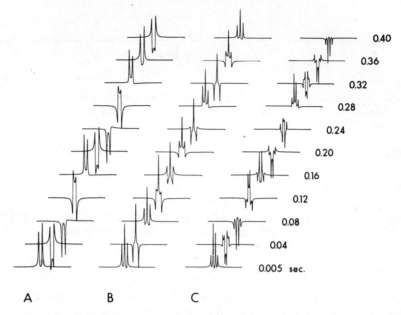

			0.40
			0.36
			0.32
			0.28
			0.24
			0.20
			0.16
			0.12
			0.08
			0.04
			0.005 sec.

A   B   C

Fig. 22. Spin–echo spectra for (A) a doublet (the methyl protons of isopropanol), (B) a triplet (the methyl protons of ethanol) and (C) a quartet (the methylene protons of ethanol). The times given are the τ_2 values in the spin–echo pulse sequence. Reprinted with permission from reference 31. Copyright 1979 American Chemical Society.

two outer lines are at a frequency of $2J$. Thus, the two outer lines are 180° out of phase with the central component at $\tau_2 = 1/(4J)$, $3/(4J)$, $5/(4J)$, etc., and in phase at $\tau_2 = 1/(2J)$, J, $3/(2J)$, etc. The outer and inner components of a quartet are modulated at frequencies of $3J$ and J, respectively. The phase modulation behavior of first-order spectra of higher multiplicity can be analyzed in a similar fashion; however, the phase modulation behavior of resonances in second-order multiplet patterns is a more complex function of J.

It is difficult to make intensity measurements as a function of τ_2 to determine T_2 values for resonances which are phase modulated. If the multiplet patterns are first order and have similar coupling constants, then τ_2 can be adjusted to give resonances all in phase. Alternatively, magnitude (power) spectra, $(U^2 + V^2)^{1/2}$, where U is the dispersion and V the absorption signal, can be used; however, there is a loss in resolution (22).

The effect of diffusion during the delay intervals on the echo intensity can be eliminated with the Carr–Purcell (23) and the Meiboom–Gill (24) modifications of the basic spin–echo pulse sequence. In the Carr–Purcell sequence, a series of 180° refocusing pulses are used: $90°–\tau–(180° – 2\tau)_n–180°–\tau–$acquisition. The 180° pulses refocus the magnetization and produce echoes at times 2τ, 4τ,

6τ, etc., after the $90°$ pulse, with alternation in sign. The effects of diffusion can be minimized by making τ sufficiently small. The major disadvantage of the Carr–Purcell method lies in the exactness and homogeneity required of the $90°$ and $180°$ pulses. In the Meiboom–Gill modification, the $90°$ pulse is applied along the x' axis, and then the train of $180°$ refocusing pulses is applied along the y' axis. This has the effect of compensating for missetting the $180°$ pulse width on even numbered echoes, from which a true measure of T_2 values can be obtained.

4.3.3. Resolution Enhancement

NMR spectra of large molecules or of multicomponent mixtures often consist of overlapping resonances. One approach to increasing the resolution is to use a spectrometer with a large magnetic field to increase the chemical shift dispersion. Shift reagents also can be used (Section 7.2.1). However, even more powerful methods based on multiple pulse experiments are available. These include methods based on differences in spin–lattice and/or spin–spin relaxation times and the recently developed two-dimensional Fourier transform (2D-FT) NMR techniques.

4.3.3.1. Based on Differences in T_1.   Overlapping resonances can be resolved with the inversion-recovery pulse sequence if their T_1 values are different. The basis of this application is that resonances of nuclei with different T_1 values pass

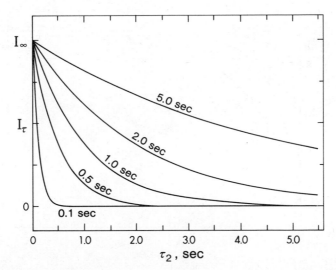

Fig. 23.   Recovery curves for nuclei having T_1 values of 0.1, 0.5, 1, 2, and 5 s following application of the $180°$ pulse in the inversion–recovery pulse sequence.

through the null point (Section 4.3.1) on their recovery curves at different times, as shown in Fig. 23. Thus, the 90° sampling pulse is timed to coincide with the null point of the resonance to be eliminated. An early application of this technique was to selectively eliminate the residual HDO resonance in ^{1}H NMR spectra of D_2O solutions by applying the 90° pulse at $\tau_1 = 0.69\ T_{1,HDO}$. This technique, called water elimination Fourier transform (WEFT) (25), may be used to observe resonances near or under the large HDO resonance provided the T_1 values are sufficiently different, or to solve the dynamic range problem (Section 7.7) when attempting to observe weak resonances in the presence of a strong solvent resonance. It is important to note that the intensity of resonances in spectra obtained by the inversion-recovery method will be reduced if they have not fully recovered when the 90° observation pulse is applied. Thus, some sort of calibration procedure must be used if quantitative information is to be obtained from the resonance intensities.

4.3.3.2. Based on Differences in T_2.

Overlapping resonances can be resolved with the spin–echo pulse sequence if spin–spin relaxation times are different (26,27). The basis of the method is that, the shorter the T_2 value (i.e., the larger the line width), the more rapidly the resonance intensity decreases as τ_2 is increased, as shown in Fig. 24.

In general, T_2 decreases as the size of the molecule increases, and thus the spin–echo pulse sequence provides a method for selectively eliminating resonances from large molecules. To illustrate, the top spectrum in Fig. 25 is the

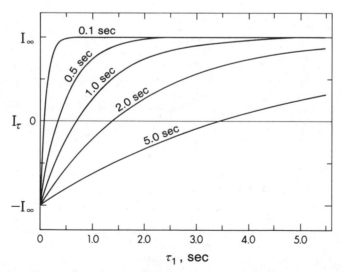

Fig. 24. The intensity of resonances of nuclei having T_2 values of 0.1, 0.5, 1, 2, and 5 s in spin–echo spectra measured as a function of τ_2.

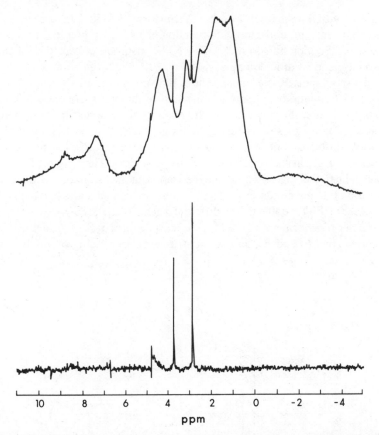

Fig. 25. 100 MHz ^{1}H-NMR spectra of a D_2O solution containing 0.003 M bovine serum albumin and 0.003 M N-methylglycine. (*top*) Spectrum obtained with the single pulse sequence; (*bottom*) spectrum obtained with the spin–echo sequence with $\tau_2 = 0.060$ s. Reprinted with permission from reference 27. Copyright 1978 American Chemical Society..

single pulse ^{1}H-NMR spectrum of a D_2O solution containing bovine serum albumin (BSA) and N-methylglycine (27). The spectrum consists of a broad envelope of resonances due to BSA with the sharp resonances from N-methylglycine superimposed on the envelope at 2.84 and 3.72 ppm. The bottom spectrum was obtained with the spin–echo pulse sequence using a τ_2 of 0.060 s. With this delay time, the transverse component of the magnetization from the ^{1}H nuclei of BSA has decayed to zero and the BSA resonances are completely eliminated. The spin–echo sequence has been used to achieve resolution enhancement in ^{1}H NMR spectra of intact human erythrocytes (28). This is discussed further in Section 8.

4.3.3.3. Two-Dimensional Fourier Transform NMR.

In the single-pulse experiment, data are collected as a function of time t_2 to give the FID. In two-dimensional Fourier transform (2DFT) NMR, data are collected as a function of two independent time domains, t_1 and t_2 (29). The t_2 time domain is the same as in the single-pulse experiment. The t_1 time domain results from an evolution period which is introduced between the start of the pulse sequence and the start of data acquisition. During the evolution period, the macroscopic magnetization is forced to undergo some prescribed motion such that its motion during the evolution period is reflected in the macroscopic magnetization at the start of acquisition, e.g., in the phase or intensity at $t_2 = 0$. The length of t_1 is incremented; thus, the macroscopic magnetization at the start of acquisition also depends on the magnitude of t_1. The 2D time-domain data array, which consists of a series of FID's collected at different values of t_1, $S(t_1, t_2)$, is subjected to two Fourier transformations, yielding a 2D spectrum, $S(F_1, F_2)$, defined by two frequency axes, F_1 and F_2, and an intensity axis. Depending on the behavior of

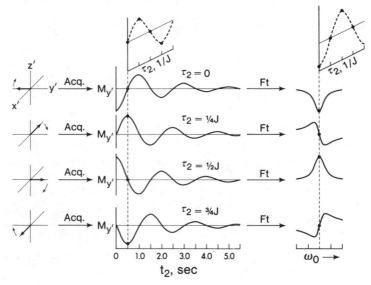

Fig. 26. A schematic representation of data collection and processing in J-resolved 2DFT NMR. The 2D data matrix, $S(t_1, t_2)$, is generated with the spin–echo pulse sequence by varying τ_2 ($\equiv t_1$). At the left is shown the magnetization for one component of a doublet, as a function of τ_2, at the start of data acquisition. The FID's generated by that magnetization are shown in the center, and the spectra obtained by Fourier transformation with respect to t_2 at the right. The spectra comprise the intermediate data matrix, $S(t_1, F_2)$, in the double Fourier transformation process. Fourier transformation with respect to t_1 (the time-domain functions in the second dimension are shown in the upper right) produces the two-dimensional data matrix, $S(F_1, F_2)$. Alternatively, the double Fourier transformation can be done in the order $S(t_1, t_2) \rightarrow S(F_1, t_2) \rightarrow S(F_1, F_2)$.

the nuclear spins during the evolution and detection periods, 2D experiments can be used to enhance resolution by separating resonances along chemical-shift and spin-coupling axes (J-resolved 2DFT NMR) and to correlate resonances linked by scalar coupling (shift-correlated 2DFT NMR). Only the homonuclear and heteronuclear J-resolved experiments will be described here.

The homonuclear J-resolved experiment is based on phase modulation in spin–echo FT NMR spectra of resonances which are split by homonuclear spin coupling (Section 4.3.2) (see, e.g., Fig. 22). In the 2D experiment, the 2D data matrix is generated by collecting a series of FID's as τ_2 in the spin–echo sequence (i.e., t_1 in the 2D experiment) is incremented (29). Generally, 2^n different τ_2 values are used. As τ_2 is incremented, the phase of the nuclear magnetization giving the various components of a multiplet pattern at $t_2 = 0$ changes periodically as a function of the coupling constant (Fig. 20), resulting in the intensity at a particular value of t_2 in the FID having a t_1 dependence. For example, the FID's obtained at four τ_2 ($\equiv t_1$) values for one component of a doublet are shown in Fig. 26. The 2D data matrix is usually processed by first doing a Fourier transformation on each of the 2^n FID's, which produces a new data matrix consisting of 2^n spin–echo spectra, differing in t_1, $S(t_1, F_2)$. This new data matrix can also be treated as a series of time-domain functions, each containing 2^n points, with the number of time-domain functions in the series equal to the number of memory addresses which define the spectrum (Fig. 26). The second step in the data processing is to Fourier transform each of these time domain functions which produces a 2D data matrix defined by two frequencies, $S(F_1, F_2)$. Since intensity in the t_2 time domain is dependent on both chemical shift and spin coupling effects, the F_2 axis contains both chemical shift and coupling constant information. However, intensity in the t_1 time domain is associated only with spin coupling, and thus the F_1 axis contains only spin coupling

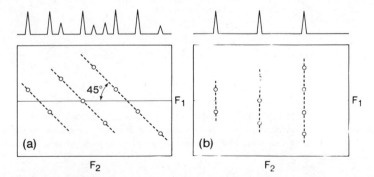

Fig. 27. Schematic representation of contour plots of the 2D data matrix, $S(F_1, F_2)$, for a doublet, triplet, and quartet (*a*) before and (*b*) after a 45° tilt. The spectra shown at the top are obtained by projection of resonance intensity onto the F_2 axis.

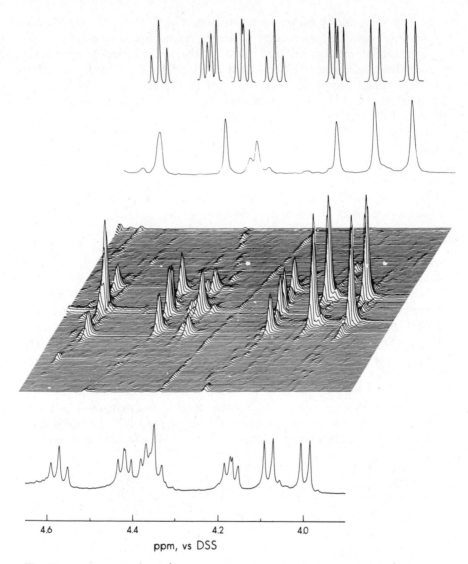

4.6   4.4   4.2   4.0

ppm, vs DSS

Fig. 28. Portions of 400 MHz ^{1}H NMR spectra of angiotensin. At the bottom is shown the one-dimensional spectrum obtained with the single-pulse sequence. In the middle is the stacked plot presentation of the J-resolved 2DFT spectrum. Above the stacked plot are shown the ^{1}H-decoupled spectra obtained by projection onto the F_2 axis and the cross sections in the $S(F_1, F_2)$ matrix at the peak maxima in the projection.

information, as shown by the contour plots (similar to topographical maps) of the $S(F_1, F_2)$ data matrix for a doublet, triplet, and quartet in Fig. 27. As this plot shows, the individual components of each multiplet lie along an axis which is tilted 45° relative to the F_1 axis (Fig.27a). Normally, the data matrix is "tilted" by 45° so that each multiplet lies along an axis parallel to F_1 and the F_2 axis contains only chemical shift information (Fig. 27b). One important consequence of this is that projection of intensity onto the F_2 axis yields an "^{1}H-decoupled" ^{1}H NMR spectrum (30). An individual multiplet pattern can be obtained from this data matrix, free of overlap from other multiplet patterns, as the cross section at the F_2 of the ^{1}H-decoupled resonance in the projection. For example, a J-resolved 2DFT NMR spectrum is shown in stacked plot form in Fig. 28. The trace behind the stacked plot is the projection onto the chemical shift axis and the multiplets at the top are the cross sections at the resonance frequencies in the projection.

It is clear that homonuclear J-resolved 2DFT NMR is a powerful method for resolving individual multiplet patterns in cluttered regions of ^{1}H NMR spectra of large molecules and of multicomponent mixtures. This method has several disadvantages, however, including: (1) spurious signals are observed for strongly coupled spin systems; (2) a large amount of time is required to collect and process the data; and (3) a large amount of data must be stored.

In contrast to homonuclear coupling, heteronuclear coupling does not cause phase modulation in spin–echo spectra because B_1 is too small to have any effect on the heteronuclear spin system. Thus, the two ^{13}C components of a ^{13}C–^{1}H doublet precess at the same frequency in the rotating frame before and after the pulse because the ^{1}H spin states are unaffected by the 180° refocusing pulse applied to the ^{13}C spin system (31,32). The components of the doublet can be

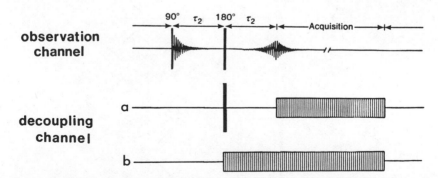

Fig. 29. Schematic representation of the pulse sequence used for spin–echo FT NMR with heteronuclear decoupling, e.g., ^{13}C{^{1}H} or ^{15}N{^{1}H}, to cause phase modulation. The spin–echo pulse sequence is applied to the observation channel while decoupling sequence (a) (the spin flip method) or (b) (the gated decoupling method) is applied to the decoupling channel.

phase modulated, however, by making the spin state of the ^{1}H nuclei different during the two delay intervals in the spin–echo pulse sequence with either of the gated decoupling sequences shown in Fig. 29. Decoupling sequence a, in which a 180° ^{1}H pulse is applied simultaneously with the 180° ^{13}C refocusing pulse, is called the spin–flip method (33). Thus, the components of the ^{13}C doublet diverge both before and after the refocusing pulse (Fig. 30) because the spin states of the ^{1}H nuclei which cause the splitting have been interchanged (Fig. 21). If the ^{1}H decoupler is off during acquisition, the frequency domain spectrum will consist of a doublet whose components are each modulated at a frequency of J. If BB ^{1}H decoupling is gated on during acquisition, the frequency domain spectrum consists of a singlet whose intensity is modulated at a frequency of J. In decoupling sequence b, BB ^{1}H decoupling is gated on during one of the two delay intervals. Thus, the components of a ^{13}C–^{1}H doublet diverge during the first delay interval, and then remain out of phase during the second delay interval. Since the components diverge during only one half of the pulse sequence, the frequency of phase or intensity modulation in the spectrum obtained, respectively, without and with BB ^{1}H decoupling during acquisition is one half that in spectra obtained by the spin–flip method. In heteronuclear J-resolved 2DFT NMR, the $S(t_1, t_2)$ data matrix is generated with one of the decoupling schemes shown in Fig. 29; thus, resonances are resolved on the basis of heteronuclear spin coupling, e.g., the F_2 axis contains ^{13}C chemical shift information and the F_1 axis ^{13}C–^{1}H spin coupling information.

There are several related 2D experiments with which resonances at different

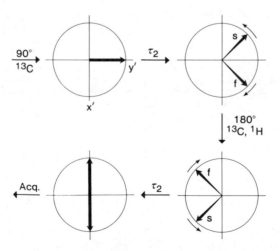

Fig. 30. Divergence of the two components of a ^{13}C doublet (^{13}C–^{1}H spin system) during the evolution of the echo in the ^{13}C{^{1}H} spin–echo experiment with decoupling scheme (a) in Fig. 29.

chemical shifts can be identified as being from the same molecule. The connectivity of resonances which are linked by dipolar coupling can be established by NOE spectroscopy (NOESY) (34), while resonances linked by scalar coupling can be established by correlated spectroscopy (COSY) or spin–echo correlated spectroscopy (SECSY) (35). These experiments have proven to be particularly useful for establishing the connectivity of resonances in ^{1}H NMR spectra of large molecules, and should prove to be equally useful in unraveling spectra of complex, multicomponent mixtures. The connectivity of ^{1}H and ^{13}C resonances can be established with heteronuclear shift correlated 2D experiments (36). ^{13}C–^{13}C connectivities in ^{13}C spectra can be established with INADEQUATE (incredible natural abundance double quantum transfer experiment) (37). Also, the INADEQUATE experiment provides a method for observing ^{13}C–^{13}C coupling with elimination of the much stronger ^{13}C–^{12}C resonance.

4.3.4. Sensitivity Enhancement

Several multiple pulse techniques for enhancement of signals from low sensitivity nuclei, e.g., ^{13}C, ^{15}N and ^{29}Si, have been devised. In the selective population inversion (SPI) (38) method one transition in the multiplet for an ^{1}H nucleus coupled to an insensitive nucleus, e.g., ^{13}C, is inverted with a selective ^{1}H pulse. This has the effect of changing populations of the levels between which ^{13}C transitions occur (Table 2). For an AX system, where A and X represent ^{13}C and ^{1}H, respectively, and $\Delta N_H = 4\Delta N_C$, the A_1 and A_2 transitions are enhanced by -3 and $+5$, respectively. For degenerate ^{1}H spin states, as in a methyl group, the ^{13}C signals have theoretical intensities of -11, -9, 15, and 13 as compared to the normal 1, 3, 3, to 1 ratio (39). Such enhancements have been

Table 2. Changes in Populations for the Selective Population Inversion Experiment for the Homonuclear and Heteronuclear^a Case

Level	Transition	Homonuclear Invert A_1	Heteronuclear Invert X_1^b (X_2)
4		$-\Delta N_A/2 - \Delta N_X/2$	$-5\Delta N_C/2\ (3\Delta N_C/2)$
3		$+\Delta N_A/2 - \Delta N_X/2$	$+5\Delta N_C/2\ (-3\Delta N_C/2)$
2		$+\Delta N_A/2 + \Delta N_X/2$	$+3\Delta N_C/2\ (-5\Delta N_C/2)$
1		$-\Delta N_A/2 + \Delta N_X/2$	$-3\Delta N_C/2\ (5\Delta N_C/2)$
	X_1	$+\Delta N_X - \Delta N_A$	
	X_2	$+\Delta N_X + \Delta N_A$	
	A_1		$-3\Delta N_C\ (5\Delta N_C)$
	A_2		$+5\Delta N_C\ (-3\Delta N_C)$

^{a}A = ^{13}C and X = ^{1}H.
$^b\Delta N_H \simeq 4\Delta N_C$.

used to observe relatively insensitive nuclei such as ^{29}Si (40) and ^{15}N (41). Because enhancements are caused mainly by changes in the ^{1}H populations, improved sensitivity may be obtained even though the observed nucleus itself is saturated (e.g., ^{13}C, ^{29}Si or ^{15}N). Also, the sensitivity gain is larger than obtained by NOE enhancements since the repetition time of the experiment is governed by ^{1}H T_1 times which are generally short. However this technique has three disadvantages: (1) the exact position of the ^{1}H resonance must be known, which in the ^{1}H spectrum is a ^{13}C satellite whose intensity is 0.5% that of the more intense ^{1}H–^{12}C resonance: (2) only a single ^{13}C resonance is enhanced since the ^{1}H inversion pulse is selective; and (3) BB ^{1}H decoupling cannot be used since the two components of the sensitivity-enhanced ^{13}C doublet are of opposite sign.

The INEPT (insensitive nuclei enhanced by polarization transfer) experiment (42) is a more generally applicable experiment which is free of some of these problems. The pulse sequence for the INEPT experiment is shown in Fig. 31, where X represents an $I = \frac{1}{2}$ nucleus (e.g., ^{13}C) which is scaler coupled to ^{1}H. The effect of this pulse sequence on the ^{1}H magnetization in the ^{13}C–H case is shown in Fig. 32. Time τ_1 is selected so that at $\tau_1/2$ after the first 90° ^{1}H pulse, the two components of the ^{1}H doublet are 90° out of phase (Fig. 32b). Simultaneous 180° pulses are applied to both the ^{1}H and ^{13}C spin systems, which has the effect of interchanging spin states and precessional frequencies (Section 4.3.2) (Fig. 32c and d). Thus, during the delay period following the 180° pulse, the

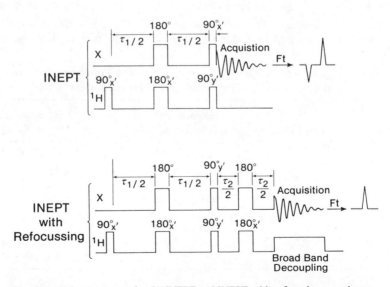

Fig. 31.   Pulse sequences for the INEPT and INEPT with refocusing experiments.

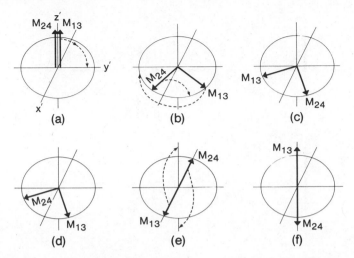

Fig. 32. The effect of the INEPT pulse sequence on the ^{1}H magnetization for a ^{13}C–^{1}H spin system. (See text for details.)

two components continue to diverge while the magnetization within each component is refocused. At time τ_1, the two components of the ^{1}H magnetization are opposed along the $\pm x'$-axes (Fig. 32e). At this time, a 90° ^{1}H pulse is applied along the y' axis, which creates ^{1}H magnetization along the $\pm z'$ axes (Fig. 32f). Thus, the net effect of the ^{1}H and ^{13}C pulses to this point has been to invert the spin population for one ^{1}H transition of the ^{13}C–^{1}H spin system. Because the ^{13}C and ^{1}H nuclei share energy levels, the energy levels for the ^{13}C nuclei now have population differences determined by γ_{1_H} (Fig. 33a), i.e., each component of the ^{13}C magnetization is larger by a factor of $\gamma_{1_H}/\gamma_{13_C}$. The effect of the ^{13}C and ^{1}H pulses up to this point has been to do what is accomplished in the SPI experiment, however, in the INEPT experiment, this is done with nonselective ^{1}H and ^{13}C pulses so that all ^{13}C–^{1}H spin systems are affected, i.e., all ^{13}C multiplets are sensitivity enhanced.

The 90° ^{13}C pulse applied simultaneously with the 90°$_y$ ^{1}H pulse serves to create ^{13}C magnetization in the $x'y'$ plane for observation (Fig. 33b). However, as in the SPI experiment, the components of a ^{13}C multiplet are out of phase with each other and, if the phase of the 90° x observation pulse is alternated (x', $-x'$, x', $-x'$, etc.), the relative intensities are $1: -1$ for doublets (^{13}CH), $1:0: -1$ for triplets (^{13}CH$_2$), and $1:1: -1: -1$ for quartets (^{13}CH$_3$). Consequently, BB ^{1}H decoupling during acquisition would cause a cancellation of the signals.

To bring the individual components of the multiplet patterns in phase, a delay is inserted between simultaneous 90° ^{1}H and ^{13}C pulses and the data acquisition (42). During this time, the two components of the ^{13}C magnetization converge

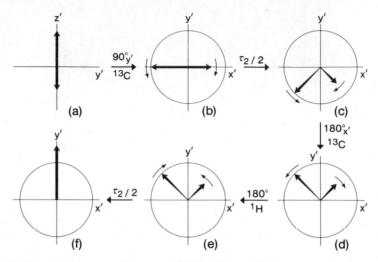

Fig. 33. Behavior of the two components of a ^{13}C doublet (^{13}C–^{1}H spin system) in the INEPT with refocusing experiment. The $90°_{y'}$ ^{13}C observation pulse is applied ($a \rightarrow b$, note change in coordinate system), followed by the first half of the refocusing period ($b \rightarrow c$), the $180°$ ^{13}C ($c \rightarrow d$), and $180°$ ^{1}H ($d \rightarrow e$) refocusing pulses, and the second half of the refocusing period ($e \rightarrow f$). (See text for details.)

due to their different precessional frequencies, but they diverge due to B_0 inhomogeneity (Fig. 33c). To reverse B_0 inhomogeneity and chemical shift effects, a 180° refocusing pulse is applied (INEPT with refocusing, Fig. 31) (43). However, it is necessary to apply simultaneously a 180° ^{1}H pulse to interchange the ^{1}H spin states so that the two ^{13}C components continue to converge (Fig. 33d and e) and refocus at time τ_2 after the 90° ^{13}C pulse (Fig. 33f). The relative phase of the components of the ^{13}C magnetization at the start of acquisition is determined by τ_2. If BB ^{1}H decoupling is used during acquisition, the intensity of singlets for ^{13}CH, ^{13}CH$_2$, and ^{13}CH$_3$ carbons is a function of τ_1 and τ_2 as given by Eqs. (48)–(50), respectively (43):

$$I_{CH} = \frac{\gamma_H}{\gamma_C} \sin(\pi J \tau_1) \sin(\pi J \tau_2) \qquad (48)$$

$$I_{CH_2} = \frac{\gamma_H}{\gamma_C} \sin(\pi J \tau_1) \sin(2\pi J \tau_2) \qquad (49)$$

$$I_{CH_3} = \frac{\gamma_H}{\gamma_C} \sin(\pi J \tau_1) [\sin(\pi J \tau_2) + \sin(3\pi J \tau_2)] \qquad (50)$$

The dependence on τ_2 is shown in Fig. 34. Maximum sensitivity enhancement

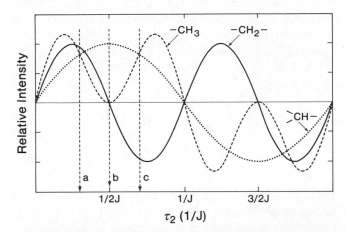

Fig. 34. The dependence predicted by Eqs. (48)–(50) for the intensity of resonances for CH, CH$_2$, and CH$_3$ type carbons in ^{13}C spectra measured by the INEPT method with refocusing and BB ^{1}H decoupling on the length of the refocusing period (τ_2). With τ_2 values of a, b, and c, resonances can be selectively enhanced, suppressed (CH$_2$ and CH$_3$), and inverted (CH$_2$), respectively.

occurs when $\tau_1 = 1/(2J)$. However, because of the different forms of the dependence on τ_2, no single τ_2 value maximizes simultaneously the intensities of resonances for all three carbon types. Selective enhancement, suppression, or inversion of resonances may be achieved, depending on carbon type and proper choice of τ_2 (Fig. 34).

The behavior shown in Fig. 34 can be used to determine the number of protons directly bonded to carbon. One would measure 3 spectra: the normal BB decoupled ^{13}C spectrum and INEPT spectra measured at the τ_2 values indicated by b and c in Fig. 34. Since resonances from quaternary carbons are absent from the INEPT spectra, quaternary carbon resonances can be identified by comparison of the BB decoupled spectrum with the INEPT spectrum measured with $\tau_2 = c$. Resonances from methine carbons can be identified from the INEPT spectrum measured with $\tau_2 = b$ while the resonances from methyl and methylene carbons can be assigned by comparison of the two INEPT spectra.

There are four problems associated with the INEPT experiment. These are:

1. It requires phase shifters and short-duration pulses on *both* observe and decoupler channels, which were not standard hardware items on spectrometers five years ago.
2. The number of pulses in the sequence requires good $\mathbf{B}_1$ and $\mathbf{B}_2$ homogeneity.
3. The amount of enhancement is dependent upon J_{CH}.
4. There is distortion in the relative intensities of coupled spectra compared

to the normal FT spectrum. The distortionless enhancement by polarization transfer (DEPT) experiment (118) provides similar information, and has the advantage that problems 2–4 are not so severe.

The DEPT pulse sequence is:

$$\text{}^1\text{H } 90°_x - \frac{1}{2J_{CH}} - 180° - \frac{1}{2J_{CH}} - \theta_y - \frac{1}{2J_{CH}} -$$

$$\text{}^{13}\text{C} \qquad\qquad 90° - \frac{1}{2J_{CH}} - 180° - \frac{1}{2J_{CH}} - \text{ACQUIRE}$$

It has two fewer pulses than INEPT, and thus is not as susceptible to pulse imperfections. For coupled spectra, all multiplets retain their characteristic intensities and the individual components of the multiplets are in phase, unlike those observed in INEPT spectra; thus, coupled subspectra may be obtained with DEPT. The functional form of the variation of peak intensities with θ for CH, CH_2, and CH_3 groups (again quaternary carbons are not observed) is identical to the equations and Fig. 34 presented for the INEPT experiment, except that θ is substituted for $\pi J \tau_2$; thus, the DEPT experiment is not as susceptible to variations in J_{CH} values. Spectral editing to obtain coupled and decoupled spectra may be accomplished with DEPT by taking spectra at different values of θ (118). Because zero and higher order coherences are created by the 90° ^{13}C pulse, simple vector diagrams fail for the DEPT experiment.

Both the INEPT and DEPT methods of sensitivity enhancement are based on the transfer of the population differences of a large γ nucleus, e.g., ^{1}H, to the energy levels for a small γ nucleus, e.g., ^{13}C, ^{15}N, and ^{29}Si, giving a theoretical sensitivity of γ_H/γ_X. Thus, the smaller is the γ_X, the greater is the sensitivity enhancement. Not only is this enhancement larger than that obtained by the NOE (Section 2.3.4.2), but there also is not the cancellation of signals as can happen with the NOE when the γ's are of opposite sign, e.g., in ^{1}H decoupled ^{15}N spectra. Another important feature of these experiments is that the rate at which the pulse sequence can be repeated for signal averaging is determined by ^{1}H T_1 values, which generally are less than those of ^{15}N, ^{29}Si, ^{77}Se, and other nuclei. Thus, the repetition rate can be faster than when the NOE is used to enhance sensitivity. In practice, the theoretical γ_H/γ_X enhancement in sensitivity is rarely achieved, owing to several factors, including incomplete inversion by the 180° pulses, spin–spin relaxation of ^{1}H magnetization during the τ_1 period, and of ^{13}C magnetization during the τ_2 period, and the use of compromise values for τ_2. Nevertheless, for some nuclei large enhancements can be achieved so that with these techniques it now is practical to do NMR on isotopically dilute, low γ nuclei, e.g., ^{15}N, at the natural abundance level.

5. MULTINUCLEAR NMR

Most elements have at least one nuclide with $I \neq 0$. However, because of low natural abundance and/or low inherent sensitivity, most were not amenable to routine NMR observation prior to the development of P/FT methods. Thus, prior to 1970, the majority of NMR measurements involved ^{1}H, with fewer measurements on ^{19}F, ^{31}P, ^{10}B, and ^{11}B. ^{13}C was the first "rare" nucleus to be studied extensively by the P/FT method. Because of the key role played by carbon in the skeletons of most molecules, ^{13}C NMR has developed to where it now is second in importance only to ^{1}H NMR (44). Since 1970, most other magnetically active nuclides also have been studied to some extent (45), although it is only with the recently developed sensitivity-enhancement methods such as INEPT that NMR measurements on some chemically important "rare" nuclides, e.g., ^{15}N, have become practical on a routine basis at the natural abundance level. The purpose of this section is to consider in more detail some of the factors which determine the sensitivity of selected nuclei, and thus the usefulness of NMR as a technique for trace analysis of compounds of these elements.

As discussed earlier (Section 2.3.4.), the inherent NMR sensitivity of an $I \neq 0$ nucleus depends on its natural abundance and its γ. Relaxation times are also of great importance in determining the actual experimental sensitivity since they determine the line width, i.e., the frequency range over which the resonance intensity is spread, and the repetition rate that can be used for signal averaging. Thus, the inherent sensitivities listed in Table 1 can only be taken as a qualitative guide to the relative sensitivities of the various nuclides. Because the relaxation times for $I > \frac{1}{2}$ nuclei are generally dominated by quadrupolar effects, it is convenient to discuss $I = \frac{1}{2}$ nuclei and $I > \frac{1}{2}$ nuclei separately.

5.1. Nuclei of Spin One-Half

There are stable $I = \frac{1}{2}$ isotopes of 24 elements. In addition to those listed in Table 1, these include ^{57}Fe, ^{77}Se, ^{89}Y, ^{103}Rh, ^{109}Ag, ^{113}Cd, ^{119}Sn, ^{125}Te, ^{129}Xe, ^{169}Tm, ^{171}Yb, ^{183}W, ^{187}Os, ^{195}Pt, ^{199}Hg, ^{205}Tl, and ^{207}Pb. Six of the $I = \frac{1}{2}$ nuclei are present at 100% natural abundance. Of these, ^{1}H, ^{19}F, and ^{31}P have a large γ and thus are sensitive to NMR detection, whereas ^{89}Y, ^{103}Rh, and ^{169}Tm have a small γ and are quite insensitive to NMR detection. There have been very few NMR measurements on the latter three, and they certainly are not suited to trace analysis by NMR. ^{205}Tl, with a natural abundance of 70.5% and a relatively large γ, is also sensitive to NMR detection.

The very high relative sensitivity of ^{1}H NMR makes ^{1}H the nucleus of choice for trace analysis. Because of the ubiquity of hydrogen, ^{1}H NMR is applicable to the majority of compounds; however, this in itself can be a problem if the sample contains a number of compounds whose resonances overlap. There are various methods by which resolved analytical resonances can be obtained for the components of interest, including the use of lanthanide shift reagents, spectrometers with large B_0, and multiple-pulse techniques. The first two are based on increasing the chemical shift dispersion, the third on using differences in relaxation times or 2DFT methods. Another problem often encountered with the use of ^1H NMR for trace analysis is that, if the sample is a solution, the solvent most likely will give strong ^1H resonance(s). This can complicate the measurement by obscuring the region of interest and by causing a dynamic range problem (Section 7.7) if the measurement is by the P/FT technique. If the sample is a solid, the problem of ^1H resonances from the solvent can be avoided by using a deuterated solvent (Section 7.1.1).

Fluorine-19 is the next most sensitive stable $I = \frac{1}{2}$ nucleus. Since there is little if any spectral interference, ^{19}F NMR is an attractive method of analysis for fluorine-containing molecules. As discussed in Section 7.2.3, a number of fluorine-containing functional group reagents have been developed with which molecules can be derivatized for analysis by ^{19}F NMR. Fluorine-19 chemical shifts cover a range of ~ 800 ppm, which corresponds to $\sim 75,000$ Hz at 2.35 T and $\sim 300,000$ Hz at 9.4 T. It generally is not possible to cover the entire chemical shift range in a single P/FT experiment, because (1) it is very difficult to obtain a uniform pulse over the entire range, i.e., $\gamma B_1 \geqslant 2\pi\Delta$, and (2) analog-to-digital conversion rates limit the spectral width, as discussed in Section 6.4. Both of these become more of a problem the larger B_0 becomes; however, it is advantageous for trace analysis to use as large a B_0 a possible for maximum sensitivity. Generally, samples do not give resonances covering the entire spectral range, in which case (1) and (2) are not a problem.

Phosphorus-31 NMR is sufficiently sensitive that it has found many applications, e.g., in measurements on intact tissues in biological chemistry (46) and analysis of organophosphorus pesticides (47). As with ^{19}F NMR, ^{31}P NMR has the advantage that there will generally be relatively few phosphorus-containing compounds in a sample, and the ^{31}P spectra will be free of the problem of overlapping resonances.

It is somewhat more difficult to do NMR on ^{13}C, ^{15}N, ^{29}Si, ^{77}Se, and ^{119}Sn because of their low natural abundance and/or small magnetogyric ratios (45). Negative magnetogyric ratios and long T_1 values add to the problem for several of the nuclei. In most compounds, these nuclei will be spin coupled to protons, often giving multiplets containing many lines. To simplify the spectrum and to increase sensitivity, it is advantageous to collapse all the resonance intensity into a single line by BB ^1H decoupling, in which case the intensities will be affected by NOEF [Eq. (19)]. Depending on the proton dipole–dipole contribution to the spin–lattice relaxation of the observed nucleus, the NOEF can cause the signal to decrease, disappear, or be negative for ^{15}N, ^{29}Si, and ^{119}Sn. Negative NOE can be avoided by using inverse gated decoupling (Section 7.3); however, the sensitivity is decreased. The INEPT pulse sequence (Section 4.3.4) provides a better solution to the negative NOE problem, while also eliminating the problem of long T_1 values since the repetition rate is governed by proton T_1's, which are generally considerably shorter than those for ^{15}N, ^{29}Si, and ^{77}Se.

Carbon-13 is a rare nucleus, but because of the importance of carbon in chemistry, ^{13}C NMR has been developed to a level where it now is possible to obtain ^{13}C spectra from micromoles of material. Carbon-13 NMR also has become a powerful tracer technique in the study of biosynthetic pathways (48). Carbon-13 spectra are usually measured with BB ^1H decoupling, which gives a series of singlets, often a well resolved singlet for each type of carbon in the molecule. One advantage of the low natural abundance of ^{13}C is that $^{13}C-^{13}C$ pairs are rare and $^{13}C-^{13}C$ couplings usually cannot be observed. Nuclear dipole–dipole relaxation involving directly bonded protons is usually the dominant relaxation mechanism, resulting in an inverse dependence of T_1 on the number of directly attached protons. Thus, the intensity of ^{13}C resonances is generally increased by the NOEF, the maximum enhancement being $\eta = 1.99$. If resonances are enhanced less than the maximum NOEF, it is necessary either to calibrate resonance intensity or to suppress the NOE with inverse gated decoupling (Section 7.3), or with shiftless relaxation reagents (Section 7.2.2), e.g., $Cr(acac)_3$, where acac is acetylacetonate, to obtain quantitative data. With the INEPT pulse sequence, the maximum theoretical sensitivity enhancement is 3.99. However, as discussed previously, this theoretical enhancement is rarely achieved. Also, since the optimum value of τ_2 in the pulse sequence depends upon the number of directly bonded protons and the coupling constant (Fig. 34), the relative intensities of resonances will not be directly equal to the relative numbers of ^{13}C nuclei. Thus, calibration procedures are necessary to obtain quantitative data from INEPT spectra.

The metals ^{113}Cd, ^{207}Pb, and ^{199}Hg all have relatively high inherent sensitivity, and NMR spectra can be obtained for all three at millimolar concentrations with P/FT techniques. The spin–lattice relaxation time of ^{199}Hg is very sensitive to its chemical bonding (49). In tetrahedral geometry, T_1 is relatively long, whereas T_1 for linear sp hybridized ^{199}Hg is short due to relaxation being dominated by the chemical shift anisotropy mechanism. For example, T_1 of $HgCl_4^{2-}$ is 6.5 s, whereas T_1 of $Hg(CH_2C_6H_5)_2$ is 0.737 s, both at 5.875 T (49). When chemical shift anisotropy dominates the relaxation, $1/T_1$ is dependent on B_0^2. Thus, T_1 for $Hg(CH_2C_6H_5)_2$ decreases to 0.0267 s when B_0 is increased to 9.40 T (49). When T_1 is short, very fast repetition rates can be used. Resonances are broad; however, resolution is generally not a problem because the chemical shift range is so large.

5.2. Nuclei of Spin Greater Than One-Half

The relaxation times of $I > \frac{1}{2}$ nuclei are generally short due to efficient relaxation by the quadrupolar mechanism resulting in broad resonances (45). The S/N ratio is reduced because the intensity is spread over a wide frequency range, however if both T_1 and T_2 are short, very fast repetition rates can be used to increase the S/N by signal averaging. Also, if T_2 is too small, intensity can be lost by spin–spin relaxation during the delay between the end of the pulse and the beginning of data acquisition. If a delay is not used, pulse breakthrough gives a strong signal at the beginning of the FID, which when Fourier transformed results in a sinusoidally rolling baseline, making the detection of very broad resonances difficult.

When relaxation is by the quadrupolar mechanism, linewidth is determined by Q, the nuclear electric quadrupole moment, and q, the electric field gradient at the nucleus. The dependence on q leads to a wide variation in $W_{1/2}$ for a particular nuclide in different compounds. When the nucleus is in an environment of cubic or higher symmetry, q is small and $W_{1/2}$ can be on the order of hertz, e.g., ^{14}N in $(CH_3)_4N^+$ and ^{35}Cl, ^{37}Cl in ClO_4^- (45). When the symmetry is less, q and $W_{1/2}$ are much larger, e.g., ^{14}N in pyridine and ^{35}Cl, ^{37}Cl in CCl_4 are on the order of kilohertz (45). Line widths for different $I > \frac{1}{2}$ nuclei increase as $|Q|$ increases. The $|Q|$ is relatively small for ^2H, ^6Li, and ^{133}Cs, and thus these nuclei tend to give relatively sharp resonances. Deuterium resonances are sufficiently sharp that stick spectra are obtained for BB ^1H-decoupled ^2H spectra of many organic molecules (45). It then is relatively easy to measure ^2H chemical

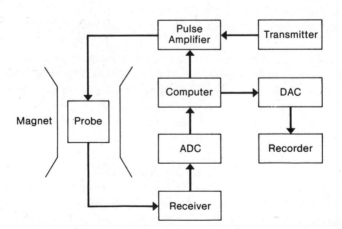

Fig. 35. Block diagram of a typical P/FT NMR spectrometer.

shifts, which parallel ^{1}H chemical shifts for the same molecule. Such measurements require spectrometers with large magnetic fields due to the low sensitivity and low frequency (e.g., 61.4 MHz at $B_0 = 9.4$ T) of ^{2}H NMR. Deuterium NMR is proving to be of considerable use in tracer studies, as discussed in Section 8.

There are many other isotopes with $I > \frac{1}{2}$; however, due to large $|Q|$ and/or low inherent sensitivity, they are not suited for detection at low concentrations.

6. INSTRUMENTATION

Since this chapter is concerned mainly with the measurement of low-intensity signals, the emphasis of this section will be on instrumentation for the P/FT experiment. The block diagram of a typical P/FT spectrometer is shown in Fig. 35. The instrumentation used in NMR is the subject of several books (13,50).

6.1. The Magnet

The magnet is the heart of any NMR spectrometer. The basic requirement is that it generate a large, uniform, and stable magnetic field. Permanent, electro-, and superconducting magnets are used, and each has its advantages and disadvantages.

The permanent magnets and electromagnets used in NMR are constructed of parallel pole pieces to give a homogeneous magnetic field. Permanent magnets are limited to fields up to about 2.1 T and to small diameter (e.g., 5 mm) sample tubes, and are generally used only in spectrometers for routine ^{1}H and ^{19}F NMR. A main advantage of electromagnets over permanent magnets is that homogeneous magnetic fields can be produced over a large enough volume to accommodate sample tubes up to 20 mm in diameter.

A superconducting magnet consists of a solenoid, made from a superconducting alloy, i.e., an alloy which has zero resistance at liquid helium temperature. The two ends of the superconducting wire in the solenoid coil are joined so that, once an electrical current is established in the solenoid, the current runs continuously so long as the solenoid is kept at liquid helium temperature. Thus, in essence, it becomes a very stable permanent magnet. To minimize the amount of liquid helium used, the solenoid is contained within a well insulated Dewar, which in turn is surrounded by a liquid nitrogen Dewar. The direction of the magnetic field from a superconducting solenoid is vertical, whereas that from both permanent and electromagnets is horizontal. The major advantage of superconducting magnets is that they can generate large magnetic fields, up to 9.4 T (^{1}H NMR at 400 MHz) with a niobium–titanium alloy and up to 11.7 T (^{1}H

NMR at 500 MHz) with a niobium–tin alloy. As mentioned previously, large magnetic fields offer two benefits for NMR: sensitivity increases as $B_0^{3/2}$ and chemical shift dispersion increases as B_0.

None of the magnets used in NMR spectrometers can, by themselves, generate a sufficiently homogeneous magnetic field for high-resolution NMR. High resolution is achieved with shim coils and by spinning the sample. The shim coils are small electric coils. By adjusting the currents in these coils, inhomogeneities in the planes parallel and perpendicular to the main magnetic field can be corrected. Spinning the sample averages out any remaining inhomogeneities in the plane perpendicular to the spin axis (the xz plane for permanent or electromagnets, and the xy plane for superconducting magnets). Spinning the sample in an inhomogeneous field results in "spinning side bands" (ssb) spaced symmetrically about each peak at the frequency of rotation. Spinning side bands can complicate the detection of resonances from trace components, but can be distinguished from real resonances by changing the spinning rate.

6.2. The Probe

The probe must (1) hold and spin the sample, (2) couple the sample to the transmitter and receiver, (3) provide a means of varying sample temperature, and (4) be easily and reproducibly placed into the magnet gap. Generally, it must also provide a field/frequency lock and a continuous source of rf irradiation, either single frequency or noise modulated, for decoupling. Requirements (1), (3), and (4) are mechanical in nature and are generally solved quite easily. Spinning the sample is accomplished by holding the sample tube in a "spinner," which is constructed of plastic or some easily machined material. The spinner should be uniform and fit snugly into the turret to prevent vibrations when spinning, which also causes ssb. The mechanical drive for spinning is usually a stream of air. Sample temperature is generally controlled by heat exchange with gas at a preset temperature.

The sample is coupled to the transmitter and receiver by means of resonant tank circuits, which usually consist of a coil and variable capacitors. Probes can be divided into two categories, depending on the number of coils used for coupling. Cross-coil probes utilize two separate coils—one for sample excitation and the other for receiving the NMR signal. Single-coil probes, as the name implies, use one coil for both functions. The advantages of the cross-coil configuration are that each coil may be optimized separately and the transmitter coil is capable of delivering a larger volume of homogeneous B_1 power to the sample, which is important in relaxation studies. However, the second advantage leads to the major drawback of crossed-coil probes, which is that B_1 power is not utilized efficiently.

Single-coil probes are mechanically easier to construct, and they deliver the transmitter power to the sample more efficiently than the cross-coil configuration. With the same transmitter power, the 90° pulse width is often a factor of 5 shorter in the single-coil arrangement, which is extremely important when observing resonances over a large frequency span. However, the homogeneity of B_1 is less.

The construction of the receiver coil is very important for the overall sensitivity of the spectrometer. For optimum performance, the receiver coil circuit is designed so that it can be tuned only over a narrow range of frequencies; to observe another nucleus, the circuit must be changed. Depending upon spectrometer design, either the entire probe must be changed or a universal probe must be used in which only the insert (receiver coil and associated glassware) need be replaced. Multinuclear probes have been designed in which the receiver coil may be tuned over a wide frequency range (e.g., from ^{31}N to ^{31}P). However, the price paid for the increased versatility is a small loss in sensitivity. Dual-frequency probes which can be computer switched between ^1H- and X-nucleus (the most common being ^{13}C) observation are commercially available. The receiver coil or the decoupling coil is double-tuned to also receive the lock signal, generally a ^2H resonance, for field/frequency stabilization.

The requirements of the receiver coil are vastly different in CW and P/FT spectrometers. In P/FT experiments, the receiver coil must first be able to accept pulse power of approximately 200 W, recover quickly from this overload, and then detect weak NMR signals from the sample with high sensitivity. In CW spectrometers, transmitter power levels seldom exceed 0.5 W.

Receiver coil size and geometry are critically related to sensitivity. Probes have been constructed to accommodate tubes having outside diameters from 2 mm (15 μL of sample) to 20 mm (10–15 mL). The larger the sample tube size, the higher the sensitivity for a constant sample concentration, as long as the homogeneity of the magnetic field over the volume of the sample remains fairly constant. If the quantity of sample is limited, higher sensitivity is achieved with probes designed for smaller-diameter tubes.

Receiver coils used with permanent magnets and electromagnets consist of a solenoid wound around a glass support column, as shown in Fig. 36a. The diameters of the glass support and sample tube are carefully chosen to give optimum filling of the receiver coil by the sample solution. In general, optimum sensitivity is achieved by using the largest sample tube which can be accommodated by a probe because the filling factor is optimized. The solenoid coil design is not used with superconducting magnets since B_1 would be colinear with B_0.

Receiver coils used with superconducting magnets are generally saddle-shaped or Helmholtz coils (Fig. 36b). With this design, samples can be changed easily. It has been shown, however, that solenoidal receiver coils placed horizontally

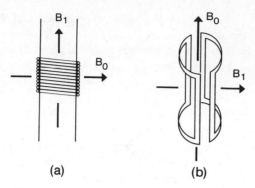

Fig. 36. Receiver coils used in NMR spectrometers. The solenoid coil design (*a*) is used with permanent magnets and electromagnets. The saddle-shaped or Helmholtz coil design (*b*) is used with superconducting solenoids.

in superconducting magnets have about 3 times the sensitivity and smaller 90° pulse widths than do comparable saddle-shaped designs, at least up to 129 MHz (51). However, this design is not generally used because the probe must be removed from the magnet each time a sample is changed. Sideways-spinning probes have been designed for superconducting systems with an increase in sensitivity (52). As the magnetic field strength and thus the resonance frequency increase, receiver coil design becomes increasingly difficult as microwave cavity technology is approached. Cavity probe design is now being used for some probes which operate in the 400–500 MHz range (53). Probes designed for NMR measurements on nuclei other than ^1H generally have a separate coil which can deliver continuously a coherent or incoherent decoupling field at the ^1H frequency.

6.3.   Radio-Frequency Hardware

6.3.1.   *Frequency Generation*

An NMR spectrometer must be capable of generating three independent fields; one for the observe channel (F_1), one for the lock channel (F_0), and one for decoupling (F_2). In older spectrometers, F_0, F_1, and F_2 generally involved the same nuclide, and audio modulation was used to generate the three frequencies for homonuclear observation, decoupling, and lock. In state-of-the-art NMR, the nuclides being observed, decoupled, and giving the lock signal may all be different, e.g. ^{13}C NMR with ^1H decoupling and an ^2H lock signal. To achieve this greater flexibility, all three frequencies are generated from the same master oscillator by appropriate multiplication, division, or mixing of the master oscillator frequency. The three frequencies may differ by only a few hertz or by many megahertz. Any drift experienced by the master oscillator is carried throughout all channels so that their proper correspondence is maintained. Alternatively, the different frequencies can be generated with frequency synthesizers.

Occasionally it is desirable to generate two rf pulses of the same frequency but of different phase, e.g., in some of the multiple pulsed experiments discussed earlier. Phase shifts of 90°, 180°, 270°, etc., can be accomplished by delay lines, transformers, or digital devices. State-of-the-art spectrometers are capable of phase shifts in smaller increments, e.g., 1° or 15°

6.3.2. The Transmitter

The transmitter amplifies the rf field and delivers it to the transmitter coil of the probe. For CW NMR, it must continuously supply a stable output of power up to 0.5 W. For P/FT NMR, it must deliver reproducibly a pulse of about 200 W to the probe in microseconds. The ideal pulse has zero rise and fall times; the B_1 power distribution for pulses with long rise times is nonuniform, which can cause phase and intensity errors in the spectrum. In P/FT NMR, the pulse amplifier should have a large bandwidth since the larger the bandwidth, the shorter the rise time.

Pulses to the transmitter are controlled by an rf switch or gate, which turns the rf field on or off at the appropriate time. When off, the gate should not allow any rf field to pass through or the rf will be amplified and passed on to the receiver preamplifier, which will generate noise in the spectrum and reduce sensitivity. Even when the gate is off, the pulse amplifier still generates noise. To isolate this from the preamplifier, crossed diodes are placed at the output, which effectively shuts off the transmitter.

6.3.3. The Receiver

The first and most critical part of the circuitry for the amplification of the NMR signal is the receiver preamplifier, because electronic noise at this first stage is very important in determining the overall sensitivity of the spectrometer. The preamplifier should be placed as close as possible to the receiver coil. It should have a low noise figure, a fast recovery from overload, and a linear response over a large bandwidth. Most preamplifiers are selectively tuned to a specific frequency, but broadband preamplifiers have been developed for use in multinuclear NMR. There is generally a slight loss in sensitivity with broadband preamplifiers.

To avoid duplicating the rest of the receiver circuit for every frequency, an intermediate frequency (IF) detection scheme is employed (Fig. 37). A frequency, ν_o + IF, is generated which, when mixed with the observation or carrier frequency, ν_o, or with the receiver NMR signal, ν_o + signal, produces the IF and IF + signal. The NMR resonance is extracted by phase-sensitive detection

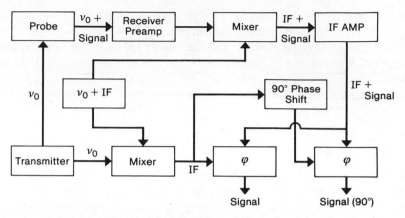

Fig. 37. Block diagram of the intermediate frequency (IF) detection scheme. (See text for details.)

of the IF + SIG frequency against the IF. In this manner, the frequency output of the mixer is the same regardless of the nucleus under investigation, and thus the main amplification is carried out at only one frequency.

Single- or double-phase-sensitive detectors may be used. If single-phase detection is used, ν_o must be positioned to the extreme right or left of the entire spectrum, or folding of signals occurs (Section 4.2.2). No signals will exist on the left or the right of the carrier, respectively, but noise will fold into the spectrum, resulting in a decrease in sensitivity. Single-side-band crystal filters can be used at the IF stage to filter out the noise on the other side of the carrier, which restores the original sensitivity.

In quadrature detection, two independent phase-sensitive detectors whose phase differs by 90° (Fig. 37) are used to detect simultaneously the real and imaginary parts of the FID. As discussed in Section 4.2.2, it is possible to distinguish signals as being higher and lower in frequency than the carrier so that the carrier may be placed in the middle of the spectrum. Thus, the transmitter needs to provide a uniform pulse power distribution over only half as large a spectral width as when single phase-detection is used, and the minimum acceptable speed of the analog-to-digital (ADC) converter is reduced by 2 (Section 6.4).

If the two detectors are not exactly 90° out of phase or if their response characteristics are not identical, then spurious peaks (quadrature images) are generated in the spectrum. To eliminate these spurious peaks, phase cycling may be used or the correction may be done via computer software.

6.3.4.   The Field/Frequency Lock

To stabilize the ratio of magnetic field and rf frequency to the precision required for the measurement of accurate chemical shifts, the spectrometer is locked to the resonance from a compound in the sample solution (internal lock) or contained in a capillary either in or out of the sample (external lock). The lock system is homonuclear or heteronuclear, depending on whether the nucleus giving the lock resonance and the nucleus being observed are the same. Most state-of-the-art spectrometers use heteronuclear ^{19}F or ^{2}H lock systems.

The lock signal is phase-shifted by 90° relative to the signal on the observe channel (phased for absorption) and is fed to the flux stabilizer as an error signal. If the error signal is zero, then the stabilizer output to the main magnetic field is zero. Any drift in the magnetic field causes the error voltage to change. The output of the stabilizer instantly corrects for this drift by increasing or decreasing the magnetic field strength.

The superconducting magnets used in high-field spectrometers are sufficiently stable that they can be operated without a lock. However, these spectrometers are routinely operated with a lock signal for shimming the magnetic field homogeneity.

6.3.5. The Spin Decoupler

The instrumentation and techniques used for spin decoupling depend on whether the observation mode is CW or P/FT and on whether the nuclei being observed and decoupled are the same (homonuclear) or different (heteronuclear).

6.3.5.1. Homonuclear Decoupling.   In the CW mode, the transmitter (F1), the decoupler (F2), and the receiver are all on simultaneously. As a result, there is a very large spike in the spectrum at the decoupler frequency, which totally obscures any resonances in that region, and beat frequencies are generally observed in multiples of 60 Hz throughout the spectrum.

In the P/FT mode, the receiver and decoupler are gated on and off to prevent receiver overload and/or computer memory overflow. Since all frequencies, including the decoupling frequency, are sampled simultaneously in the P/FT mode, the large decoupler signal would be present in the FID if F2 were on during data acquisition. This large signal would make it difficult to detect the much smaller resonance signals, a dynamic range problem, and would also overload the receiver and cause memory overflow when doing signal averaging. RF power levels in the range 0.005–0.02 W are typically used for $^{1}H\{^{1}H\}$ decoupling.

6.3.5.2. Heteronuclear Decoupling.   Heteronuclear decoupling usually refers to decoupling ^{1}H while observing another nucleus, e.g., ^{13}C or ^{15}N. Thus, F2 can be on during acquisition since the F1 and F2 frequencies are quite different.

Carbon-13 spectra are routinely measured with BB ^{1}H decoupling, i.e., all

^{1}H nuclei are irradiated simultaneously and continuously by F2. Since ^{1}H resonances cover a range of ~ 10 ppm, this requires that F2 be able to cover a wide frequency range, e.g., 1000 and 4000 Hz when B_0 is 2.35 and 9.4 T, respectively. Because the efficiency with which F2 decouples over the entire ^{1}H frequency range can determine the resolution and sensitivity of ^{13}C resonances, a great deal of effort has been directed towards developing efficient methods for BB ^{1}H decoupling (54,55). Radio-frequency power levels on the order of 1–10 W are generally used, with pseudorandom noise modulation to distribute the power over a wide range of frequencies. This corresponds to irradiation with a band of frequencies, whose power falls off at the band limits. A decoupler bandwidth of ± 5 ppm (^{1}H) is generally satisfactory. Achieving this obviously becomes more difficult the stronger B_0 becomes. The decoupler bandwidth can be increased by increasing the decoupler power, but this has the deleterious effect of excessive sample heating, especially with samples in high-dielectric solvents and ionic solutions.

Recently, a new BB decoupling scheme has been devised in which the ^{1}H nuclei are subjected to a series of 180° pulses at a rate which is fast compared to J_{CH} (56). With this technique, called composite pulse decoupling, a decoupler bandwidth of 8 kHz was obtained versus 3 kHz with normal broadband decoupling at the same power level.

For assignment purposes, the technique of coherent single frequency off-resonance decoupling (SFORD) is often used (Section 7.3) (57). Typically, decoupler powers in the 0.1–1 W range are used.

6.4.   The Interface

The FID's coming from the receiver are continuous, i.e., analog, in nature. They are passed through a low-pass audio filter, to eliminate noise which has been folded in from outside the spectral width, and then converted to digital form. This is accomplished by an analog-to-digital converter, which samples the FID instantaneously at a series of discrete, equally spaced times. Upon instruction by the computer, the ADC is connected to the receiver. The signal from the receiver charges a capacitor to a potential determined by the magnitude of the signal. The charge on the capacitor is converted to a digital number, which is fed to an address in the computer memory. This entire process should be very rapid, of the order of 10 μs, so that large frequency ranges can be sampled. The input voltage to the ADC is normally set to utilize the full range of the ADC without clipping the signal, which leads to distortion and generates harmonics. The three main parameters that determine ADC performance are the maximum input voltage, its speed of conversion, and its resolution in representing the signal as a binary number. The ADC's most commonly used in commercial spectrometers are of 12-bit resolution, with 16-bit ADC's available as accessories from some manufacturers. The analog signal must be sampled digitally at least twice per cycle of the highest frequency component or, for a particular spectral width, the sampling rate must be twice the spectral width. Typically, with a 12-bit ADC, spectral widths of 100 kHz are easily attainable. With a 16-bit ADC, the maximum spectral width is generally less because the sampling time is longer.

The interface also contains a digital-to-analog converter (DAC) to convert the digitized spectrum in the computer to analog form for plotting on an xy recorder or digital plotter or displaying on an oscilloscope.

6.5. The Computer

The role of the computer in NMR spectroscopy has changed from just a simple computer of average transients (CAT) to collecting and processing accumulated data and controlling a number of spectrometer functions, e.g., receiver gain, carrier and decoupler frequency, lock, homogeneity, and sample temperature.

6.5.1. Hardware

The minicomputers used in most NMR spectrometers have a 16- to 24-bit word-length. The wordlength is important in determining the maximum number of FID's which can be signal averaged, as discussed in Section 7.4. Some computers are capable of acquiring data in double precision. Also important is the number of words contained in core memory, which determines the computer (digital) resolution (Fig. 12). Computers used in NMR spectrometers typically have up to 256K of memory, part of which, 16K, might be used by the program. Computers with much larger memory are starting to be used in NMR spectrometers.

After the data are collected, the time-domain FID is Fourier transformed to give the frequency-domain spectrum. The Fourier transform is done by the CPU. In general, Fourier transformation requires N^2 (N = number of words) multiplications. However, the Cooley–Tukey algorithm (58) reduces the number of multiplications to $N \ln(N)$, which drastically reduces the time required for Fourier transformation. The only restriction is that N be a power of 2. The transformation is essentially performed in place using only a few additional scratch locations, with the spectrum replacing the original data. This reduces the amount of core necessary, but the original data is destroyed.

6.5.2. Software

The computer program (software) is written to perform several functions, including control of the spectrometer during data acquisition, processing of the data, and finally plotting of the spectrum. To perform these functions, the software employs two modes of user control over the operation of the system, parameters and commands, both of which consist of alphanumeric characters. Subroutines within the program carry out specific tasks and where necessary are called by the executive which oversees all computer operations.

The acquisition subroutines initiate collection of data by first turning off the receiver and turning on the pulse. Then the receiver and ADC are turned on to detect and sample the FID. Acquisition parameters include the length and phase of the pulse, the number of pulses, spectral width, number of points sampled (size of FID), time between pulses, and data routing.

Once the FID has been collected, it is processed with the processing software. The first step might involve the multiplication of the FID by one of a variety of functions to enhance the desired information in the final spectrum. The FID is the superposition of a number of exponentially decaying sinusoidal signals; the intensity information is in the beginning of the FID, and linewidth information is in the rate of decay. The latter part of the FID contains mostly noise, but it is necessary for high resolution. If the FID is truncated, i.e., if data collection is stopped, before it decays to zero, then Fourier transformation leads to (sin x)/x wiggle beats on either side of the main peak. These can be removed by multiplication of the FID by a trapezoidal function which goes to zero at the longest time in the collection period (apodization).

The processing software also generally contains functions with which the resolution or sensitivity can be enhanced (Fig. 38) (59). For example, the FID can be multiplied by a decreasing or increasing exponential function or by a Gaussian function (transformation from a Lorentzian to a Gaussian lineshape). Multiplication by the decreasing exponential preferentially favors the beginning of the FID, which improves sensitivity at the expense of resolution (Fig. 38C); whereas multiplication by an increasing exponential improves resolution at the expense of sensitivity. A Gaussian function may improve either sensitivity or resolution (Fig. 38A), or combinations thereof, depending on the location of its maximum in the FID.

The final steps after Fourier transformation are phase correction, plotting, and integration. Two types of phase corrections are necessary: zero order, which are independent of frequency, and first order, which show a first-order dependence on frequency. Zero- and first-order phase corrections are easily performed via external phase knobs. Once the spectral region of interest has been selected, the appropriate data are routed through the DAC to the recorder or scope.

Integration of resonances to determine intensities involves summing over digital values rather than over an analog signal as in CW NMR. The accuracy of results obtained from the integrals depends in part on the accuracy of the data collected, as determined for example by the repetition time, the pulse power distribution, and the NOE. Further precautions must be taken during acquisition and data manipulation. If the data are not represented accurately, e.g., if each resonance is not defined by enough points, then the integrals will not give a true

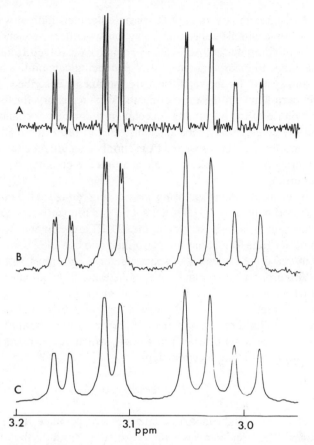

Fig. 38. The effect of multiplication of the FID by software functions on resolution and sensitivity. Spectrum *B* was obtained by Fourier transformation of a 32K FID collected over a spectral width of 3623 Hz. The digital resolution is 0.221 Hz per point. Spectrum *A* was obtained by doing a Gaussian multiplication on the same FID, to increase resolution, before Fourier transformation. Note the appearance of long-range couplings in the resonances on the right. Spectrum *C* was obtained by multiplying the same FID by an exponentially decreasing function, to increase sensitivity, before Fourier transformation. The resonances are the AB part of the ABX pattern for the $-CH_2CH-$ protons of histidine.

picture of the relative proportions of chemically distinct nuclei. If the resonance of interest is narrower than the computer resolution, its integral will be lower than its true value, and more data points are needed. Also, if a spectrum contains a broad resonance and a sharp resonance, multiplication of the FID by any form of resolution enhancement function or a delay before data acquisition will preferentially favor the sharper resonance. In fact, if the delay is large enough, the broad resonance may have decayed to zero by the time acquisition begins.

The above discussion covers only the basic computer and software requirements needed for a simple relaxation delay–pulse–collect sequence and data processing. Both computer and software systems have evolved from the simple one-pulse sequence to incorporate the pulse-sequencing capabilities necessary to prepare the spin system for the sophisticated experiments described earlier, e.g., the 2D FT experiments. In these experiments, it is necessary for the computer to control the phase of the transmitter, decoupler, and receiver channels, to gate them on and off at precisely timed intervals, and to route the data to specific locations. User interaction has changed from simply executing certain commands via alphanumeric characters, executed one at a time, to creating experiments via a microprogram.

In many experiments, data processing and plotting represent the major fraction of the total experiment time, e.g., if the solute is sufficiently concentrated that little signal averaging is necessary or in the 2D FT experiments where a large number of Fourier transformations are performed and complex data matrices are plotted. To increase the efficiency of spectrometer usage, most software is now capable of "foreground/background" operation, e.g., the computer can control the spectrometer and collect data in one region of memory (a background operation) while it is plotting data from a second region of memory and the user is processing data in a third region of memory (a foreground operation). Processing time can be decreased with array processors, which are offered as optional hardware on state-of-the-art spectrometers.

6.5.3. *Peripherals*

Computer peripherals might include a teletype, a paper tape reader, and mass storage devices. The user interacts with the software through the teletype. The teletype also echoes commands and parameters entered by the user and provides the user with a hard copy of peak printout (chemical shift, intensity, etc.) and parameters. The paper tape reader is used to input programs. Storage devices may be random access floppy or hard discs or serial access magnetic tape units.

7. EXPERIMENTAL TECHNIQUES

7.1. Sample Preparation

Some of the factors to be considered when preparing a sample are the state of the sample, the concentrations of the compounds of interest, possible impurities, and the field/frequency lock signal. Both liquid and solid samples are generally run as solutions. Gaseous samples are characterized by long T_1 values and are not run routinely. In many cases, the sample will be obtained as a solution. If the compounds of interest are present at a sufficiently high concentration and if the solvent resonances do not obscure the region of interest, the sample can be run directly. If the concentration is too low, it may be possible to use one of the methods described in Section 7.7 for the detection of weak signals in the presence of strong signals, or it may be necessary to work up the sample using preconcentration and separation techniques such as solvent extraction. This is a common problem when doing ^{1}H NMR, but much less of a problem when measuring NMR spectra for other nuclei, e.g., ^{31}P and ^{19}F.

7.1.1.   The Solvent

A pure, chemically inert solvent capable of dissolving the sample over the desired concentration and temperature ranges and having no resonances in the spectral region of interest must be used. Because of the dynamic range problem in P/FT NMR when the spectrum contains both strong and weak signals, most solvents used in ^{1}H NMR are deuterated, e.g., $CDCl_3$, CD_3COCD_3, C_6D_6, CD_3SOCD_3, D_2O, CD_3OD, and CD_3CN. These solvents have the added benefit of providing an ^{2}H resonance for the field/frequency lock. Deuterated solvents are also used routinely in ^{13}C NMR for the ^{2}H lock signal. In ^{13}C NMR, it generally is not necessary to use solvents depleted in ^{13}C because carbons bonded to deuterium have long T_1 values, and thus their resonances are partially saturated at the repetition rates normally used. For most ^{1}H NMR measurements, deuteration at the 99.7% level is sufficient; however, for measurements at trace levels, it may be necessary to use 100% deuterated solvents.

7.1.2.   Paramagnetic Impurities

Contaminants, either in particulate or solution form, can have a deleterious effect on the spectrum. A particulate of magnetic material can cause the resolution to deteriorate as it moves in the receiver coil region. Particulates can be removed by filtering the solution through a cotton plug in a disposable pipette. Dissolved paramagnetic species, including transition metal ions and molecular oxygen, can shorten relaxation times by the electron–nuclear dipolar mechanism. Dissolved transition metal ions can be removed by equilibrating the sample with a chelating resin. Molecular oxygen can be removed by degassing the solvent, e.g., by taking it through several freeze–pump–thaw cycles or by bubbling an inert gas (N_2, He, Ar) through it.

7.1.3.   Sample Cells

Sample cells are normally cylindrical tubes made from Pyrex glass. Tubes come in a variety of sizes, with outside diameters ranging from 1 to 35 mm. The usual practice is to use tubes of the maximum diameter which can be accommodated by the probe to optimize the filling factor. Tubes must be perfectly straight and the glass must be of a uniform thickness to minimize spinning sidebands (ssb). If it is necessary to use an external reference or an external lock signal, they can be contained in a coaxially aligned cell, e.g., a 2–3-mm o.d. sealed capillary. Care must be used to ensure that these are placed concentrically to prevent ssb. Small coils and microcells have been developed for samples of limited size (60), as have egg-shaped and spherical microcells. The advantage of these cells is that the sample is concentrated in the detection region, resulting in higher sensitivity over standard cells by as much as a factor of 5. The full advantage of the microcell is realized only when it is properly positioned and has a good filling factor. One problem is that the homogeneity changes dramatically, and these cells are more difficult to shim.

The NMR tube should be filled to a height which is above the top of the receiver coil for maximum resolution. Generally, one tries to fill all sample tubes to approximately the same height, because shim settings are different for different sample heights. When sample volume or concentration is not a problem, adding more solution or solvent generally results in better resolution. The larger the diameter of the tube, the greater the risk of vortexing when spinning the sample, which will degrade the resolution. Vortex plugs can be used to prevent vortexing.

7.1.4. Chemical Shift References

Resonance frequencies are normally measured relative to the resonance of a standard or reference compound. The reference compound should be chemically inert and soluble in the solvent over the temperature range to be used, and should give a sharp resonance, generally at one end of the spectral region with no interfering resonances. Both internal references (those dissolved in the sample solution) and external references (physically separated from the sample) are used. Sample cells for external references were discussed in the previous section.

Tetramethylsilane (TMS) is universally accepted as the reference for ^{1}H and ^{13}C NMR because it is inert, has a resonance "upfield" from most resonances, and has a low boiling point, making it easy to remove. Several disadvantages of TMS are that its resonance does shift in solutions of aromatic compounds and it is not suitable for aqueous solutions or high-temperature studies. Hexamethyldisiloxane (HMDS) has been used for high-temperature work, and the water-soluble sodium salt 2,2-dimethyl-2-silapentane-5-sulfonate (DDS) or tetramethylammonium salts are used for aqueous solutions.

7.2. Chemical Reagents for NMR

The measurement and interpretation of NMR spectra can often be aided by adding chemical reagents to the sample. For example, faster repetition rates can be used by adding a reagent which reduces T_1 values; resolution can be increased by adding a reagent which increases chemical shift dispersion; and functional group analysis can be done on complex mixtures by forming derivatives which contain a nuclide not originally present in the sample. The solvent also can be chosen to modify the NMR spectrum.

7.2.1. Shift Reagents

The overlap of resonances is a common problem in ^{1}H NMR spectroscopy. Often, resonances can be spread apart by adding certain lanthanide ions or lanthanide complexes to the solution (61,62).

The mechanism by which the lanthanide-induced shifts (LIS) are produced involves the formation of a Lewis acid–base adduct by reaction of the lanthanide shift reagent (LSR) with a nonbonded pair of electrons on the molecule, e.g., a nonbonded pair of electrons on a carbonyl, alcohol, or ether oxygen. The nuclei of the molecule in this form are shielded or deshielded by the unpaired electrons of the paramagnetic lanthanide ion. Thus, the effectiveness of a particular LSR depends both on its strength as a Lewis acid toward functional groups of the molecule, and on the paramagnetic properties of the LSR.

The paramagnetic lanthanides shield or deshield nuclei by (1) the contact shift mechanism in which unpaired electron density is transferred to the nucleus as a result of delocalization into molecular orbitals or spin polarization, and (2) the dipolar (pseudocontact) mechanism, in which nuclei experience magnetic fields generated by the unpaired electrons on the lanthanide. Thus, dipolar shifts result from through-space interactions. Lanthanide-induced shifts in ^{1}H NMR spectra are caused mainly by the dipolar mechanism (63), whereas ^{13}C resonances can be shifted by both the contact and dipolar mechanisms (64). The magnitude of the dipolar contribution Δ_i to the LIS for a particular resonance depends on the distance between the nucleus and the lanthanide ion, r_i, which has the effect of spreading the resonances apart.

Generally, molecules exchange rapidly between their free and LSR-complexed forms, so that exchange-averaged resonances are observed. Thus, the magnitude of the observed LIS depends on the mole fraction which is complexed. LSR can be added until the necessary resolution has been achieved, e.g., until a well resolved analytical resonance is obtained for each component of interest in a multicomponent mixture. The spectra in Fig. 39 provide an example of the use of LSR to obtain analytical resonances for each component of a mixture. By adding a LSR, analytical resonances are obtained for each alcohol in a mixture containing methanol, ethanol, n-propanol, n-butanol, n-pentanol, and n-hexanol, on a low field spectrometer operating at 60 MHz (65). By combining LSR and high field spectrometers, it should be possible to analyze quite complicated mixtures by NMR.

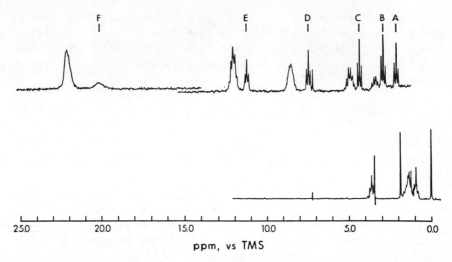

Fig. 39. The effect of Eu(DPM)₃ on the 60 MHz ¹H NMR spectrum of CDCl₃ solution containing methanol, ethanol, *n*-propanol, *n*-butanol, *n*-pentanol, and *n*-hexanol. (*lower*) Spectrum of CDCl₃ solution. (*upper*) Same solution plus Eu(DPM)₃. The resonances identified are for the methyl protons of *n*-hexanol (A), *n*-pentanol (B), *n*-butanol (C), *n*-propanol (D), ethanol (E), and methanol (F). Reprinted with permission from reference 65. Copyright 1971 American Chemical Society.

A variety of lanthanide derivatives have been developed for use as LSR. Complexes of Eu(III), Pr(III), Yb(III), Ho(III), and Dy(III) are the most effective by inducing large shifts without excessive broadening of the resonances. Generally, Eu(III) and Yb(III) complexes cause a decrease in shielding, while the others cause an increase in shielding. The aquated ions themselves or their EDTA chelates have been used for aqueous samples. Neutral chelates formed with β-diketone ligands are used for nonaqueous samples. These include the chelates tris(dipivaloylmethanato)Ln(III), Ln(DPM)₃ (R = R′ = *t*-butyl) and tris-(1,1,1,2,2,3,3-heptafluoro-7,7-dimethyl-4,6-octanedionato)Ln(III),   Ln(FOD)₃

(R = *t*-butyl, R′ = *n*-C₃F₇). LSR formed from fluorinated β-diketones generally are stronger Lewis acids. β-diketone chelates are effective as shift reagents for molecules containing a variety of functional groups, including alcohol, carbonyl, ether, ester, nitro, sulfoxide, and amino groups. They are not effective, however, for carbon–carbon double bonds.

Recently, binuclear shift reagents containing Ln(III) and Ag(I) ions have been developed as shift reagents for aromatic and olefinic compounds (66); the substrate binds to Ag(I), which holds it close to the lanthanide. To illustrate their application, Wenzel and Sievers have shown that by adding a Yb(III)–Ag(I) reagent to unleaded gasoline, analytical resonances can be resolved for toluene, o-, m- and p-xylene, durene, mesitylene, 1,2,3- and 1,2,4-trimethylbenzene (66).

7.2.2.   Relaxation Reagents

Paramagnetic species may increase nuclear relaxation rates, as discussed in Section 2.3.3. Complexation with the substrate is not necessary; however, if there is binding, those nuclei closest to the binding site will be the most affected. Paramagnetic relaxation reagents (PARR), which either do not complex the substrate or complex it very weakly, are useful for reducing relaxation times throughout the entire molecule. Since T_2^* is almost always less than T_1, PARR can be added to reduce T_1 with no effect on line width. Such reductions in T_1 are important when doing quantitative analysis by P/FT NMR, because faster repetition rates can be used and thus signal averaging to increase the S/N ratio is more efficient. PARR are most useful with nuclides having long T_1 values, e.g., ^{13}C, ^{29}Si, and ^{15}N, which can have T_1 values longer than 20 s.

PARR are also useful for quenching the NOE (67), which can distort resonance intensities, e.g., in ^{13}C NMR spectra measured with BB ^{1}H decoupling. NOE's occur when T_1 relaxation is dominated by the nuclear dipole–dipole mechanism. The PARR effectively short circuits the relaxation through the electron–nuclear dipole–dipole mechanism, thus quenching the NOE.

Care must be exercised when using PARR to suppress NOE's in quantitative NMR measurements. The assumptions made are that the PARR has reduced all T_1 values and that NOE's have been suppressed completely. However, it has been shown (68) that, in medium and large organic molecules, the suppression of the NOE by PARR may not be complete due to the efficiency of the ^{13}C–^{1}H dipolar mechanism, even at high concentrations of the PARR.

The PARR most commonly used in organic solvents include Cr(acac)$_3$ and Fe(acac)$_3$ (69). Cr(III), Fe(III), and Gd(III) complexes of diethylenetriamine-pentaacetic acid and triethylenetetraaminehexaacetic acid have been used as aqueous PARR (70). The water soluble 4-hydroxy- and 4-amino-2,2,6,6-tetra-methylpiperidinooxy organic free radicals also have been used as aqueous PARR (71).

7.2.3.   Fluorine-Containing Reagents

Another approach to the problem of obtaining analytical resonances for the compounds of interest is to form derivatives with reagents which give well resolved signals. Reagents have been developed for ^{1}H NMR, however their application is limited due to the inherently small ^{1}H chemical shift range and potential overlap with sample resonances. Fluorine-containing derivatives are generally much more useful because the ^{19}F chemical shift range is considerably larger, chemical shifts are more sensitive to small changes in molecular structure, and there are usually no background signals. Since ^{19}F is almost as sensitive as ^{1}H to NMR observation, this provides a very attractive approach to trace analysis by NMR.

There are several ^{19}F reagents for forming derivatives with functional groups containing active hydrogen, e.g., OH, SH, and NH. Manatt described the use of trifluoroacetic anhydride to form trifluoroacetyl derivatives of alcohols and phenols (72). The trifluoroacetyl ^{19}F resonance is a singlet. Since each hydrogen is replaced by 3 equivalent fluorine atoms, there is effectively an increase in sensitivity. Dorn and co-workers have proposed trifluoroacetyl chloride (TFAC) as an alternative reagent for forming the same derivatives (73),

$$CF_3 \overset{\overset{\displaystyle O}{\|}}{C} Cl + RXH \rightarrow CF_3 \overset{\overset{\displaystyle O}{\|}}{C} XR + HCl$$

where X is O, S, or N. The derivatives are prepared *in situ* by bubbling reagent gas (B.P., $-18°$C) in a solution of the sample, e.g., in chloroform or tetrahydrofuran. Trifluoroacetic acid, which is formed by reaction with water, can be removed by pouring the sample over anhydrous K_2CO_3. Dorn and co-workers present chemical shift and yield data for some 50 trifluoroacetyl derivatives of alcohols, phenols, thiols, and primary and secondary amines (73). The chemical shifts depend on the nature of the heteroatom (O, N, or S) and the number and type of groups on the carbon attached to the heteroatom. The compound, $\alpha\alpha\alpha$-trifluoroacetophenone was used as an internal intensity standard for quantitative analysis. Jung et al. have shown that natural products such as polyols, glycosides, and mixtures of terpenols and steroids can be easily and rapidly analyzed by ^{19}F NMR as their trifluoroacetyl derivatives (74).

Other ^{19}F reagents include hexafluoroacetone (HFA) (75), 2,2,2-trifluorodiazoethane (76) and trifluoromethanesulfonyl chloride (77). HFA forms adducts with alcohols, phenols, thiols, and other functional groups which contain active hydrogen (75). These derivatives potentially offer even higher sensitivity than trifluoroacetyl derivatives since they contain 6 equivalent fluorine atoms. The ^{19}F chemical shifts are very sensitive to the nature of RXH, providing both qualitative and quantitative information. A particularly interesting application is the determination of trace water in organic solvents (75b). This is difficult to do by ^{1}H NMR because of the dynamic range problem, exchange with other active hydrogen, and resonance overlap. The ^{19}F method is reported to have high sensitivity, few interferences, and, perhaps most important when compared to the Karl Fischer method, it can be used to quantitate simultaneously trace water and alcohols, etc., as well as to characterize the alcohols.

In view of the high sensitivity of ^{19}F NMR, its wide chemical shift range, and the lack of interfering background resonances, it would seem that ^{19}F NMR with fluorine-containing derivatives will become an even more important method of trace analysis. At present, its range of application is limited only by the range of derivatives which have been developed.

7.2.4. Miscellaneous Reagents

It often is difficult in ^{1}H NMR to determine the position of resonances from exchangeable protons, e.g., NH and OH protons. In some solvents, e.g., DMSO-d_6, proton exchange with residual H_2O in the solvent is slow so that relatively sharp resonances are observed, often split by coupling to hydrogens on adjacent carbons. The assignment of the resonance to an exchangeable proton can be confirmed by exchanging it for deuterium, e.g., by adding D_2O or CD_3OD. A reduction in intensity or disappearance of the resonance after exchange verifies the presence of exchangeable protons.

Trichloroacetylisocyanate (TCAI) can be used as an *in situ* reagent for the determination of hydroxyl and amino groups by ^{1}H (78) and ^{13}C NMR (79). TCAI reacts with alcohols and amines to form a urethane or a substituted urea:

$$RXH + O{=}C{=}N-\overset{\displaystyle O}{\overset{\displaystyle \|}{C}}-CCl_3 + RX\overset{\displaystyle O}{\overset{\displaystyle \|}{C}}-\overset{\displaystyle H_A}{\overset{\displaystyle |}{N}}-\overset{\displaystyle O}{\overset{\displaystyle \|}{C}}-CCl_3$$

where X is O or N. Proton H_A gives resonances in the 7.5–8.5 ppm region. Its chemical shift is sufficiently sensitive to the nature of RX that information can be obtained about the number and type of different alcohols and amines. The shifts of the protons on the carbinol carbon depend on whether the alcohol is primary or secondary. Carbon-13 chemical shifts of TCAI derivatives also can be used to distinguish primary, secondary and tertiary alcohols. Often, TCAI-induced shifts are used to spread apart resonances in regions of spectral congestion.

7.3. Assignment of Resonances

It sometimes is possible to assign resonances on the basis of their chemical shifts, intensities, and spin–spin coupling patterns. The chemical environment of a nucleus can be identified from the chemical shift, using the extensive chemical shift–structure correlations in the literature. The number of nuclei giving the resonance can be determined from its intensity, and their neighbors can be identified from the spin–spin coupling patterns. However, when dealing with large molecules or multicomponent mixtures, additional information may be necessary to make the assignments. Spin–lattice relaxation times and the magnitude of the NOE are often useful, as are the results of experiments with lanthanide shift reagents and paramagnetic relaxation reagents and spin decoupling experiments.

The shifts induced by LSR are useful for assignment purposes since the magnitude of the shift decreases with distance from the site of complexation. Thus, if the molecule contains only one binding site, or if it has several sites but of differing Lewis basicity towards the shift reagent, the shift reagent not only spreads the resonances apart, but it also sorts the resonances according to their distance from the site of complexation. Relaxation or broadening reagents with Lewis acidity toward groups in the molecule, e.g., $Gd(DPM)_3$ and $Gd(FOD)_3$, can be used to obtain similar information. Relaxation reagents of this type complex the molecule, and thus selectively decrease relaxation times and broaden resonances, the effect decreasing with distance from the site of complexation.

Resonances which are connected by spin–spin coupling can be identified by several experimental techniques. The simplest and most widely used is selective decoupling, either homo- or heteronuclear (Section 2.3.2.2), with which connectivities of resonances can be established. Hall and Sanders have outlined a general strategy for the analysis of ^{1}H-NMR spectra of complex molecules which uses 2D FT J-resolved spectroscopy to resolve and analyze all the resonances, and relaxation rates, NOE and decoupling experiments to assign them (80).

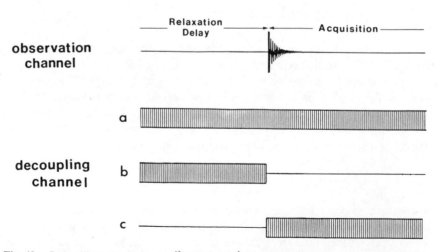

Fig. 40. Decoupling schemes used in ^{13}C NMR. BB ^{1}H decoupling (*a*) on throughout the relaxation delay and acquisition periods, (*b*) gated off during acquisition, and (*c*) gated on during acquisition. (See text for details.)

These same techniques are useful for assigning NMR spectra of complicated, multicomponent mixtures.

Carbon-13 spectra are normally measured with BB ¹H decoupling on throughout the experiment [decoupling scheme (*a*) in Fig. 40]. This has the effect of increasing sensitivity by up to 8 due to multiplet collapse and NOE. However, ¹H coupling information, which is potentially useful for assignment purposes, is lost and intensities no longer are directly proportional to the number of nuclei [spectrum (*C*) in Fig. 41]. ¹H-coupled ¹³C spectra can be measured while retaining the sensitivity enhancement due to the NOE by gating the BB ¹H decoupler (81) according to scheme (*b*) in Fig. 40. The decoupler is on during the delay period before the ¹³C pulse to build up the NOE. Immediately before the pulse, it is gated off, and ¹H coupling is restored. With decoupling scheme (*c*) (inverse gated decoupling), the decoupler is gated on only during acquisition giving BB ¹H decoupled ¹³C spectra with the NOE suppressed (82).

The first step in the assignment of a ¹³C spectrum is to determine the number of protons attached to each carbon. In some cases, this can be determined from the fully coupled spectrum measured with gated decoupling sequence (*b*) in Fig. 40. Often, however, the ¹H-coupled resonances overlap so that it is difficult to make unequivocal assignments [spectrum (*A*) in Fig. 41]. The SFORD technique (Section 6.3.5) can be used to reduce the apparent magnitude of $J_{CH'}$ which

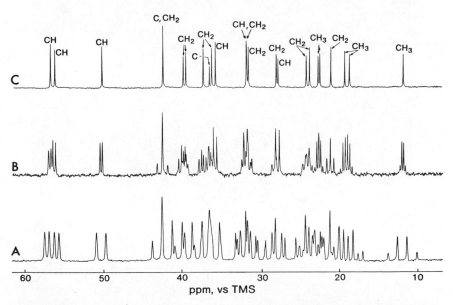

Fig. 41. The high field portion of the (*A*) fully coupled, (*B*) off resonance ¹H decoupled, and (*C*) BB ¹H decoupled 100.6 MHz ¹³C spectrum of cholesterol. Reprinted with permission from reference 32.

has the effect of bringing resonances in each multiplet together while retaining the characteristic doublet, triplet and quartet patterns for CH, CH_2, and CH_3 carbons. This simplifies the spectrum, making assignments more straightforward in many cases. For example, the SFORD spectrum for cholesterol is shown in Fig. 41B. The number of attached protons can easily be determined for most of the resonances from the reduced multiplet patterns. However, there are ambiguities in the 24 and 32 ppm regions. There are also two drawbacks of the SFORD method which should be noted: (1) multiplicities may be distorted by second order spin–spin coupling effects in the ^1H spectrum (83) and (2) the power levels used in broadband and SFORD are different, which may cause sample temperature and thus chemical shifts to be different (84).

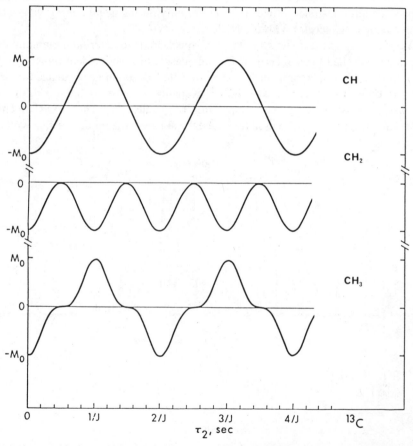

Fig. 42. Calculated intensity modulation of the singlets in SEFT spectra obtained with gated decoupling for doublets, triplets, and quartets. The horizontal axis gives τ_2 in units of J. Carbon-13 NMR spectra are measured with decoupling scheme (*b*) in Fig. 29.

An alternative method for determining the number of attached protons is based on intensity modulation in ^{13}C spectra measured by the spin–echo Fourier transform method with gated BB ^{1}H decoupling (31,32). The pulse sequence is shown in Fig. 29. ^{1}H decoupling is gated so that the precession frequencies of the different components of a ^{13}C multiplet are different during the two delay periods, the result is that they are out of phase with each other at the start of acquisition. This can be accomplished by gating the decoupler on during one of the delay periods [scheme (*b*) in Fig. 29] or by applying a broadband 180° ^{1}H pulse simultaneously with the 180° ^{13}C pulse [scheme (*a*) in Fig. 29]. With both schemes, the components of a ^{13}C multiplet are out of phase at the end of the evolution period (Fig. 30). If BB ^{1}H decoupling is gated on during acquisition, the ^{13}C resonances appear as singlets, but their intensity varies with τ_2 (Fig. 42), the exact intensity variation depending on whether there are 0, 1, 2, or 3 protons attached to the carbon as shown in Fig. 43. The CH and CH$_3$ resonances are differentiated by a spin–echo difference experiment (*B* in Fig. 43) and quaternary carbons are identified by doing the experiment with $\tau_2 = 1/(2J)$(*C* in Fig. 43).

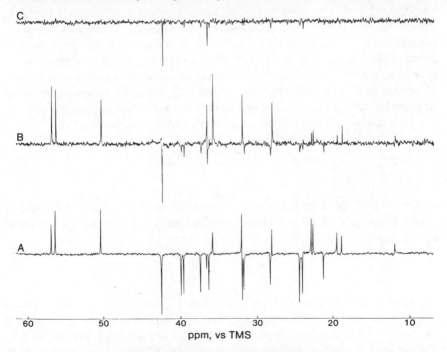

Fig. 43. The high field portion of the ^{13}C spectra of cholesterol measured by the spin–echo method with gated ^{1}H decoupling (scheme *b* in Fig. 29). *A* was measured with $\tau_2 = 0.008$ s; *B* is the difference spectrum obtained by subtracting the FID obtained with $\tau_2 = 0.0032$ s from that obtained with $\tau_2 = 0.0048$ s; *C* was measured with $\tau_2 = 0.004$ s. Reprinted with permission from reference 32.

With this method, the first step in the assignment of ^{13}C spectra of large molecules can be done easily (31,32). It also should prove to be of considerable use in the analysis of the ^{13}C spectra of mixtures, e.g., petroleum samples. This experiment, which has recently been modified so that signal averaging can be done more efficiently, has been given the name APT for attached proton test (85).

It should be mentioned that similar information can be obtained by the INEPT or DEPT experiment (Fig. 34); however, these experiments are more difficult to perform and more demanding of the spectrometer (86).

7.4. Signal Averaging

As discussed in Section 4.2.3, the S/N ratio can be increased by a factor of $\sqrt{N}$ by signal averaging N spectra in CW NMR or N FID's (scans) in P/FT NMR. There is a practical limit to N beyond which the increase in S/N is small for the time invested. The computer word length also limits the number of scans which can be averaged in P/FT NMR. Usually, the resolution of the ADC is less than the computer word length. As successive scans are added to memory, the number of bits used in each computer word increases. Eventually, the accumulated FID will completely fill all the bits in those words that hold the data points at the beginning of the FID. Further accumulation causes memory overflow, i.e., the FID is clipped. This causes distortion in resonance intensities and a decrease in S/N because every part of the FID contains information about all frequencies. It also causes harmonics to be generated. The total number of scans N which can be accumulated before memory overflow is given by (87)

$$N^{1/2} = \frac{-1 + [1 + 4S(S+1)\ 2^{w-d}]^{1/2}}{2S} \tag{51}$$

where S is the S/N of the largest signal in the spectrum, and d and w the number of bits in the ADC and the computer word, respectively. If S is large compared to one, Eq. (51) reduces to

$$N = 2^{w-d} \tag{52}$$

A practical consequence is that, for spectra which contain only very weak signals, e.g., ^{13}C spectra, the ADC resolution should be kept to a minimum to increase the number of accumulations possible, and thus the potential increase in S/N. The software in use today generally has provisions for preventing memory overflow by dividing down the number of bits when memory is almost full. In order to maintain statistical significance, both memory and successive ADC readings are divided down by the same factor.

If qualitative rather than quantitative information is to be derived from the spectrum, a flip angle and repetition rate are used which will give the largest increase in S/N in a given period of time. The repetition rate is determined on the basis of T_2^*, and then a flip angle is chosen on the basis of the repetition rate and the estimated value for the longest T_1 of the nuclei of interest (Section 4.2.3). When using these conditions, the relative intensities are no longer equal to the relative numbers of nuclei because the magnetization does not return to equilibrium between pulses, i.e., the resonances are partially saturated.

7.5. Quantitative P/FT NMR

If quantitative information is to be derived from the spectrum, pulse and acquisition conditions must be chosen that are consistent with the maximum tolerable error level. To properly define the resonances, the FID should be sampled for at least $3-5T_2^*$. Also, if T_2^* values for different resonances are quite different, the delay between the end of the pulse and the start of data acquisition must be sufficiently short that differential T_2^* relaxation does not occur.

Pulse conditions in quantitative P/FT NMR generally are a compromise between those which give maximum sensitivity enhancement and those which give equilibrium resonance intensities, i.e., relative resonance intensities which are exactly equal to relative numbers of nuclei (88). If sensitivity is not a problem, the conditions often used are $\alpha = 90°$ and $t_r = 5$ times the longest T_1 of the nuclei of interest. Under these conditions, the intensity of the resonance for the nuclei having the longest T_1 is 99.3% of its equilibrium intensity. The intensity of resonances for nuclei whose T_1 values are shorter will be closer to the equilibrium intensity, e.g., the intensity of resonances for nuclei having T_1 values 0.75 and 0.5 of the longest T_1 will be 99.9 and 100% of their equilibrium intensity. The larger the range of T_1 values, the larger the range of intensity differences for the same number of nuclei. The maximum range of intensity differences due to incomplete relaxation with $t_r = 5T_1$ is 0.7%.

If sensitivity is a problem, other compromise pulse conditions can be selected which give some sensitivity enhancement while keeping intensity errors due to incomplete relaxation at an acceptable level. The resonance which is most saturated will have a fractional intensity (i.e., the fraction of its equilibrium intensity) given by (89)

$$f = \frac{1 - E}{1 - E \cos \alpha} \tag{53}$$

where $E = \exp(-t_r/T_1)$. The maximum percentage error in resonance intensity ϵ for these pulse conditions is given by

$$\epsilon = \frac{100\,E(1 - \cos \alpha)}{1 - E \cos \alpha} \tag{54}$$

The maximum error is plotted for $\alpha = 30°$, $60°$, and $90°$ as a function of t_r in Fig. 44 (89). In most experiments, a limit can be set on the maximum error tolerable. For a given t_r, the maximum pulse angle which can be used within these error limits is given by (89)

$$\alpha = \cos^{-1} \frac{\epsilon - 100E}{(\epsilon - 100)E} \tag{55}$$

where E is for the longest T_1 of the nuclei of interest. This may not be the optimum pulse angle for maximum sensitivity enhancement. The values of t_r and α at which maximum sensitivity is obtained at various error levels are: $t_r = 4.5T_1$, $\alpha = 84°$ (1% error level); $t_r = 3.5T_1$, $\alpha = 81°$ (2.5%); $t_r = 2.5T_1$, $\alpha = 79°$ (5%); $t_r = 2.0T_1$, $\alpha = 73°$ (10%); $t_r = 1.2T_1$, $\alpha = 65°$ (20%); and $t_r = 0.7T_1$, $\alpha = 55°$ (30%) where T_1 is the maximum T_1 of the nuclei of interest (89). The differences in relative intensity within a spectrum will generally be less than the maximum error used in Eq. (55) because the range of T_1 values generally will be limited. Thus, all resonances will be saturated to some extent.

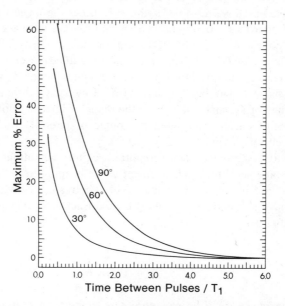

Fig. 44. The maximum range of errors in resonance intensities with flip angles of 30°, 60°, and 90° as a function of the time between pulses (repetition time) in units of the longest T_1 of the nuclei of interest (89).

Other potential sources of error include digital resolution and the delay between the end of the pulse and the start of data acquisition. Both are discussed earlier in this chapter. The selection of optimum conditions and sources of error in quantitative analysis by P/FT NMR has been considered by several authors (71,88–91).

7.6. Quantitative Measurements

The integrated intensity (area) of a resonance is related to concentration by:

$$I = kC \tag{56}$$

where k is a constant whose magnitude depends on the nuclide and the measurement conditions. For spectra measured by CW NMR, k is the same for all resonances in the spectrum if B_1 is low enough that saturation does not occur. For k to be the same for all resonances in spectra measured by the P/FT method, the repetition rate must be slow enough for complete T_1 relaxation between pulses and, if spin decoupling is used, pulse conditions must be chosen to eliminate the NOE. When spectra are measured under these conditions, relative concentrations can be obtained directly from relative peak areas with the equation

$$\frac{C_A}{C_B} = \frac{I_A n_B}{I_B n_A} \tag{57}$$

where C_A is the concentration of A, I_A the integrated intensity of a resonance from A, and n_A the number of chemically equivalent nuclei giving the resonance.

Resonance intensities will be in error if there is overlap with ^{13}C satellites from other resonances. All resonances for ^1H directly bonded to ^{12}C will be symmetrically flanked by satellite resonances from ^1H bonded to the 1.1% of naturally occurring carbon which is ^{13}C. Satellites are generally 60–70 Hz to high and low frequency from each ^1H–^{12}C resonance. This splitting has two effects of importance in quantitative measurements: (1) the actual resonance intensity is too low by 1.1% and (2) the satellites may overlap with other resonances, causing their integrated intensities to be too large. The intensities of all resonances for ^1H bonded to carbon will be affected equally by (1) and this effect can be ignored if the resonances used for quantitative measurements are all for carbon-bonded ^1H nuclei. However, the different resonances will be affected differently by (2). If the ^{13}C satellite from resonance B overlaps resonance A, the integrated intensity for A can be corrected by

$$I_{A,C} = I_A - 0.0055 I_B \tag{58}$$

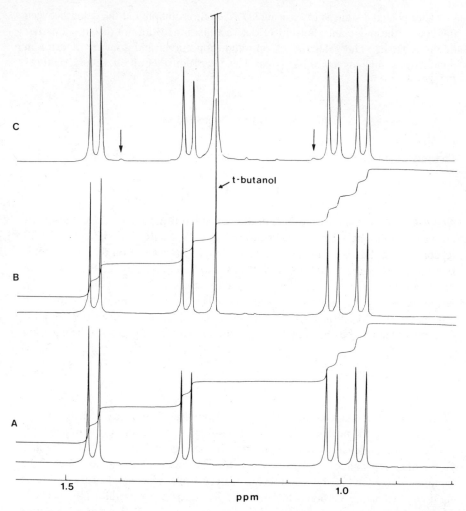

Fig. 45. The 360 MHz ^{1}H-NMR spectra of the methyl region of (A) the amino acid mixture in Fig. 3; (B), (C) after the addition of t-butanol internal reference. Integral traces are shown for (A) and (B). (See text for details.)

where $I_{A,C}$ is the corrected integral for A. To illustrate, the methyl region of the ^{1}H NMR spectrum in Fig. 3 is expanded in Fig. 45A. This spectrum was measured by P/FT NMR with $t_r = 8T_1$; thus, k in Eq. (56) is the same for the different resonances. Also shown is the integral for this region. The heights of the steps in the integral trace are directly proportional to the integrated intensities of the resonances. The integrals for the alanine, threonine and valine resonances are 53.0, 36.8, and 84.0 mm, respectively. The alanine and threonine doublets are separated by 60 Hz. The corrected integrals for alanine and threonine are 52.8

and 36.5, respectively, from which the relative concentrations for alanine:threonine:valine are calculated to be 1.00:0.692:0.796. When the integrals are similar in intensity, as here, neglect of ^{13}C satellites causes little error. For example, relative concentrations calculated directly from the uncorrected integrals are 1.00:0.694:0.794.

Absolute concentrations can be obtained by calibrating the response of the spectrometer. If the analyte is available in pure form, a calibration curve can be prepared by measuring the intensity of its resonance(s) as a function of concentration. The concentration of the unknown can then be read from the calibration curve. For this method to work, the calibration curve and the unknown must be measured using exactly the same instrument conditions. Also, identical tubes must be used throughout.

The standard addition method is more convenient when the pure analyte is available. In this method, two solutions are prepared (1) which contains only the unknown and (2) which contains the unknown plus a known amount of pure material. If the unknown is diluted by the same amount in both solutions, the concentration of the unknown, C_x, in solution (1) is given by

$$C_x = \frac{I_x C_s}{I_{x+s} - I_x} \tag{59}$$

where I_x and I_{x+s} are the integrated intensity for an analytical resonance in solutions (1) and (2), respectively, and C_s is the known concentration of the standard in (2). The absolute concentration is then calculated from C_x and the dilution factor.

Absolute concentrations can also be obtained by using other pure compounds as intensity standards when measurements are made under conditions that k in Eq. (56) is the same for all resonances in the spectrum. Care should be taken to select an intensity standard which has a high molecular weight to minimize weighing errors, is readily available in pure form, and gives an NMR spectrum consisting of only a few resonances, preferably singlets, which do not overlap any resonances from the sample. The intensity reference either can be added to the sample solution (internal reference) or contained in a separate, smaller tube which is positioned coaxially in the sample tube (external reference) (92). In the latter case, spectra are obtained simultaneously for sample and reference, and concentrations can be obtained from calibration curves of intensity ratio (analyte:standard) versus concentration of analyte. If an internal reference is used, it must not react with any of the components of the sample. Absolute concentrations can be obtained from the relative concentrations of analyte and standard, calculated from integrated intensities using Eq. (57), and the absolute concentration of the internal standard. Maximum accuracy is obtained for intensity ratios near unity. To illustrate, spectrum B in Fig. 45 was obtained after adding

t-butanol to the amino acid mixture, the final concentration in the NMR tube being 0.00116 M. The integrals are 48.4, 34.0, 26.9, and 77.0 mm for alanine, threonine, t-butanol, and valine, respectively. For highest accuracy, the alanine integral must be corrected for overlap by ^{13}C satellites from threonine and t-butanol, the threonine integral must be corrected for overlap by a ^{13}C satellite from alanine, and the valine integral must be corrected for overlap by a t-butanol satellite. The corrected intensities are 48.1, 33.7, 26.9, and 76.9, from which the alanine, threonine and valine concentrations are calculated to be 0.00622, 0.00436, and 0.00497 M, respectively. Concentrations calculated directly from the integrals are 0.00627, 0.00441, and 0.00499 M.

Errors due to neglect of overlap with ^{13}C satellites increase as the relative intensity of resonances giving the satellites increases. For example, the t-butanol concentration is increased in spectrum C in Fig. 45, as is the intensity of the ^{13}C satellites. This can be a major source of error when doing trace analysis if there are strong resonances 50–70 Hz away from the weak resonances. In some cases, it is convenient to use ^{13}C satellites, e.g., from the solvent, as an internal intensity reference in trace analysis (93).

Spinning side bands can have the same affect as ^{13}C satellites on integrated intensities. For quantitative measurements, it is essential that ssb be reduced to negligible intensity. Alternatively, if resonances are sufficiently separated, the integral can be measured with the sample not spinning.

Resonance intensities in spectra measured with spin decoupling, e.g., ^{13}C or ^{15}N spectra measured with BB ^{1}H decoupling, are generally affected by the NOE; thus k in Eq. (56) can be different for each resonance in the spectrum. The NOE can be eliminated by using inverse gated decoupling with a sufficiently long time between pulses [scheme (c) in Fig. 40] or they can be suppressed with relaxation reagents (Section 7.2.2). However, for low concentration samples the sensitivity may be so low that, unless the resonance intensity is enhanced by the NOE, the time required to obtain a sufficiently high S/N for precise determination of resonance intensities would be prohibitively long. In such cases, quantitative information can still be obtained by calibrating the resonance intensities for known concentrations of the compounds of interest relative to those for a known amount of a reference compound (94). Calibration factors which relate the signal intensities to that of the reference are obtained for each compound. Concentration ratios in mixtures run under exactly the same conditions can be obtained by using three calibration factors. Results obtained by this procedure will be in error if T_1 values are not identical for sample and standard.

7.7. The Dynamic Range Problem

The observation of a very weak resonance in the presence of a strong resonance by P/FT NMR is difficult due to the limited dynamic range of the ADC. This problem occurs routinely with samples in H_2O. The H_2O resonance is in the middle of the spectrum and memory overflow [Eq. (52)] prevents long term averaging. If the divide down technique is used to prevent memory overflow, S/N for the weak signals decreases because significant bits are lost. The final S/N value S_f for a weak resonance of intensity S_w in the presence of a large resonance of intensity S_L is given by (87)

$$S_f = 2^{(w-d)/2} \times S_w \tag{60}$$

provided that $S_L/S_w \leq 2^{d+1}$, where d and w are the number of bits in the ADC and the computer word, respectively. However, if $S_L/S_w > 2^{d+1}$, then the weak resonance will never be observed no matter how many accumulations are taken. In the first case, S_w is buried in the noise but noise is able to trigger the least significant bit (lsb) in the digitizer. S_w is always present and coherent, whereas noise is random; thus, S_w will be pulled out from the noise. In the second case, because the strong signal dominates, noise cannot trigger the lsb, and S_w is lost.

An obvious solution to the dynamic range problem is to increase the bit size of the ADC. Figure 46 shows the increase in S/N for a weak signal in water solution when the resolution of the ADC is increased from 12 to 16 bits. Memory overflow due to the strong signal can be avoided with the technique of block averaging. Several FID's are accumulated, Fourier transformed, and stored in memory. This is repeated, with coaddition of the frequency domain spectra until a sufficiently strong signal is obtained for the weak peak. This technique takes advantage of the fact that the large resonance in the frequency domain spectrum can overflow memory without affecting the weak resonances.

Several experimental solutions to the dynamic range problem have been developed. The water eliminated FT (WEFT) technique (25), which is based on differences in T_1 values for the H_2O and solute resonances, was discussed in Section 4.3.3.1. Notch filters have been used to selectively filter out large resonances (95). Solvent resonances can be reduced in intensity by presaturation

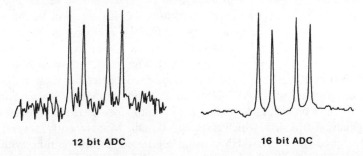

12 bit ADC   **16 bit ADC**

Fig. 46. The effect of digitizer resolution on S/N for weak resonances in the presence of a strong solvent resonance. Four FID's were coherently added to obtain each spectrum, which is for the methyl protons of 0.01 M valine in 90% H_2O/10% D_2O solvent. Measured at 360 MHz.

with an intense selective pulse from the decoupler before the observation pulse is applied (96). An ingenious technique of tailored excitation involves generation of the excitation frequencies by the computer and omitting those frequencies near the large resonance (97). Among the most widely used techniques are those involving soft pulses, i.e., pulses of relatively low power and correspondingly longer length (98). A long 90° pulse of length τ has a null in its effectiveness for flipping nuclei at frequencies approximately $\pm 1/\tau$ away from the carrier frequency. Thus, by careful positioning of the carrier, resonances over a range of frequencies can be excited selectively while leaving the large water resonance unperturbed. A modified soft-pulse technique developed by Redfield et al. has the advantage that the null in its effectiveness for flipping nuclei occurs over a broader frequency range (99). The Redfield method uses a soft pulse divided into five segments of relative lengths 2:1:4:1:2. The phase of the second and fourth segments is shifted by 180° relative to that of the other segments. With the 2-1-4-1-2 pulse sequence, water resonance suppression of 100–1000 can be achieved.

One consequence of the use of a soft pulse is that the flip angle, and thus the resonance intensity, depends on distance from the carrier frequency. Also, spectra require a large first-order phase correction, which leads to a rolling baseline. A disadvantage of the Redfield method and several of the other experimental solutions to the dynamic range problem is that resonances near the large solvent resonance are lost.

The dynamic range problem can be avoided by using rapid scan FT NMR (Section 3.3). The sweep window is adjusted to cover only the region of interest, thus avoiding the strong solvent resonances.

7.8.   Liquid Chromatography–P/FT NMR

The analysis of complex mixtures by NMR often involves first the separation of the components of the mixture, e.g., by high performance liquid chromatography (HPLC), followed by measurement of their NMR spectra. The recent coupling of HPLC and P/FT NMR increases the efficiency of the analysis of mixtures by combining these two steps (100,101). The first coupling of HPLC and P/FT NMR used stopped flow conditions, i.e., the eluent flow was stopped while pulsing the NMR spectrometer (101f). The stopped flow method has also been studied by Bayer et al. (101d), who reported a detection limit in the higher nanomole range using ^{1}H P/FT NMR at 250 and 400 MHz. Stopping the flow had little effect on chromatographic peak width since broadening is dependent on longitudinal diffusion, which is small in liquids. More recently, several papers have described HPLC–P/FT NMR using continuous flow conditions with detection limits similar to or lower than those of the refractive index detector (100,101).

In addition to the usual considerations for measuring spectra by the P/FT method, (1) the chromatographic system must be designed so that the residence time of the solute in the magnetic field before it enters the rf coil will be at least three times the longest T_1 of the compounds of interest so the longitudinal magnetization will reach at least 95% of its equilibrium value and (2) the rf coil/ sample cell must be designed so that the residence time, τ, of the solute in the rf coil is sufficiently short to avoid additional resonance broadening. Under flow conditions $T^*_{2\text{eff(flow)}}$ is given by

$$\frac{1}{T^*_{2\text{eff(flow)}}} = \frac{1}{T^*_{2eff(static)}} + \frac{1}{\tau} \tag{61}$$

where $T^*_{2\text{eff(static)}}$ is the effective T_2 under static conditions. With electromagnet spectrometers, the first condition has been met by passing the eluent through a coil located in the magnetic field just before the sample cell (100b). With superconducting solenoids, this condition is met by coupling the HPLC to the NMR probe through the bottom of the solenoid (100a). Since solenoid magnets have large fields at the base of the Dewar, full magnetization is achieved as the solute travels to the sample cell if the flow rate is not too fast. The second condition is met by using a sufficiently large cell volume [e.g., 120 μL (100a)] and a sufficiently slow flow rate (e.g., 1 mL or less/min). With high field spectrometers, injection volumes down to 10 μL have been used (100a). Line widths as small as 0.5 Hz have been achieved (101d).

Another consideration is the solvent, which must have satisfactory properties both for HPLC and P/FT NMR. Resonances from ^{1}H nuclei in the solvent may cause dynamic range problems as well as overlap with resonances of interest. Freon 113 (1,1,2-trifluorotrichloroethane), CS_2, CCl_4, and C_2Cl_4 have been proposed as solvents (100b), as have the deuterated solvents generally used in P/ FT NMR. Separations can be achieved with as little as 15 mL of solvent (100a), making it practical to use deuterated solvents. Also, normal protonated solvents have been used with spectrometers having an ADC and computer with more bits (101d) and with solvent suppression by presaturation (101b). The use of fluorinated derivatives and ^{19}F NMR has been proposed (100b). An advantage of the superconducting solenoid spectrometers is that the magnetic field is sufficiently stable that no lock signal is required.

Several instrument configurations have been used. Generally the system includes a conventional LC detector (refractive index or UV) prior to the NMR cell. P/ FT NMR data can be acquired continuously, e.g., (1) a fixed number of FID's are coadded in computer memory, (2) stored on disk, and (3) then back to (1) while the previously accumulated FID is Fourier transformed in another part of memory and plotted. Typically, successive NMR spectra can be accumulated in 30–60 s. Alternatively, the beginning and end of the data acquisition sequence can be timed according to peaks in the chromatograph from the other detector.

Compared to other HPLC detectors, NMR provides considerably more information with which the identity of the molecule can be established. It is less sensitive than some detectors; however, with high-field spectrometers, the sensitivity is similar to that of the refractive index detector (100a). At present, directly coupled HPLC–P/FT NMR instrumentation is not commercially available.

8. APPLICATION TO TRACE ANALYSIS

NMR is suited to both qualitative and quantitative organic analysis; the resonance frequencies and spin–spin coupling patterns provide qualitative information and their intensities provide quantitative information. Until recently, however, NMR was generally not considered for use in trace analysis because of its relatively low sensitivity. As described earlier in this chapter, that has changed as a result of recent developments in instrumentation and techniques.

As with other spectroscopic methods of organic trace analysis, it may be necessary to work up the sample prior to the NMR measurement. The work up procedure for a particular analysis will depend on the sample matrix, the nucleus used in the NMR measurement, the concentration of the sought for constituent, and whether qualitative or quantitative information is to be obtained. In this section, examples are given to illustrate the application of NMR to organic trace analysis. It seems likely that NMR will become an increasingly important technique for organic trace analysis as more laboratories become equipped with high sensitivity, high field P/FT NMR spectrometers.

8.1.   Qualitative Organic Analysis

NMR has been widely used in conjunction with other spectroscopic methods to identify trace constituents in a variety of sample types. The main problem generally is to isolate the various constituents in reasonably pure and sufficiently concentrated form for the NMR measurement. A typical procedure might involve:

1. Extraction of the analytes from the sample matrix with an organic solvent.
2. Cleanup of the extract by column chromatography or thin-layer chromatography to remove nonanalyte organic compounds.
3. Concentration of analytes by reducing the volume.
4. Isolation of the individual analyte compounds by chromatography.

A large amount of qualitative information can be obtained from NMR spectra of the isolated compounds. In the most straightforward case, the unknown can be identified by matching its spectra with those of known compounds. If this is not possible, the structural formula of a compound can often be determined from a detailed analysis of its ^{1}H and ^{13}C spectra, using the techniques described earlier in this chapter together with the extensive chemical shift correlations in the literature.

8.2.   Quantitative Organic Analysis

Quantitative organic analysis by NMR is based on the measurement of resonance intensities. If measurement conditions are carefully chosen, relative concentrations can be obtained directly from the relative intensities of resonances in the spectrum (Section 7.6). Absolute concentrations can be obtained by comparing analyte resonance intensities to those of a standard, either added to the sample or in an external capillary. If the above measurement conditions are not met, as is generally the case when measuring NMR spectra of very dilute solutions, quantitative results can be obtained by using one of the calibration procedures described in Section 7.6.

8.2.1.   Environmental Samples

Becconsall (102) has described an analysis scheme for the identification and determination of trace organic pollutants in water by ^{1}H NMR. The sample is preconcentrated by extraction into carbon tetrachloride. Water is removed from the extract by shaking it with 0.5–1.0 times its volume of D_2O, removing the D_2O, and then adding Linde 4A molecular sieves. To achieve large increases in analyte concentrations, sample to CCl_4 volume ratios of 40:1 to 200:1 were used. Silicone stopcock grease was added to the solvent to serve both as a chemical shift reference and an intensity reference. The silicone grease resonance was calibrated for intensity measurements by carrying standard alkane solutions in water through the same solvent extraction procedure. The method can be used to determine all hydrogen-containing compounds that are efficiently extracted by the solvent, including hydrocarbons, halogenated hydrocarbons, and organochlorine pesticides. The ^{1}H NMR measurements were made by the P/FT method at 100 MHz. Alkane concentrations in the range 0.1 to 2.4 mg L^{-1} were determined, and the detection limit was estimated to be 15 μg L^{-1} of alkane when an extraction ratio of 200:1 and 90 min of signal averaging were used. With higher sensitivity high field spectrometers, the detection limit is probably 1 to 2 orders of magnitude lower. Results were in agreement with those obtained by infrared measurements on CCl_4 extracts. Becconsall noted that separation and preconcentration schemes used for preparing samples for measurement by IR will generally be applicable to sample preparation for ^{1}H NMR measurement since it is advantageous to eliminate water in both cases.

The P/FT spectrometer used by Becconsall was equipped with an external lock facility for field/frequency stabilization, so that it was not necessary to add a deuterated compound to obtain a lock signal. The addition of a deuterated lock compound is to be avoided if possible when doing trace analysis since it inevitably will give large ^{1}H resonances, due to residual hydrogen, which will obscure regions of the spectrum. Most spectrometers are not equipped with an external lock facility; however, the superconducting magnets used in high field spectrometers are generally sufficiently stable that these spectrometers can be operated without a lock for field/frequency stabilization.

Phosphorus-31 P/FT NMR has been used to determine phosphorus-containing compounds, including inorganic and organic phosphates, phosphonates, and thiophosphates, in wastewater and related aqueous systems (47,71,103). The procedures generally employ a relaxation reagent to allow the use of faster repetition rates for more efficient signal averaging, since the ^{31}P T_1 values in these compounds can be long and cover a wide range (typically 1–25 s). Gurley and Ritchie used several Fe(III) complexes, including Fe(III)–HCl, Fe(III)–acac, and Fe(III)-EDTA, as relaxation reagents for ortho, pyro, and tripolyphosphate (47,103). These were not entirely satisfactory, however, because of broadening of some ^{31}P signals due to specific complexation. Stanislawski and Van Wazer found the water-soluble 4-hydroxy- and 4-amino-2,2,6,6-tetramethylpiperidino-oxy free radicals to be more satisfactory with the same inorganic phosphates, presumably because there is no significant chemical interaction between the free radicals and the phosphates (71). Gurley and Ritchie used the relaxation agent Gd(FOD)$_3$ with a mixture containing malathion, parathion, triethylthiophosphate, and triphenylphosphate (47). The use of Gd(FOD)$_3$ requires that the ^{31}P compounds be extracted into an organic solvent since it decomposes in aqueous solution. Quantitative results have been obtained by comparison of resonance intensities with those of KH$_2$PO$_4$ or tetraethylammonium phosphate as an external reference in solution in a concentric tube.

The above examples of determination of phosphorus compounds in environmental samples by ^{31}P P/FT NMR have all used low field, electromagnet spectrometers. With these instruments, detection limits in the 10–20 ppm (phosphorus) range have been achieved by signal averaging for several hours. Detection limits should be considerably lower with state-of-the-art high-field spectrometers. The ^{31}P method is particularly suited to the analysis of environmental samples since the spectra are assignable to specific phosphorus compounds, little or no sample preparation is usually necessary because the sample will contain few other phosphorus compounds, and the ^{31}P chemical shift range is large. It may be necessary to treat the sample with EDTA or some other strong complexing agent if it contains transition metals, e.g., Fe(III), at a sufficiently high concentration to cause broadening of resonances, or it may be necessary to preconcentrate the sample if the ^{31}P compounds are present at levels below the detection limit.

The paper by Stanislawski and Van Wazer (71) is recommended for a good discussion of the instrumental factors which must be considered when attempting to do quantitative analysis by ^{31}P P/FT NMR. In their procedure, they used a 30° pulse to permit the use of a rapid repetition rate and to minimize the effect of pulse power dropoff.

8.2.2. Biological and Clinical Samples

NMR has become an important method for making measurements on compounds at low levels in biological systems (46). Because it is totally noninvasive and nondestructive, NMR measurements can be made directly on intact, living systems. Applications have included the study of metabolic pathways and the measurement of intracellular pH on organisms as diverse as bacteria and humans (46a). NMR spectra can also be obtained directly from biological fluids, e.g., blood plasma and urine (104). With state-of-the-art high-sensitivity high-dispersion spectrometers, well resolved resonances are obtained for a multitude of components present at low levels in these fluids, providing an excellent demonstration of the use of NMR for the analysis of complex, multicomponent mixtures.

The basic requirement for obtaining NMR spectra for molecules in biological samples, as for other samples, is that they produce signals which are intense enough to be distiguished from noise and narrow enough to be resolved from other signals. The first requirement is related to concentration, the second to mobility. Many of the small molecules in cellular systems are freely mobile, and thus give narrow resonances. In fact, this dependence on mobility makes it possible to observe selectively resonances from the small molecules in the presence of the macromolecules of the cell and the cell membrane.

Most studies on intact cells and tissues have been by ^{31}P NMR. ^{31}P spectra are simple and easy to interpret, and some of the most important biological compounds, e.g., adenosine di- and triphosphate (ADP and ATP), inorganic phosphate (P_i), phosphocreatine (PCr), and others involved in energy metabolism, contain phosphorus and are present at concentrations (0.2 mM and above) which can be detected by ^{31}P NMR. The first such studies were by Moon and Richards (105), who detected signals from 2,3-diphosphoglycerate, P_i and ATP in high resolution ^{31}P spectra of red blood cells. They made use of the pH dependence of the chemical shift of P_i to determine the intracellular pH (pH$_i$) of intact red blood cells. P_i exists mainly as HPO_4^{2-} and $H_2PO_4^{-}$ in the physiological pH region. The chemical shifts of the two species are different. However, because of rapid proton exchange a single ^{31}P resonance is observed, the chemical shift of which is determined by the fraction of P_i in each form. Thus, the chemical shift of the averaged resonance is pH dependent, and serves as a pH indicator. This dependence has since been used to measure pH$_i$ of a variety of biological systems (46). It should be mentioned that the entire ^{31}P spectrum is obtained so that, in the majority of these studies, pH$_i$ is determined from the chemical shift of P_i while information about the metabolism of other phosphorus-containing compounds is obtained simultaneously; in other words, the NMR spectrometer is being used as much more than a pH meter.

NMR probes have been designed so that biological samples can be kept in good physiological condition, e.g., muscle, perfused hearts, livers, and kidneys (46). For measurements on whole animals, including humans, rf coils have been designed which are placed on the surface (106), e.g., of an arm (107,108). Phosphorus-31 spectra are obtained from a region of approximately 1 mL volume immediately in front of the coil. Phosphorus-31 spectra of muscle contain resonances for ATP, PCr, and P_i from which the pH_i has been estimated. The chemical shifts of the three resonances for ATP are sensitive to complexation, and indicate that in muscle the ATP is mainly complexed by Mg^{2+}. The instrumentation is sufficiently sensitive that time-course studies are possible, e.g., to monitor changes in metabolite levels following exercise or cessation of blood flow. This has found clinical application. As an example, Radda et al. made ^{31}P measurements before, during, and after exercise on the forearm of an individual known to have mitochondrial NADH-coenzyme Q reductase deficiency (108). Before exercise, the $PCr:P_i$ ratio was found to be lower than normal while the pH_i was normal (7.04). During exercise (opening and closing the fist every 2 s), the PCr decreased and P_i increased more rapidly than in normal controls. By the second minute of exercise, 65–75% of the PCr had been consumed and pH_i had dropped to 6.56. After exercise, the recovery of PCr was much slower than in normal controls, which was concluded to be consistent with a decreased rate of oxidative metabolism. The use of NMR to study energy metabolism in this way is a recent development, and can be expected to become an important method of clinical diagnosis.

Carbon-13 NMR also has been used to study metabolic pathways. Because of its low sensitivity and low natural abundance, it is necessary to use ^{13}C enriched samples, e.g., ^{13}C-enriched glucose. However, this has the advantage that the label can be used as a tracer to provide information about the metabolic pathway.

The use of ^1H NMR has also been applied to biological systems. Although ^1H NMR is more sensitive than ^{31}P or ^{13}C NMR, it has not been applied as widely because the spectra normally consist of a very large number of overlapping resonances. The measurement of ^1H spectra for molecules present at low concentrations is further complicated by the much more intense water resonance. The water resonance problem can be solved by replacing the H_2O with D_2O, or by using solvent suppression techniques (Section 7.7). The problem of overlapping resonances can sometimes be solved by using multiple pulse techniques. To illustrate, the ^1H NMR spectrum obtained by the spin-echo FT method for intact human red blood cells which had been washed with D_2O solution is shown in Fig. 47 (31). The broad envelope of interfering resonances due to hemoglobin protons is eliminated with the spin-echo method (28). As described in

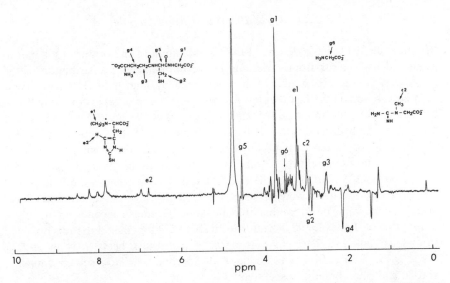

Fig. 47. The 400 MHz ^{1}H spin–echo Fourier transform NMR spectrum of intact packed human erythrocytes which had been washed with a D_2O solution of isotonic saline–glucose solution. A delay of 0.060 s was used between pulses in the spin–echo pulse sequence. (See text for details.)

Section 4.3.3, the spin–spin relaxation time becomes a resolution parameter. Since spin–spin relaxation times decrease as molecular size increases, the hemoglobin resonances can be selectively eliminated by proper choice of delay time in the spin–echo pulse sequence (28,109). Resonances in ^{1}H spin–echo FT NMR spectra of red cells have been assigned to a number of intracellular compounds, including glutathione, ergothioneine, creatine, glucose, glycine, alanine, valine, lactic acid, and choline (28,110). Although these compounds are all present at the millimolar or submillimolar level, it took only approximately 1 min to obtain the spectrum in Fig. 47.

High-resolution ^{1}H NMR spectra can also be obtained for biological fluids, including blood plasma and urine (104). These measurements require a minimum of sample preparation, e.g., urine can be run directly while the plasma is simply separated from the cells. These types of measurements are so recent that their clinical utility is yet to be established. It seems likely that the relative intensity patterns for the array of diverse compounds which are detected will also be of some use. Also, it should be possible to increase the selectivity and sensitivity of the measurement with the extraction techniques which are commonly used with other instrumental methods. With relatively simple sample workup, it should be possible to achieve detection limits approaching the micromolar level.

8.2.3. *Agricultural and Food Samples*

Martin et al. (111) have described the identification of the origin of ethanol in grain and fruit spirits from variations in the intramolecular distribution of the trace level isotope ^2H in ethanol as determined by high field ^2H NMR. Their method is based on the fact that the percentage of ^2H in organic compounds can be slightly different from that of Standard Mean Ocean Water (SMOW) in which 0.015% of the hydrogen is the ^2H isotope. For example, they found that, in the ethyl groups of 22 representative organic compounds, the proportion of ^2H in the methylene group relative to that in the methyl group (defined as R) varied from 1.695 $((C_2H_5)_2N\text{-}CH_2CO_2Et)$ to 2.558 (C_2H_5Br), as compared to a statistical proportion of 2 if there were no selective enrichment or depletion. They have found that the value of R for ethanol in alcoholic beverages depends on the origin of the ethanol. For example, R varies from about 2.2 for ethanol in rums obtained from sugar cane to 2.7 for ethanols extracted from sugar beets. R is about 2.3–2.35, 2.45, 2.55, and 2.6 for ethanol in bourbon whiskey, scotch whiskey, vodka, and plum brandy, respectively.

The total ^2H content can be measured by comparing the intensity of the ^2H resonances of ethanol to the intensity of a reference contained in a coaxial cell, whose ^2H content can be measured with respect to SMOW. The total ^2H content, but not the relative ^2H content of different chemical sites in the molecule, can also be determined by mass spectroscopy. The relative ^2H contents of the different chemical sites also cannot be obtained by ^1H NMR because the variation in percentage of ^2H is so small as to make an undetectable difference in the intensities of the ^1H resonances. It seems likely that, with the increasing availability of multinuclear high field NMR instruments, quantitative determination of the intramolecular distribution of the trace level isotope ^2H will become a powerful analytical method for the study of chemical and biosynthetic mechanisms and for determining the origin of natural products (111,112). However, because of the variation in ^2H abundance, care must be exercised in determining molecular concentrations by comparison of ^2H resonance intensities.

An interesting application of ^{13}C NMR is the determination of 4-hydroxy-L-proline (**1**), a nonessential amino acid, in meat protein (113). Collagen, the principal protein in connective tissue, is the only protein which contains significant levels of **1**; thus, it is the accepted indicator of connective tissue in meat and of meat quality. O'Neill et al. (113) described a procedure in which meat samples are freeze dried, defatted, hydrolyzed in 6 M HCl, and then concentrated. β-Hydroxy-α-methylphenethylammonium chloride (**2**) is then added as an internal intensity standard, and the ^{13}C spectrum of the sample–standard concentrate is measured at 80°C. The concentration of **1** is determined by comparison of the intensity of its C-4 resonance with the intensity of the β-C resonance of **2**. These two resonances are resolved from the other ^{13}C resonances, but are separated from each other by only 2.9 ppm, so that errors from different amounts of excitation due to uneven B_1 power distribution are small. The sample is concentrated to a relatively high viscosity to shorten the ^{13}C T_1 values (to ~0.1 s for C-4 of **1** and ~0.08 s for the β-C of **2**) so that fast repetition rates can be used [e.g., O'Neill et al. used a repetition time of 0.85 s (113)]. To eliminate errors due to different responses to **1** and **2**, the method can be calibrated by adding **1** to the concentrate or poly(4-hydroxyl-L-proline) to the freeze-dried sample. The method was shown to be comparable in accuracy and precision to the standard colorimetric method which requires essentially the same sample pretreatment (113b). The standard deviation was 0.16% at the 3% level of **1** (w/w of dried, defatted sample) by ^{13}C NMR versus 0.10% at the 1.4% level by colorimetry. The ^{13}C method requires more instrument time (~30 min of signal averaging) than the colorimetric method, but it has several advantages, including the additional information which can be obtained simultaneously about the other amino acids in the protein hydrolysate.

Other applications of NMR to the analysis of agricultural and food samples have included the determination of inositol hexaphosphate (phytate) in food stuffs by ^{31}P NMR (114) and of sucrose in sugar beet juices by ^{1}H NMR (115), the identification of dihydrotutin and dihydrohyenancin as trace components of a range of toxic honeydew honeys (116), and the determination of the carbohydrates glucose, fructose, myo-inositol, pinitol, and sequoyitol in *Pinus radiata* wood tissue and needles (91).

9. SUMMARY

Although the advantages of NMR for chemical analysis have long been recognized—qualitative and quantitative information can be obtained for multicomponent samples and the method is nondestructive—it has not been widely used for trace analysis due to its low inherent sensitivity. However, parallel developments in instrumentation and techniques have increased the sensitivity of NMR by over 300-fold over the past 20 years (117), making it now possible to obtain ^{1}H and ^{13}C NMR spectra from nanomole and micromole quantities of material, respectively. As a result, state-of-the-art NMR is well suited to organic trace analysis, as the examples described in this chapter illustrate, and it is to be expected that it will become an important tool for trace chemical analysis in the future. Because it is noninvasive, NMR is expected to play a particularly important role in biomedical analysis, both for routine screening and for diagnosis through *in situ* measurements.

10.   ADDENDUM

Since the writing of this chapter, several important new experimental techniques have been developed. Of particular importance to trace analysis by ^{1}H NMR are the new multiple-pulse methods for suppressing or eliminating the solvent resonance (119,120). In each of these methods, a series of closely spaced pulses is used to flip the nuclei giving the resonances of interest while returning the magnetization from the solvent to the z axis at the start of acquisition. In one group of methods (119), the carrier is set at the solvent resonance frequency. In its simplest form (the so-called $1\bar{1}$ method), a series of 2 pulses of the same length but shifted in phase by 180° is used. The delay between the 2 pulses is chosen to give maximum magnetization in the xy plane for the spectral region of interest. More pulses can be used with lengths in the ratio of binomial coefficients and phases alternating between 0° and 180°. Solvent resonance suppression of more than 1000-fold can be achieved with the $1\bar{3}3\bar{1}$ pulse sequence (119b). In another method (120), the carrier is set in the spectral region of interest, and two 45° pulses of the same phase are used. The time delay between the two pulses is chosen so that the magnetization giving the solvent resonance has precessed through 180° in the rotating frame. It should be noted that, with this method, two solvent resonances can be eliminated by proper choice of carrier frequency and delay time between the pulses.

The poor sensitivity of NMR experiments has always been a concern to NMR spectroscopists, and the solution to this problem has been approached through improvements in probe design and data handling. In the first case the improvements in probe design have lead to a substantial increase in sensitivity of commercially available spectrometers. In a recent article (121), the receiver coil and preamplifier of a high-resolution system were cooled to liquid He temperatures, reducing thermal noise and thus dramatically increasing sensitivity. In the second case, the maximum entropy method (MEM) uses a mathematical approach to enhance the information content within the FID (122), and improves not only the S/N ratio, but also the resolution of NMR spectra. The drawback to this method is the requirement of a mainframe computer.

Two-dimensional NMR experiments have continued to evolve. These experiments and other modern pulse methods are described in an excellent review article (123). Another useful paper considers the question of optimal conditions for obtaining quantitative ^{13}C NMR data efficiently (124). There has been continued development of HPLC/NMR techniques (125) and the field has been reviewed recently (126). The applications of NMR in biology and medicine have continued to grow at a rapid pace, in particular imaging by NMR techniques and *in vivo* NMR spectroscopy (127). Chemical applications of NMR imaging have been reviewed (128).

ACKNOWLEDGMENTS

We thank Annabelle Wiseman for her skill and patience in typing the many versions of this manuscript.

REFERENCES

1. J. A. Pople, W. G. Schneider, and H. J. Bernstein, *High-Resolution Nuclear Magnetic Resonance*, McGraw-Hill, New York, 1959.
2. J. W. Emsley, J. Feeney, and L. H. Sutcliffe, *High Resolution Nuclear Magnetic Resonance Spectroscopy*, Pergamon, New York, 1965.
3. E. D. Becker, *High Resolution NMR—Theory and Chemical Applications*, 2nd ed., Academic, New York, 1980.
4. E. W. Garbisch, Jr., *J. Chem. Educ.*, **45**, 311, 402, 480 (1968).
5. W. A. Anderson and R. Freeman, *J. Chem. Phys.*, **37**, 85 (1962).
6. J. H. Noggle and R. E. Schirmer, *The Nuclear Overhauser Effect: Chemical Applications*, Academic, New York, 1971.
7. A. Abragam, *The Principles of Nuclear Magnetism*, Oxford, New York, 1961.
8. T. C. Farrar and E. D. Becker, *Pulse and Fourier Transform NMR—Introduction to Theory and Methods*, Academic, New York, 1971.
9. L. M. Jackman and F. A. Cotton, Eds., *Dynamic Nuclear Magnetic Resonance Spectroscopy*, Academic, New York, 1975.
10. R. K. Gupta, J. A. Ferretti, and E. D. Becker, *J. Mag. Res.*, **13**, 275 (1974).
11. Reference 3, p. 260.
12. D. Shaw, *Fourier Transform NMR Spectroscopy*, Elsevier, Amsterdam, 1976.
13. M. L. Martin, J.-J. Delpuech, and G. J. Martin, *Practical NMR Spectroscopy*, Heyden, Philadelphia, 1980.
14. T. C. Farrar, *Am. Lab.*, March, 22 (1972).
15. J. D. Ellett, Jr., M. G. Gibby, U. Haeberlen, L. M. Huber, M. Mehring, A. Pines, and J. S. Waugh, *Advan. Magn. Reson.*, **5**, 117 (1971).
16. R. R. Ernst and W. A. Anderson, *Rev. Sci. Instrum.*, **37**, 93 (1966).
17. F. W. Wehrli, *Advan. Mol. Relax. Processes*, **6**, 139 (1974).
18. R. L. Vold, J. S. Waugh, M. P. Klein, and D. E. Phelps, *J. Chem. Phys.*, **48**, 3831 (1968).

19. E. L. Hahn, *Phys. Rev.*, **80,** 580 (1950).
20. E. L. Hahn and D. E. Maxwell, *Phys. Rev.*, **88,** 1070 (1952).
21. (a) R. Freeman and H. D. W. Hill, *J. Chem. Phys.*, **54,** 301 (1971); (b) R. Freeman and H. D. W. Hill, Chapter 5 in Ref. 9.
22. A. C. McLaughlin, G. G. McDonald, and J. S.Leigh, Jr., *J. Mag. Res.*, **11,** 107 (1973).
23. H. Y. Carr and E. M. Purcell, *Phys. Rev.*, **94,** 630 (1954).
24. S. Meiboom and D. Gill, *Rev. Sci. Instrum.*, **29,** 688 (1958).
25. S. L. Patt and B. D. Sykes, *J. Chem. Phys.*, **56,** 3182 (1972).
26. I. D. Campbell, C. M. Dobson, R. J. P. Williams, and P. E. Wright, *FEBS Lett.*, **57,** 96 (1975).
27. D. L. Rabenstein, *Anal. Chem.*, **50,** 1265A (1978).
28. F. F. Brown, I. D. Campbell, P. W. Kuchel, and D. L. Rabenstein, *FEBS Lett.*, **82,** 12 (1977).
29. R. Freeman and G. Λ. Morris, *Bull. Mag. Res.*, **1,** 5 (1979).
30. W. P. Aue, J. Karhan, and R. R. Ernst, *J. Chem. Phys.*, **64,** 4226 (1976).
31. D. L. Rabenstein and T. T. Nakashima, *Anal. Chem.*, **51,** 1465A (1979).
32. D. W. Brown, T. T. Nakashima, and D. L. Rabenstein, *J. Mag. Res.*, **45,** 302 (1981).
33. G. Bodenhuasen, R. Freeman, R. Niedermeyer, and D. L. Turner, *J. Mag. Res.*, **26,** 133 (1977).
34. S. Macura and R. R. Ernst, *Mol. Phys.*, **41,** 95 (1980).
35. K. Nagayama, K. Wüthrich, and R. R. Ernst, *Biochem. Biophys. Res. Comm.*, **90,** 305 (1979).
36. (a) A. Bax and G. A. Morris, *J. Mag. Res.*, **42,** 501 (1981); (b) A. A. Maudsley and R. R. Ernst, *Chem. Phys. Lett.*, **50,** 368 (1977); (c) R. Freeman and G. A. Morris, *J. C. S. Chem. Comm.*, 684 (1978).
37. A. Bax, R. Freeman and T. A. Frenkiel, *J. Am. Chem. Soc.*, **103,** 2102 (1981).
38. K. G. R. Pachler and P. L. Wessels, *J. Mag. Res.*, **12,** 337 (1973).
39. H. J. Jakobsen, S. Aa. Linde, and S. Sørensen, *J. Mag. Res.*, **15,** 385 (1974).
40. S. Aa. Linde, H. J. Jakobsen, and B. J. Kimber, *J. Am. Chem. Soc.*, **97,** 3219 (1975).
41. H. J. Jakobsen and W. S. Brey, *J. Am. Chem. Soc.*, **101,** 774 (1979).
42. G. A. Morris and R. Freeman, *J. Am. Chem. Soc.*, **101,** 760 (1979).
43. D. P. Burum and R. R. Ernst, *J. Mag. Res.*, **39,** 163 (1980).
44. (a) J. B. Stothers, *Carbon-13 NMR Spectroscopy*, Academic, New York, 1972; (b) G. C. Levy, R. L. Lichter, and G. L. Nelson, *Carbon-13 Nuclear Magnetic Resonance Spectroscopy*, Wiley, New York, 1980; (c) F. W. Wehrli and T. Wirthlin, *Interpretation of Carbon-13 NMR Spectra*, Heyden, London, 1976.
45. R. K. Harris and B. E. Mann, Eds., *NMR and the Periodic Table*, Academic, New York, 1978.
46. (a) D. G. Gadian, *Nuclear Magnetic Resonance and Its Applications to Living Systems*, Oxford Univ. Press, Oxford, 1982; (b) P. W. Kuchel, *Crit. Rev. Anal. Chem.*, **12,** 154 (1981); (c) G. K. Radda and P. J. Seeley, *Ann. Rev. Physiol.*, **41,** 749 (1979).
47. T. W. Gurley and W. M. Ritchey, *Anal. Chem.*, **48,** 1137 (1976).

48. (a) F. W. Wehrli and T. Nishida, *Prog. Chem. Org. Nat. Prod.*, **36**, 1 (1979); (b) J. C. Vederas, *Can. J. Chem.*, **60**, 1637 (1982).

49. R. E. Wasylishen, R. E. Lenkinski, and C. Rodger, *Can. J. Chem.*, **60**, 2113 (1982).

50. E. Fukushima and S. B. W. Roeder, *Experimental Pulse NMR. A Nuts and Bolts Approach*, Addison-Wesley, Reading, Mass., 1981.

51. D. I. Hoult and R. E. Richards, *J. Mag. Res.*, **24**, 71 (1976).

52. E. Oldfield and M. Meadows, *J. Mag. Res.*, **31**, 327 (1978).

53. H. J. Schneider and P. Dullenkopf, *Rev. Sci. Instrum.*, **48**, 68 (1977).

54. J. B. Grutzner and R. E. Santini, *J. Mag. Res.*, **19**, 173 (1975).

55. V. J. Basus, P. D. Ellis, H. D. W. Hill, and J. S. Waugh, *J. Mag.Res.*, **35**, 19 (1979).

56. M. H. Levitt and R. Freeman, *J. Mag. Res.*, **43**, 502 (1981).

57. H. J.Reich, M. Jautelat, M. T. Messe, F. J. Weigert, and J. D. Roberts, *J. Am. Chem. Soc.*, **91**, 7445 (1969).

58. J. W. Cooley and J. W. Tukey, *Math. Comput.*, **19**, 297 (1965).

59. (a) I. D. Campbell, C. M. Dobson, R. J. P. Williams, and A. V. Xavier, *J. Mag. Res.*, **11**, 172 (1973); (b) A. G. Ferrige and J. C. Lindon, *J. Mag. Res.*, **31**, 337 (1978).

60. J. N. Shoolery, *Topics in Carbon-13 NMR Spectroscopy*, **3**, 28 (1979).

61. R. E. Sievers, Ed., *Nuclear Magnetic Resonance Shift Reagents*, Academic, New York, 1973.

62. W. DeW. Horrocks, in *NMR of Paramagnetic Molecules*, G. N. LaMar, W. DeW. Horrocks, and R. H. Holm, Eds., Academic, New York, 1973.

63. A. F. Cockerill, G. L. O. Davies, R. C. Harden, and D. M. Rackman, *Chem. Rev.*, **73**, 553 (1973).

64. G. E. Hawkes, C. Marzin, S. R. Johns, and J. D. Roberts, *J. Am. Chem. Soc.*, **95**, 1661 (1973).

65. D. L. Rabenstein, *Anal. Chem.*, **43**, 1599 (1971).

66. T. J. Wenzel and R. E. Sievers, *Anal. Chem.*, **54**, 1602 (1982).

67. G. N. LaMar, *J. Am. Chem. Soc.*, **93**, 1040 (1971).

68. G. C. Levy and U. Edlund, *J. Am. Chem. Soc.*, **97**, 4482 (1975).

69. G. C. Levy and R. Komorski, *J. Am. Chem. Soc.*, **96**, 678 (1974).

70. T. J. Wenzel, M. E. Ashley, and R. E. Sievers, *Anal. Chem.*, **54**, 615 (1982).

71. D. A. Stanislawski and J. R. Van Wazer, *Anal. Chem.*, **52**, 96 (1980).

72. S. L. Manatt, *J. Am. Chem. Soc.*, **88**, 1323 (1966).

73. P. Sleevi, T. E. Glass, and H. C. Dorn, *Anal. Chem.*, **51**, 1931 (1979).

74. G. Jung, W. Voelter, and E. Breitmaier, *Mikrochimica Acta*, 850 (1970).

75. (a) F. F.-L. Ho, *Anal. Chem.*, **46**, 496 (1974). (b) F.F.-L. Ho and R. R. Kohler, *Anal. Chem.*, **46**, 1302 (1974).

76. K. L. Koller and H. C. Dorn, *Anal. Chem.*, **54**, 529 (1982).

77. F. F. Shue and T. F. Yen, *Anal. Chem.*, **54**, 1641 (1982).

78. V. W. Goodlett, *Anal. Chem.*, **37**, 431 (1965).

79. A. K. Bose and P. R. Srinivasan, *Tetrahedron*, **31**, 3025 (1975).

80. L. D. Hall and J. K. M. Sanders, *J. Am. Chem. Soc.*, **102**, 5703 (1980).

81. O. A. Gansow and W. Schittenhelm, *J. Am. Chem. Soc.*, **93**, 4294 (1971).

82. R. Freeman, H. D. W. Hill, and R. Kaptein, *J. Mag. Res.*, **7**, 327 (1972).
83. J. B. Grutzner, *J. C. S. Chem. Comm.*, 64, (1974).
84. H.-J. Schneider and W. Freitag, *J. Am. Chem. Soc.*, **98**, 478 (1976).
85. S. L. Patt and J. N. Shoolery, *J. Mag. Res.*, **46**, 535 (1982).
86. D. M. Doddrell and D. T. Pegg, *J. Am. Chem. Soc.*, **102**, 6388 (1980).
87. J. W. Cooper, in *Topics in Carbon-13 NMR Spectroscopy*, Vol. 2, G. C. Levy, Ed., Wiley, New York, 1976.
88. D. J. Cookson and B. E. Smith, *Anal. Chem.*, **54**, 2591 (1982).
89. D. L. Rabenstein, *J. Chem. Educ.*, **61**, 909 (1984).
90. (a) B. Thiault and M. Mersseman, *Org. Mag. Res.*, **7**, 575 (1975); (b) B. Thiault and M. Mersseman, *Org. Mag. Res.*, **8**, 28 (1976).
91. J. W. Blunt and M. H. G. Munro, *Aust. J. Chem.*, **29**, 975 (1976).
92. K. Hatada, Y. Terawaki, H. Okuda, K. Nagata, and H. Yuki, *Anal. Chem.*, **41**, 1518 (1969).
93. F. F. Caserio, Jr., *Anal. Chem.*, **38**, 1802 (1966).
94. T. D. Alger, M. Solum, D. M. Grant, G. D. Silcox, and R. J. Pugmire, *Anal. Chem.*, **53**, 2299 (1981).
95. A. G. Marshall, T. Marcus, and J. Sallos, *J. Mag. Res.*, **35**, 227 (1979).
96. J. P. Jesson, P. Meakin, and G. Kneissel, *J. Am. Chem. Soc.*, **95**, 618 (1973).
97. B. L. Tomlinson and H. D. W. Hill, *J. Chem. Phys.*, **59**, 1775 (1973).
98. A. G. Redfield and R. K. Gupta, *J. Chem. Phys.*, **54**, 1418 (1971).
99. A. G. Redfield, S. D. Kunz, and E. K. Ralph, *J. Mag. Res.*, **19**, 114 (1975).
100. (a) J. F. Haw, T. E. Glass, and H. C. Dorn, *Anal. Chem.*, **53**, 2327 (1981); (b) J. F. Haw, T. E. Glass, D. W. Hausler, E. Motell, and H. C. Dorn, *Anal. Chem.*, **52**, 1135 (1980); (c) J. F. Haw, T. E. Glass, and H. C. Dorn, *Anal. Chem.*, **53**, 2332 (1981).
101. (a) J. Buddrus and H. Herzog, *Org. Mag. Res.*, **13**, 153 (1980); (b) J. Buddrus, H. Herzog, and J. W. Cooper, *J. Mag. Res.*, **42**, 453 (1981); (c) J. Buddrus and H. Herzog, *Org. Mag. Res.*, **15**, 211 (1981); (d) E. Bayer, K. Albert, M. Nieder, E. Grom, G. Wolff, and M. Rindlisbacher, *Anal. Chem.*, **54**, 1747 (1982); (e) E. Bayer, K. Albert, M. Nieder, E. Grom, and T. Keller, *J. Chromatog.*, **186**, 497 (1979); (f) N. Watanabe and E. Niki, *Proc. Japan. Acad. Ser B*, **54**, 194 (1978).
102. J. K. Becconsall, *Analyst*, **103**, 1233 (1978).
103. T. W. Gurley and W. M. Ritchey, *Anal. Chem.*, **47**, 1444 (1975).
104. J.K. Nicholson, M. P. O'Flynn, P. J. Sadler, A. F. Macleod, S. M. Juul, and P. H. Sönksen, *Biochem. J.*, **217**, 365 (1984).
105. R. B. Moon and J. H. Richards, *J. Biol. Chem.*, **248**, 7276 (1973).
106. J. J. H. Ackerman, T. H. Grove, G. C. Wong, D. G. Gadian, and G. K. Radda, *Nature*, **283**, 167 (1980).
107. B. D. Ross, G. K. Radda, D. G. Gadian, G. Rocker, M. Esiri, and J. Falconer-Smith, *New Eng. J. Med.*, **304**, 1338 (1981).
108. G. K. Radda, P. J. Bore, D. G. Gadian, B. D. Ross, P. Styles, D. J. Taylor, and J. Morgan-Hughes, *Nature*, **295**, 608 (1982).
109. D. L. Rabenstein and A. A. Isab, *J. Mag. Res.*, **36**, 281 (1979).

110. D. L. Rabenstein, A. A. Isab, and D. W. Brown, *J. Mag. Res.*, **41**, 361 (1980).
111. (a) G. J. Martin, M. L. Martin, F. Mabon, and M.-J. Michon, *Anal. Chem.*, **54**, 2380 (1982); (b) G. J. Martin, M. L. Martin, F. Mabon, and M.-J. Michon, *J. Agric. Food Chem.*, **31**, 311 (1983).
112. (a) M. J. Garson and J. Staunton, *Chem. Soc. Rev.*, 539 (1979); (b) G. J. Martin and M. L. Martin, *Tetrahedron Lett.*, **36**, 3525 (1981).
113. (a) M. L. Jozefowicz, I. K. O'Neill, and H. J. Prosser, *Anal. Chem.*, **49**, 1140 (1977); (b) I. K. O'Neill, M. L. Trimble, and J. C. Casey, *Meat Sci.*, **3**, 223 (1979).
114. I. K. O'Neill, M. Sargent, and M. L. Trimble, *Anal. Chem.*, **52**, 1288 (1980).
115. D. W. Lowman and G. E. Maciel, *Anal. Chem.*, **51**, 85 (1979).
116. J. W. Blunt, M. H. G. Munro, and W. H. Swallow, *Aust. J. Chem.*, **32**, 1339 (1979).
117. G. C. Levy and D. J. Craik, *Science*, **214**, 291 (1981).
118. (a) D. T. Pegg, D. M. Doddrell, and M. R. Bendall, *J. Chem. Phys.*, **77**, 2745 (1982); (b) D. M. Doddrell, D. T. Pegg, and M. R. Bendall, *J. Mag. Res.*, **48**, 323 (1982).
119. (a) P. Plateau and M. Gueron, *J. Am. Chem. Soc.*, **104**, 7310 (1982). (b) P. J. Hore, *J. Mag. Res.*, **55**, 283 (1983).
120. G. M. Clore, B. J. Kimber, and A. M. Gronenborn, *J. Mag. Res.*, **54**, 170 (1983).
121. P. Styles, N. F. Soffe, C. A. Scott, D. A. Cragg, F. Row, D. J. White, and P. C. J. White, *J. Mag. Res.*, **60**, 397 (1984).
122. S. Sibisi, J. Skilling, R. G. Bereton, E. D. Laue, and J. Staunton, *Nature*, **311**, 446 (1984).
123. R. Benn and H. Günther, *Angew. Chem. Int. Ed. Engl.*, **22**, 350 (1983).
124. D. J. Cookson and B. E. Smith, *J. Mag. Res.*, **57**, 355 (1984).
125. D. A. Laude, Jr. and C. L. Wilkins, *Anal. Chem.*, **56**, 2471 (1984).
126. H. C. Dorn, *Anal. Chem.*, **56**, 747A (1984).
127. T. F. Budinger and P. C. Lauterbur, *Science*, **226**, 288 (1984).
128. S. L. Smith, *Anal. Chem.*, **57**, 595A (1985).

INDEX